2024 37th International Vacuum Nanoelectronics Conference (IVNC 2024)

AA001154

Brno, Czech Republic
15-19 July 2024

IEEE Catalog Number: CFP24VAC-POD
ISBN: 979-8-3503-7977-8

**Copyright © 2024 by the Institute of Electrical and Electronics Engineers, Inc.
All Rights Reserved**

Copyright and Reprint Permissions: Abstracting is permitted with credit to the source. Libraries are permitted to photocopy beyond the limit of U.S. copyright law for private use of patrons those articles in this volume that carry a code at the bottom of the first page, provided the per-copy fee indicated in the code is paid through Copyright Clearance Center, 222 Rosewood Drive, Danvers, MA 01923.

For other copying, reprint or republication permission, write to IEEE Copyrights Manager, IEEE Service Center, 445 Hoes Lane, Piscataway, NJ 08854. All rights reserved.

*** This is a print representation of what appears in the IEEE Digital Library. Some format issues inherent in the e-media version may also appear in this print version.

IEEE Catalog Number: CFP24VAC-POD
ISBN (Print-On-Demand): 979-8-3503-7977-8
ISBN (Online):   979-8-3503-7976-1
ISSN: 2164-2370

Additional Copies of This Publication Are Available From:

Curran Associates, Inc
57 Morehouse Lane
Red Hook, NY 12571 USA
Phone: (845) 758-0400
Fax: (845) 758-2633
E-mail: curran@proceedings.com
Web: www.proceedings.com

TABLE OF CONTENTS

3D Printed, Quadrupole Mass Filter with High Filter Resolution for Detecting Carbon-13 Isotopes 1
Colin C. Eckhoff, Hyeonseok Kim, Luke J. Metzler, Randall E. Pedder, Luis Fernando Velásquez-García

A Carbon Nanotube Cold Cathode Microwave Electron Gun 3
Jiaxin Li, Yu Zhang, Yanlin Ke, Ningsheng Xu, Shaozhi Deng

A Fully Chip-Scale Integrated X-Ray Source 5
Pawel Urbanski, Tomasz Grzebyk

A Gold Needle Tip Array Ultrafast Electron Source with High Beam Quality 7
Leon Bruckner, Constantin Nauk, Philip Dienstbier, Constanze Gerner, Bastian Lohrl, Timo Paschen, Peter Hommelhoff

A MEMS-Type Ionization Vaccum Sensor Based on an On-Chip Thermionic Electron Source 9
Yanqing Zhao, L. Zhiwei, Dengzhu Guo, Xianlong Wei

A Method of Shaping a Carbon Nanotube Pillar Emitter 11
Zhaoying Xu, Jun Jiang, Yu Zhang, Yuan Huang, Shaozhi Deng

A Nano-Focus X-Ray Source with Nanoneedle Cold Cathode by Simulation 13
Junhao Zhong, Shuai Tang, Mingkai Gou, Yan Shen, Yu Zhang, Yifeng Huang, Juncong She, Jun Chen, Ningsheng Xu, Shaozhi Deng

A Nanoscale Vacuum Transistor Based on Vertical Silicon Dioxide Tunneling Junctions 15
H. Yidan, L. Zhiwei, Xianlong Wei

A Photo-Electric Co-Excited Self-Focusing Nano-Cold-Cathode Electron Gun Device 17
X. Pengbin, Yan Shen, Dong Han, Xiaoyu Qin, Yu Zhang, X. Ningsheng, Shaozhi Deng

A Self-Consistent Combined Laplace-Schroedinger Solution of Field Emission from the
Hemisphere on a Post Model 19
G. C. Kokkorakis, D. Karaoulanis, J. A. Roumeliotis, J. P. Xanthakis

Advancing Debris-Free EUV Light Source Technology with Reduced Visible Light Effect by Cold
Cathode Electron Beam (C-Beam) Irradiation 21
Umesh Balaso Apugade, Bishwa Chandra Adhikari, Iksu Kim, Kyu Chang Park

Alternative Definition of the Apex Field Enhancement Factor for a Regular Array of
Electrostatically Interacting Post Emitters 23
Thiago A. De Assis, Richard G. Forbes, Fernando F. Dall'Agnol

An In-Depth Analysis of the Impact of Vacuum Conditions on the Field Emission Characteristics of
a Cold Cathode C-Beam 25
Ketan R. Bhotkar, Jaydip Sawant, Ravindra Patil, Kyu Chang Park

Area Scaling of Dense Si Field Emitter Arrays 27
Shabnam Ghotbi, Saeed Mohammadi

Beam Divergence and Interference Characteristics of Field Emission Beam: Effects of Transverse
Momentum and Source Spatial Coherence 29
Soheil Hajibaba, Soichiro Tsujino

Coherent Particle Acceleration on a Nanophotonic Chip 31
Leon Bruckner, Tomas Chlouba, Roy Shiloh, Stefanie Kraus, Julian Litzel, Peter Hommelhoff

Cold Cathode Coherent-Structure Flat Panel X-Ray Source Using Microarray Metal Target 33
*Guicai Qi, Qi Liu, Song Kang, Zhuoran Ou, Wangjiang Wu, Yuan Xu, Linghong Zhou,
Juncong She, Shaozhi Deng, Ningsheng Xu, Jun Chen*

Compact and Low-Cost Scanning Electron Microscope 35
Casimir Kuzyk, Alexander Dimitrakopoulos, R. Fabian Pease, Alireza Nojeh

Compact Modeling Approach of Field Emitter Arrays 36
Youngjin Shin, Nedeljko Karaulac, Winston Chern, Akintunde I. Akinwande

Conductively Coated 3D Printed Emitters for Electron Devices 38
Daniel Burda, Mario Kandra, Mohammad Allaham, Alexandr Knápek, Milan Matejka

Contact Interface of Graphene Sensors 40
*Patrik Staron, Robert Macku, Petr Sedlák, Nikola Papež, Ramazanov Shihkgasan, Farid
Orudzhev, Mohammed A. Al-Anber, Dinara Sobola*

Controlled Synthesis of Tungsten Oxide Nanowires Prepared by Thermal Oxidation for Application
in Cold Cathode Flat Panel X-Ray Source 42
Qi Liu, Zufang Lin, Song Kang, Guofu Zhang, Shaozhi Deng, X. Ningsheng, Jun Chen

Design of Setup for Laser Induced Plasma Etching 44
Lukas Silhan, Jan Novotny, Tomas Plichta, Jan Jezek, Ondrej Vaculik, Mojmir Sery

Evaluation of Virtual Source Size Measurement System for a Field Emission Electron Gun 46
Erina Kawamoto, Soichiro Matsunga

Electron Emission Energy Spread Analysis Under Plasmon Resonance 48
Yuyue Ding, Yinyao Chen, Tao Cui, Zheyu Song, Yan Shen, Shaozhi Deng

Electron Emission Properties of Planar-Type Electron Emission Devices Based on a Graphene/H-
BN/Si Heterostructure Fabricated by Inductively Coupled Plasma-Enhanced Chemical Vapor
Deposition 50
Katsuhisa Murakami, Hiromasa Murata, Masayoshi Nagao

Electron-Source Investigations with Discharge-Resistant CNT Field Emitter Cathodes 52
Wolfram Knapp

Enhanced EUV Lighting with Focusing Electrode Adapted C-Beam Irradiation Technique 54
Iksu Kim, Umesh Balaso Apugade, Kyu Chang Park

Enhanced Field Emission Properties of Titanium Nitride Coated ZnO Nanowires 56
*Xinran Li, Guofu Zhang, Zhuoran Ou, Zhipeng Zhang, Shaozhi Deng, Ningsheng Xu, Jun
Chen*

Enhancing X-Ray Emission of Cold Cathode Flat-Panel X-Ray Source by Pulsed Driving 58
Ruowen Fan, Song Kang, Guofu Zhang, Juncong She, Shaozhi Deng, Jun Chen

Exploring the Future of Field Electron Emission Theory 60
Richard G. Forbes

Extended Definitions of Fermi Level and Fermi Energy and How Lattice Expansion Affects Fermi
Energy 62
Richard G. Forbes

Fabrication of Double-Gate Zinc Oxide Nanowire Field Emitter Arrays for Achieving High Current............. 64
Zhuoran Ou, Chengyun Wang, Guofu Zhang, Xinran Li, Hai Ou, Juncong She, Shaozhi Deng, Jun Chen

Field Electron Emission Characteristics of Uncoated Alumel Tips.. 66
Marwan S. Mousa, Enas A. Arrasheed, Adel M. Abuamr, Ammar Al Soud, Dinara Sobola

Field Emission Characteristics from Vertical Few-Layer Graphene Growth on Graphite Substrate................. 68
Yiming Huang, Shuai Tang, Mingkai Gou, Haonan Zhao, Yan Shen, Yu Zhang, Juncong She, Jun Chen, Shaozhi Deng

Field Emission X-Ray System for Online Conveyor Belt Imaging.. 70
Rui Zhou, Jiaqi Wang, Zhemiao Xie, John T. W. Yeow

From Deformation to Performance: WO$_3$-Coated Tungsten Emitters Created by Anodization 72
Zuzana Košelová, Daniel Burda, Mohammad M. Allaham, Zdenka Fohlerová, Alexandr Knápek

Glass-Extraction Electrode for Field Emission Applications 74
Aleksandra M. Buchta, Alexander Kassner, Folke Dencker, Marc C. Wurz

HfN and Hf Spindt-Type FEA Fabrication Using Triode Reactive High Power Pulsed Magnetron
Sputtering 76
Hiromasa Murata, Shun Kondo, Md. Suruz Mian, Katsuhisa Murakami, Takeo Nakano, Masayoshi Nagao

High Performance Paper-CNT Field Emitters.. 78
Michal Krysztof, Piotr Szyszka, Pawel Urbanski, Tomasz Grzebyk

High Quantum Efficiency Graphene-Oxide-Semiconductor Electron Source Without Negative
Electron Affinity Surface.. 80
Hidetaka Shimawaki, Masayoshi Nagao, Katsuhisa Murakami

Characterization and Analysis of Field Electron Emission from Copper Tips 82
Marwan S. Mousa, Adel M. Abuamr, Ammar Al Soud, Alexandr Knápek, Dinara Sobola

Characterization of Electron Emission from Broad Area Composite Emitter 84
Ammar Al Soud, Marwan S. Mousa, Aseel Aljabarat, Adel M. Abuamr, Ahmad Telfah, Alexandr Knapek, Enas A. Arrasheed, Dinara Sobola

Improved Method for Determining the Distribution of FEA Currents by Optical CMOS Sensors.................... 86
Matthias Hausladen, Andreas Schels, Philipp Buchner, Mathias Bartl, Ali Asgharzade, Simon Edler, Dominik Wohlfartsstätter, Michael Bachmann, Rupert Schreiner

Improved Performance of Planar Vacuum Field Emission Transistors Via Angled Gates and
Extended Anode 88
Zelin Yu, Zhenpeng Wang, Zhuoya Zhu, Mei Xiao, Wei Lei, Xiaobing Zhang

Improving the Performance of Field Electron Emission from Carbon Fiber Emitters with an Epoxy
Resin 478 Coating Layer 90
Issam Trrad, Marwan S. Mousa, Ahmad M D Jaber, Adel M. Abuamr, Ali F. Alqaisi, Alexandr Knápek

Improving the Performance of Nanoscale Vacuum Channel Transistors Via Composite Gate 92
Xin Zhai, Zhuoya Zhu, Mei Xiao, Wei Lei, Xiaobing Zhang

In Situ Observation of Nanoprotrusion Growth on a Carbon Coated Tungsten Nanotip Under Field 94
Guodong Meng, L. Yimeng, Roni Aleksi Koitermaa, Veronika Zadin, Yonghong Cheng, Andreas Kyritsakis

Lateral Glow Discharge Ion Source for the Integrated MEMS Quadrupole Mass Spectrometer 96
Piotr Szyszka

MEMS-Based Vacuum Analytical Instruments for Space Exploration 98
Tomasz Grzebyk, Piotr Szyszka, Michal Krysztof, Pawel Urbanski, Marcin Bialas, Pawel Knapkiewicz, Jan Dziuban

Miniaturized Rubidium Source for Generating Vapor Phase Atoms for Magneto Optical Traps 100
Jannik Koch, Leonard Frank Diekmann, Alexander Kassner, Folke Dencker, Marc Christopher Wurz

Modeling of Horizontal Integrated Silicon Field Emitter for On-Chip Free Electron Interactions 102
Goulven Rouillé, Catherine Weng, Anthony Ayari, Sorin-Mihai Perisanu, Sylvain Combrié, Laurent Gangloff, Xavier Checoury

Modelling of X-Ray Diffraction on Multilayer Objects 104
Artur Ovcharenko, Serhii Lebedynskyi, Oleksandr Lebed

Narrow Energy Spread Electron Emission from Si-Tip with Ultra-Thin Diamond Like Carbon
Coating 106
Wen Zeng, Yang Chen, Wenqi Feng, Yifeng Huang, Runze Zhan, Jun Chen, Shaozhi Deng, Ningsheng Xu, Juncong She

Noise Characterization of Graphene Sensors 108
Patrik Staron, Robert Macku, Petr Sedlák, Nikola Papež, Ramazanov Shihkgasan, Farid Orudzhev, Mohammed A. Al-Anber, Dinara Sobola

Numerial Simulation of a Vacuum Cold Cathode X-Ray Detector Driven by a Dual-Gate Thin Film
Transistor 110
Zhongbin Pu, Zhipeng Zhang, Jiaquan Kong, Juncong She, Shaozhi Deng, Jun Chen

On-Chip Integrated Si-Tip Field Electron Emission Vacuum Transistor with Saturated Output
Characteristics 112
Zhen Wang, Yuan Huang, Yang Chen, Yifeng Huang, Jun Chen, Ningsheng Xu, Shaozhi Deng, Juncong She

Optimization of High-Performance Single Island Carbon Nanotube Electron Beam (C-Beam) for
Microscopy Application 114
Ravindra Patil, Aniket Karande, Ketan Bhotkar, Kyu Chang Park

Optimization of Sputtering Condition for TiN-Coated Si-FEA 116
Hiromasa Murata, Katsuhisa Murakami, Masayoshi Nagao

Origin of the Slope-Intercept Linear Relationship in Field Emission 118
Anthony Ayari, Pascal Vincent, Sorin Perisanu, Philippe Poncharal, Stephen T. Purcell

Outgassing and Leak Rates of Bonding Technologies for Quantum Systems 120
Verena Velthaus, Jakob Buchheim

Photoelectron Emission from Molybdenum Disulfide/Hexagonal Boron Nitride/Graphene
Heterostructure 122
Guichen Song, Shaozhi Deng, Jun Chen

Photoresponse of Field Emission Current from FAPbI$_3$ Perovskite Film 124
    Bin Wen, Zhuoran Ou, Guofu Zhang, Manni Chen, Juncong She, Shaozhi Deng, Jun Chen

Planar On-Chip Auto-Ponderomotive Devices for Electron Beam Control 126
    Franz Schmidt-Kaler, Michael Seidling, Robert Zimmermann, Nils Bode, Fabian Bammes,
    Lars Radtke, Peter Hommelhoff

Planar-Gate Zinc Oxide Nanowire Cold Cathode for Line-Coded Flat Panel X-Ray Source 128
    Junhang Xie, Qi Liu, Guofu Zhang, Song Kang, Shaozhi Deng, Ningsheng Xu, Jun Chen

Portable Scanning and Acquisition System for Miniature SEM 130
    Marcin Bialas

Potential and Charge Density at the Surface of a Field Emitter 132
    Chris Edgcombe, Janis Huns

Scanning Enhancement of STM-Tungsten Probes by Applying Colloidal Graphite Coatings 134
    Mohammad M. Allaham, Zuzana Koselova, Daniel Burda, Alexandr Knapek

Scattering Matrix Approach to Electron Transmission Probability Through a Flat Semiconductor /
Vacuum Interface 136
    Nathaniel Hernandez, Marc Cahay, James Hart, Jonathan O'Mara, Jonathan Ludwick,
    Dennis E. Walker, Tyson Back, Harris Hall

Silicon Field Emitter Arrays for Vacuum Integrated Circuits 138
    Nedeljko Karaulac, Akintunde I. Akinwande

Silicon Nanowire Field Emitters with Integrated Extraction Gates Using Benzocyclobutene as an
Insulator 140
    Philipp Buchner, Alexander Kaiser, Matthias Hausladen, Mathias Bartl, Michael Bachmann,
    Rupert Schreiner

Simulating Vacuum Arc Initiation by Coupling Emission, Heating, and Plasma Processes 142
    Roni Aleksi Koitermaa, Tauno Tiirats, Veronika Zadin, Flyura Djurabekova, Andreas
    Kyritsakis

Simulation of the Electrical Properties of a Graphene Monolayer Field Effect Transistor 144
    Ammar Al Soud, Ahmad M D Assa'D Jaber, Vladimir Holcman, Petr Sedlak, Dinara Sobola

Simulations and Investigations of Silicon Nanowire Field Emitters 146
    Mathias Bartl, Philipp Buchner, Matthias Hausladen, Ali Asgharzadehkhorasani, Michael
    Bachmann, Rupert Schreiner

Single Column Multiple Electron Beam Imaging from N-Type Silicon 148
    Jáchym Podstránský, Matthias Hausladen, Jakub Zlámal, Alexandr Knápek, Rupert Schreiner

Slow Dynamics in Localized Heating of Carbon Nanotube Forests 150
    Mokter M. Chowdhury, Kevin Voon, Jeff F. Young, George A. Sawatzky, Alireza Nojeh

Study of ALD Grown Multilayers Exhibiting Vacancy Induced Conductivity for Electron Emitters 152
    Daniel Burda, Mohammad Allaham, Alexandr Knápek, Marwan S. Mousa

Study of High Compression Ratio Cold Cathode Electron Gun 154
    Qianqian Tang, Mengjie Li, Junjie Zhang, Mei Xiao, Xiaobing Zhang

Temperature Dependence of the Field Emission Characteristics of AlGaN/GaN Nanoscale Vacuum Diodes 156

Nathaniel Hernandez, Marc Cahay, James Hart, Jonathan O'Mara, Jonathan Ludwick, Dennis E. Walker, Tyson Back, Harris Hall

The Relation Between the Electron Backscattering Coefficients and Elastic Peak Intensity for C, Si, Cu, Ag, and Au at Low Energy Range 158

Ahmad M. D. Jaber, Mohamed M. El-Gomati, Marwan S. Mousa, Alexandr Knápek

Theoretical Analysis of Electron Emission from Hafnium Carbide Tip 160

Toshiaki Kusunoki, Noriaki Arai

Theoretical Analysis of the Nanovoids Influence on the Field Emission 162

Serhii Lebedynskyi, Yuliia Lebedynska, Roman Kholodov

Two-Stage Amplifier Using Field Emitter Array-Based Vacuum Transistor and Its Application to a Wien Bridge Oscillator 164

Ryosuke Hori, Tomoaki Osumi, Masayoshi Nagao, Hiromasa Murata, Yasuhito Gotoh

Uniform Current Supply in Gated P-Type Si-Tips for Achieving High-Performance Field Electron Emitter Array 166

Yang Chen, Yifeng Huang, Jun Chen, Shaozhi Deng, Ningsheng Xu, Juncong She

Vacuum Flat-Panel Solar Blind Ultraviolet Detectors Using PIN-Structure Photocathodes Formed by ZnO Nanowires on NiO-Ga$_2$O$_3$ 168

Dunhan Mo, Zhipeng Zhang, Zhuoran Ou, Juncong She, Shaozhi Deng, Jun Chen

Developing Cold Cathode Flat Panel X-Ray Source Module for Portal X-Ray Imaging System 170

Haonan Wei, Qi Liu, Zhuoran Ou, Song Kang, Guofu Zhang, Zhipeng Zhang, Shaozhi Deng, Ningsheng Xu, Jun Chen

Author Index

TECHNICAL DIGEST

2024 37th International Vacuum Nanoelectronics Conference (IVNC)

Brno Exhibition Centre, Czech Republic

15-19 July 2024

IVNC 2024 Conference Chairs

General Chair
Alexandr Knápek — Institute of Scientific Instruments of the CAS

School Chair
Alexandr Knápek — Institute of Scientific Instruments of the CAS

Treasurer of IVNC 2024
Ľubomír Martinický — Centre of Administration and Operations of the CAS

Event Planner
Veronika Novotná — Institute of Scientific Instruments of the CAS

Publications Coordinator
Veronika Novotná — Institute of Scientific Instruments of the CAS

Webmaster Chair
Veronika Novotná — Institute of Scientific Instruments of the CAS

IVNC International Steering Committee

Akintunde I. Akinwande	Massachusetts Institute of Technology, USA
Heinz Busta	Industry Representative (retired), USA
Marc Cahay	University of Cincinnati, USA
Shaozhi Deng	Sun Yat-Sen University, CHINA
Jan Dziuban	Wroclaw University of Technology, POLAND
Yasuhito Gotoh (ISC Secretary)	Kyoto University, JAPAN
Christopher Holland (Chair)	SRI International, USA
Charles Hunt	University of California, Davis, USA
Hans W.P. Koops	Hawilko Gmbh, GERMANY
Cheol Jin Lee	Korea University and Luminax Co., Ltd., KOREA
Hidenori Mimura	Shizuoka University, JAPAN
Alireza Nojeh	University of British Columbia, CANADA
Kyu Chang Park	Kyung Hee University, KOREA
Stephen Purcell	Université Claude Bernard Lyon 1, FRANCE
Rupert Schreiner	OTH Regensburg, GERMANY
Jonathan Shaw	Naval Research Laboratory, USA
Soichiro Tsujino	Paul Scherrer Institut, SWITZERLAND
Luis Fernando Velásquez-García	Massachusetts Institute of Technology, USA
Ningsheng Xu	Fudan University, CHINA

IVNC Technical Committee 2024

Heinz Busta	Industry Representative (retired), USA
Jan Dziuban	Wroclaw University of Technology, POLAND
Richard Forbes	University of Surrey, UK
Mimura Hidenori	Shizuoka University, JAPAN
Wolfram Knapp	Industry Representative (retired), GERMANY
Michał Krysztof	Wroclaw University of Technology, POLAND
Alireza Nojeh	University of British Columbia, CANADA
Jonathan Shaw	Naval Research Laboratory, USA
Rupert Schreiner	OTH Regensburg, GERMANY
Luis Fernando Velásques-García	Massachusetts Institute of Technology, USA
John P Xanthakis	National Technical University of Athens, GREECE
Ningsheng Xu	Fudan University, CHINA

WELCOME

Greetings and welcome to IVNC 2024 - the 37th International Conference on Vacuum nanoelectronics, held for the first time in the Czech Republic, specifically in the second largest city of Brno, which is world famous in the field of electron optics and electron microscopy. IVNC 2024 is held under the auspices of the Mayor of Brno, JUDr. Markéta Vaňková and is organized by the Institute of Scientific Instruments and the Centre of Administration and Operations (both part of the Czech Academy of Sciences), which is also the main financial sponsor of the conference. Technical co-sponsorship is provided by the IEEE Electron Devices Society.

The International Vacuum Nanoelectronics Conference (IVNC) series is dedicated to the science and technology of micro- and nanodevices that utilize vacuum and high electric field phenomena. The first IVNC was held in 1988 in Williamsburg, VA and was chaired by C. A. Spindt (SRI) and H. F. Gray (NRL). Initially focused on field emission, displays, and power tubes, IVNC evolved into the world's preeminent conference series on electron sources and their applications, and then expanded to include the fundamentals, fabrication, and applications of other vacuum and high electric field phenomena such as plasmas, liquid ionization, gas ionization, X-rays, mass spectrometry, space propulsion, vacuum quantum systems, and harsh electronics. The IVNC alternates between North America, Europe and Asia and has nearly equal participation from these continents. Since its inception, the IVNC series has provided a forum for researchers to discuss the latest research results and promote progress toward commercialization.

In 2024, oral and poster sessions will be held, complemented by social activities that offer ample opportunities for interaction and networking. This year's technical program will feature 3 plenaries and 3 invited lectures from leading experts in the fields of electron microscopy, electron optics, and advanced micro- and nanoelectronics. After careful consideration by a twelve-member ad hoc committee composed of members from the International Steering Committee of IVNC and representatives from IEEE (EDS and MEMS Society), a total of 95 contributions were selected for presentation, including 44 oral presentations and 51 posters. The proceedings will be visible and accessible via IEEE Xplore after the conference concludes.

This conference would not be possible without the generous contributions of time, effort, and financial support from all our colleagues, sponsors, and other supporters. We thank the ad hoc Technical Program Committee for their intensive efforts in reviewing abstracts and the International Steering Committee for their advice and support. We are grateful to Lubomir Martinicky and his team from the Centre of Administration and Operations at the Czech Academy of Sciences in Prague. We would also like to extend our thanks to the IEEE Electron Devices Society for their technical co-sponsorship, without which it would not be possible to publish the abstracts in IEEE Xplore®. We hope you find our conference enjoyable and fruitful, that you enjoy your stay in Brno, and that you will be happy to return to us in the future.

Alexandr Knápek

Chair IVNC 2024

3D Printed, Quadrupole Mass Filter with High Filter Resolution for Detecting Carbon-13 Isotopes

Colin C. Eckhoff[1], Hyeonseok Kim[2], Luke J. Metzler[3], Randall E. Pedder[3], and Luis Fernando Velásquez-García[4,*]

[1] *Massachusetts Institute of Technology, Department of Electrical Engineering and Computer Science, Cambridge, MA, USA*
[2] *Massachusetts Institute of Technology, Department of Mechanical Engineering, Cambridge, MA, USA*
[3] *Ardara Technologies LP, Ardara, PA, USA*
[4] *Massachusetts Institute of Technology, Microsystems Technology Laboratories, Cambridge, MA, USA*
* Corresponding author: Velasquez@alum.mit.edu

Abstract—**We report the design, fabrication, and characterization of additively manufactured quadrupole mass filters with sufficient resolution and sensitivity to detect carbon-13 isotopes. This is a new milestone for 3D-printed quadrupoles.**

Keywords—3D-printing, isotope identification, m+1 peak recognition, quadrupole mass filter, quadrupole mass spectrometry

I. INTRODUCTION

Quadrupole mass spectrometry is an important scientific technique for analytical chemistry. Its relevance in many sectors of research and industry motivates the development of smaller, less expensive hardware [1]. 3D printing allows the creation of complex structures with optimized materials and geometries [2],[3]. Additive manufacturing is a promising approach towards the miniaturization and cost reduction of quadrupole mass filters (QMFs). We have demonstrated that additively manufactured QMFs are capable of basic mass spectrometry [4],[5]; in this work, we show that 3D-printed quadrupole mass filters are capable of isotope discrimination.

II. PART DESIGN

QMFs have four conductive surfaces driven at RF voltages to create the electrical fields necessary for mass filtering [6]. The ideal QMF electrodes are hyperbolic (Figure 1). Due to cost and simplicity, the QMF rods are usually implemented as stain-less steel circular rods aligned parallel to one another by means of dielectric collars. To avoid the fastening procedures usually utilized to build QMFs, we 3D print the part monolithically and selectively metallize the rods. This way, the rods are precisely pre-aligned during production. Electrical isolation between the rods is attained by selectively metallizing the rods via electroless plating. Each rod is connected to the QMF overstructure via posts that enable the easy application of maskant to attain electrically isolated rods.

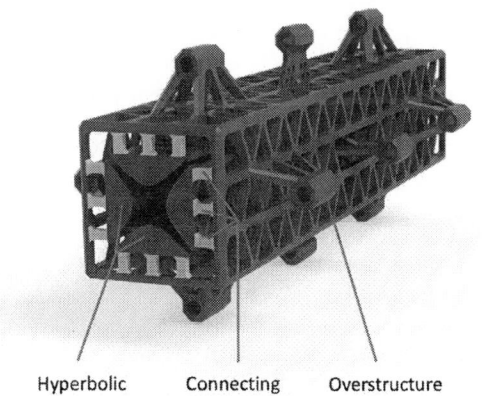

Hyperbolic rods   Connecting posts   Overstructure

Figure 1. Render of our 3D-printed QMF design.

The design of the quadrupole is skeletonized to maximize the strength-to-weight ratio of the part while minimizing printable feedstock consumption. The opposite rod separation distance is equal to 6 mm—a 50% reduction from our prior work [4],[5], attaining a larger mass range. The length of the quadrupole is 125 mm—limited by the capability of the printer employed.

III. MATERIALS & METHODS

A. Printing

The QMF is printed via digital light processing of glass-ceramic resin, as this printing approach and material has demonstrated compatibility with high vacuum environments [4],[5],[7]. The printing technique can resolve with high precision complex structures (Figure 2).

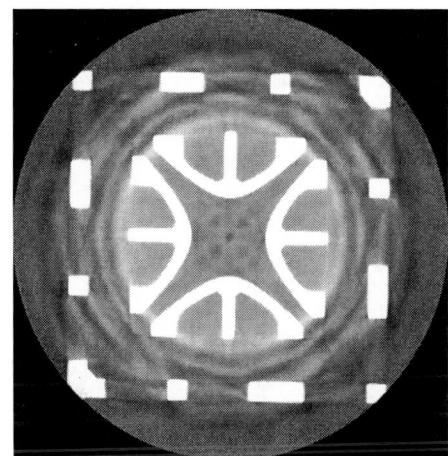

Figure 2. CT scan of midsection of 3D-printed quadrupole, showing precisely aligned rod surfaces. The separation between opposite hyperbolic rods is equal to 6 mm.

B. Plating

Electroless nickel-boron plating was employed to metallize the rods after printing. Relative to other plating methods, electroless plating produces a more uniform film [8]. Nickel-boron is sufficiently conductive and corrosion resistant for this application.

IV. EXPERIMENTAL RESULTS

A. Setup

The QMF was mounted in series with a commercial ion source and an electron multiplier, then installed into a vacuum chamber. Pressures as low as 6×10^{-9} Torr were achieved, evidencing the low outgassing rate of the 3D-printed QMF. Argon and perfluorotributylamine (PFTBA) were leaked into the vacuum chamber during the characterization experiments.

979-8-3503-7977-8/24 $31.00 © 2024 IEEE

The ion source is an electron impact ionizer via thermionic emission. The QMF was operated at 2.4 MHz.

B. Experimental Data

We conducted mass spectrometry of argon. We expect to see two major peaks at 40 Da and 20 Da, corresponding to Ar^+ and Ar^{2+}, respectively. Both of these peaks show up in the mass spectra with clarity, and their relative amplitudes are similar to the references in the NIST database, i.e., the peak at 20 Da is less than one-fifth of the peak at 40 Da (Figure 3). Next, we introduced PFTBA into the chamber to get a more thorough characterization of our QMFs. Unlike argon, which has two significant peaks, PFTBA has many, and they are spread throughout the low hundreds of Da in mass range. Our results show these characteristic peaks (Figure 4).

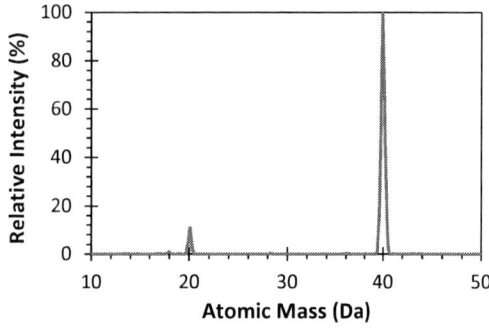

Figure 3. Argon mass scan with 3D-printed quadrupole.

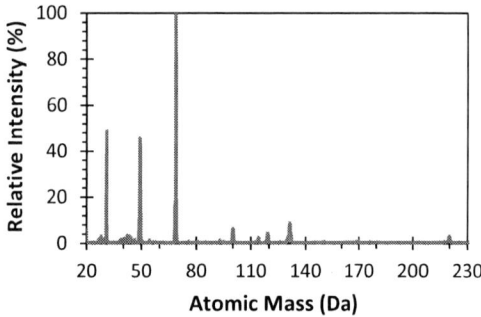

Figure 4. PFTBA scan with 3D-printed quadrupole.

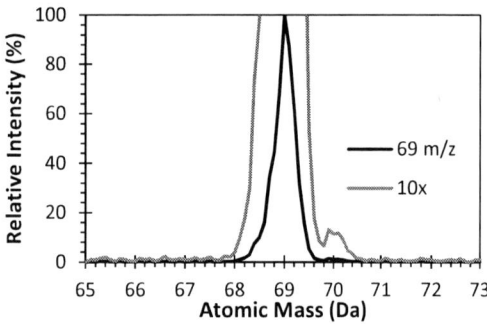

Figure 5. Close-up of the 69 Da peak from PFTBA, showing carbon-13 isotope to its right at 70 Da.

We also examined the ability of our QMFs to detect trace peaks. PFTBA's distinctive peak at 69 Da comes from a molecular fragment with one carbon atom. Carbon-13 makes up about 1.1% of naturally occurring carbon, and it is only 1 Da heavier than carbon-12 (the most common carbon isotope).

When the 69 Da fragment is made of carbon-13, the peak is at 70 Da. This is a tiny peak at 70 Da—about 1.1% the height of the 69 Da peak. It's so small, the data need to be magnified a tenfold to clearly observe it. The carbon-13 peak (also known as an M+1 peak) is visible and clearly resolved (Figure 5). Resolution of carbon-13 peaks have always been a rigorous benchmark of mass spectrometers. The impressive resolution from our QMFs is underscored by their competitive resolution-transmission characteristics (Table 1). Even in high-resolution scanning modes, relative transmission remains high enough for trace sensitivity.

Resolution (FWHM)	Relative Transmission
2.00 Da	97.92%
0.70 Da	43.57%

Table 1. Resolution-transmission specifications for 69 Da.

V. DISCUSSION

Detecting isotopes with high resolution and sensitivity is a necessary requirement for practical mass spectrometry. By revealing the coveted carbon-13 peak, this work demonstrates a great advance forward in the feasibility of alternatively-manufactured, analytical grade QMFs. Further refinements and innovations to additive manufacturing, and the fabrication of parts thereby, will continue to push the performance boundaries of inexpensive, lightweight, and compact mass spectrometers.

ACKNOWLEDGMENT

This work was sponsored by the Empiriko Corporation (Newton, MA, USA).

REFERENCES

[1] R. R. A. Syms, "Status and future trends of the miniaturization of mass spectrometry," 2015 28th IEEE International Conference on Micro Electro Mechanical Systems (MEMS), Estoril, Portugal, 2015, pp. 134-139, doi: 10.1109/MEMSYS.2015.7050904.

[2] A. L. Beckwith, L. F. Velásquez-García, and J. T. Borenstein, "Microfluidic model for evaluation of immune checkpoint inhibitors in human tumors," Advanced Healthcare Materials, vol. 8, no. 11, 1900289, June 2019, doi: 10.1002/adhm.201900289

[3] A. L. Beckwith, J. Borenstein, and L. F. Velásquez - García, "Tunable plant-based biomaterials via in vitro cell culture using a Zinnia Elegans model," Journal of Cleaner Production, vol. 288, 125571, March 2021, doi: 10.1016/j.jclepro.2020.125571

[4] C. C. Eckhoff, N. K. Lubinsky, L. J. Metzler, R. E. Pedder, and L. F. Velásquez–García, "Low-cost, compact quadrupole mass filters with unity mass resolution via ceramic resin vat photopolymerization," Advanced Science, vol. 11, no. 9, Dec. 2023, doi: 10.1002/advs.202307665.

[5] C. C. Eckhoff, N. K. Lubinsky, R. E. Pedder and L. F. Velasquez–Garcia, "Miniature, monolithic, fully additively manufactured glass-ceramic quadrupole mass filters for point-of-care mass spectrometry," 2023 IEEE 36th International Vacuum Nanoelectronics Conference (IVNC), Cambridge, MA, USA, 2023, pp. 204-206, doi: 10.1109/IVNC57695.2023.10188968.

[6] P. H. Dawson, Quadrupole Mass Spectrometry and its Applications, Elsevier, Amsterdam, 1976

[7] J. Izquierdo-Reyes, Z. Bigelow, N. K. Lubinsky, L. F. Velásquez-García, "Compact retarding potential analyzers enabled by glass-ceramic vat polymerization for CubeSat and laboratory plasma diagnostics," Additive Manufacturing, vol. 58, 103034, October 2022. Doi: 10.1016/j.addma.2022.103034

[8] J. Sudagar, J. Lian, W. Sha, "Electroless nickel, alloy, composite and nano coatings – A critical review," J. Alloys Compd., vol. 571, pp. 183–204, September 2013, doi: 10.1016/j.jallcom.2013.03.1

A Carbon Nanotube Cold Cathode Microwave Electron Gun

Jiaxin Li, Yu Zhang*, Yanlin Ke, Ningsheng Xu, Shaozhi Deng

State Key Laboratory of Optoelectronic Materials and Technologies, Guangdong Province Key Laboratory of Display Material and Technology, School of Electronics and Information Technology, Sun Yat-sen University,
Guangzhou, People's Republic of China
stszhyu@mail.sysu.edu.cn

Abstract— **Microwave cold cathode electron gun is an advanced gun type for achieving high current and current density which is very helpful for high frequency microwave vacuum electron devices. To improve the efficiency of microwave generated field emission, a novel cylindrical reentrant resonant cavity with a tapered gradient circular truncated cone and an input coaxial line was designed to minimize the area of the coupling ring and increase the electromagnetic energy transferred to the cavity. Simulation results showed that when applying an input microwave power of 200 W, the cathode surface electric field was up to 3.65×10^6 V/m, with a maximum field emission current of 1.14 A and a corresponding current density of 9.07 A/cm^2.**

Keywords—microwave electron gun, field emission, resonant cavity

I. Introduction

Microwave driven electron emission is a good way to achieve high current and high current density electron beam [1]. Microwave cavity is the related structure to generate a radio frequency electric field on the cathode surface to drive on electron emission. As the frequency increase, the size dimension of cavity shrinks, thus the cathode emission area reduces accordingly. Thus, the required electron current density for VED increases. Compare to thermionic cathode, field emission cold cathode got the advantage of small size, easy for microfabrication, room temperature working and direct modulation under microwave [2]. The field emission current density reaches as large as 10^7 A/cm^2. Therefore, it fits the requirement of high frequency microwave electron gun and has applications on high power vacuum electronics devices (VED), such as x-ray free-electron lasers, terahertz radiation source and synchrotron [3,4,5]. In this paper, we proposed a new design of carbon nanotube (CNT) cold cathode microwave electron gun using a resonant cavity aiming for achieving ampere scale current and ampere per centimeter square scale current density under radio frequency electric field modulation.

II. Experimental and Results

The proposed microwave cold cathode electron gun consists of three parts: a cold cathode, a cavity, and an anode collector. The cavity is a cylindrical reentrant resonant cavity which's model is shown in Fig. 1. The cathode is located on a coaxial column in the center of the cavity, and the top of the cavity has an anode plate. According to the theory of reentrant resonator, when the cylindrical resonator works in TM$_{010}$ mode, the electric field in the cavity is concentrated in the gap between the cathode and the anode while the magnetic field is located at the bottom of the cavity around the column. The anode

is used to collect electrons and a voltage bias can also be applied on it to accelerate the electrons, thereby increasing the output current. The key parameters of the resonator include the size and the coupling structure of the cavity which are related to the maximum electric field intensity for driving field emission.

The coupling structure is used for exchange energy from power source to the cavity. A suitable coupling structure can input the microwave power to the cavity and stimulate a larger radio frequency electric field in high efficiency. The coupling ring is the most popular structure for TM$_{010}$ reentrant resonator. The greater the magnetic induction intensity in the area where the coupling ring is formed, the greater the excited electric field will be. According to the distribution of the electromagnetic field in the reentrant resonator in Fig. 1, the strongest distribution of the magnetic field is at the bottom of the cavity around the coaxial column, so this location is most suitable for placing the coupling ring.

The dimension of the coupling ring radius r_L and the embedding depth l_L effects on the electric field intensity at the surface of the cold cathode. The simulation results in Fig. 2 showed that increase l_L and r_L, the electric field intensity decreases accordingly. It revealed that the area of the coupling ring should be reduced as much as possible to increase the electromagnetic energy transferred to the cavity. On the other side, the resonant frequency is inversely proportional to the l_L and r_L According to this rule, the resonant cavity is designed as shown in Fig. 3(a). The resonant frequency of the cavity is f=6.06 GHz, r_L=1 mm, l_L=1.6 mm, and the maximum electric field intensity on the cathode surface is 2.00×10^5 V/m when the input power is 1.0 W.

However, the technique problem is that the fabrication and assembly of a coupled rings with size around 1-2 mm in the cavity is difficult. To overcome this problem, a new resonator cavity structure with a tapered gradient circular truncated cone and an input coaxial line was proposed as shown in Fig. 3(b). The side of the circular truncated cone, the inner wall of the cavity and the coaxial line form a new coupling ring structure. The area of the coupling ring is 0.36 mm^2 which is 13.25 times smaller than that in Fig. 3(a). This structure reduces the fabrication difficulty and shrinks the area of the coupling ring.

Another improvement to increase the electric field in this structure is on the anode plate. A cylinder extends down at the middle of the anode plate with the same radius of the cathode was added. It shortens the gap between anode and cathode, then increase the intensity of the excitation electric field.

The simulation result based on this optimized cavity structure are shown in Fig. 4. The resonant frequency is 6.08 GHz. Applying the input power of 1.0 W, the maximum electric field obtained increases by 2.3 times, reaching 6.63×10^5 V/m which is 3.3 times larger than that in Fig. 3(a). Increasing the input power to 200 W, when the resonation reach a stable state, the generated electric field on the cathode surface can be up to 3.65×10^6 V/m (Fig. 4(a)), which is fully sufficient to drive on field emission. By applying cold cathode field emission parameters of carbon nanotubes with a diameter of 4 mm on this model, the maximum emission current reach 1.14 A (Fig. 4(b)), and the corresponding current density is 9.07 A/cm^2. Due to the high radio frequency and short pulse width, the Joule heating effect is weakened, thus the carbon nanotube cathode can sustain such a high current and current density which is the most distinct advantage of microwave electron gun compare to the direct-current electron gun.

III. CONCLUSION

We designed a new structure of a cold cathode microwave electron gun using a cylindrical reentrant resonant cavity. A circular truncated cone structure and an input coaxial line were designed to reduce the area of the coupling ring and increase the efficiency of input power transferred to the cavity. The simulation results showed that under a input microwave power of 200 W, the surface electric field of the cathode can be as high as 3.65×10^6 V/m, which is sufficient to drive on field emission. The microwave generated field emission current reached 1.14 A with the corresponding current density of 9.07 A/cm^2.

ACKNOWLEDGMENT

This work was supported by the National Natural Science Foundation of China (grant no. 62274188, U22A2020), the National Key Basic Research Program of China (grant no. 2019YFA0210201), the Guangdong Basic and Applied Basic Research Foundation (grant no. 2023A1515011876), the Science and Technology Department of Guangdong Province and the Fundamental Research Funds for the Central Universities.

REFERENCES

[1] Teo. K et al, "Microwave devices: Carbon nanotubes as cold cathodes," Nature, 437(7061), 2005.
[2] Sergey V. Baryshev et al, "Planar ultrananocrystalline diamond field emitter in accelerator radio frequency electron injector: Performance metrics," Applied Physics Letters, 105(20), 2014.
[3] F. Floreani et al, "Concept of a miniaturised free-electron laser with field emission source," NIM-A, 483(1-2), 2002.
[4] S. V. Kutsaev et al, "Thermionic microwave gun for terahertz and synchrotron light source," Rev. Sci. Instrum, 91(4), 2020.
[5] Y. Xing et al, "A Cold-Cathode Microwave and Terahertz Radiation Source: Experimental Realization @10's GHz and Computational Design @THz," IEEE Electron Device Letters, 40(9), 2019.

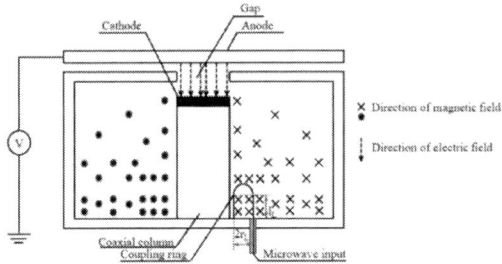

Figure 1 Schematic of a reentrant resonator and its TM$_{010}$ mode field distribution

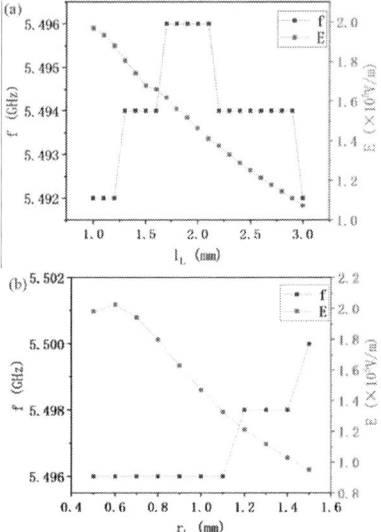

Figure 2 (a) Relationship of l$_L$ on the resonant frequency and electric field intensity, (b) relationship of r$_L$ on resonant frequency and electric field intensity.

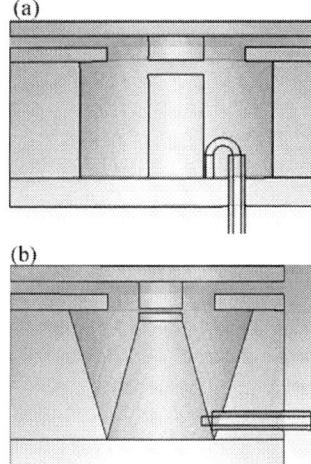

Figure 3 (a) Cavity structure with a coaxial cathode column and a coupling ring, (b) cavity structure with a tapered gradient circular truncated cone cathode and a coupling coaxial line.

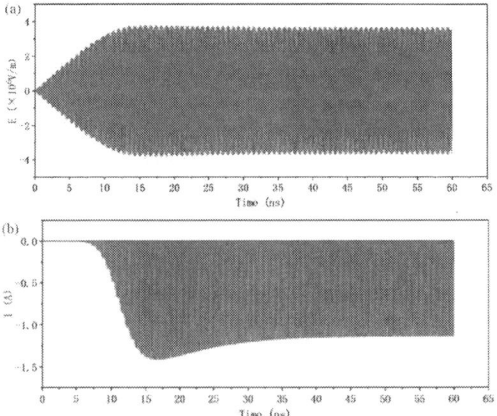

Figure 4 The evolution of microwave drive on field emission in 60 ns, (a) the electric field intensity on the carbon nanotube cathode surface, (b) the current emitted from the carbon nanotube cathode.

A Fully Chip-Scale Integrated X-Ray Source

Paweł Urbański , Tomasz Grzebyk

Faculty of Electronics, Photonics and Microsystems, Wrocław University of Science and Technology
Wrocław, Poland
pawel.urbanski@pwr.edu.pl

Abstract— In today's dynamic world of technological innovation, miniaturization and integration are key drivers of progress. In medicine, industry, materials science and many other fields, the use of innovative micro electro-mechanical systems (MEMS) solutions is becoming more and more common. In this context, we present the pioneering achievement of creating the world's first X-ray source made entirely in MEMS technology. Traditional X-ray sources are usually large, complex, and expensive. However, because of advances in the field of MEMS, it has become possible to create ultra-compact and efficient counterparts. We have overcome the existing problems with hermetic sealing, high vacuum stabilization and risk of electric short-circuits and now present the results obtained for a stand-alone chip-scale device. This achievement, due to the small size of the instrument and possibility of batch fabrication, could be a key step to enable faster access to advanced diagnostic technologies in healthcare and industry.

Keywords— Integration, MEMS (Micro Electro Mechanical Systems), Miniaturisation, X-ray source, Field electron source

I. INTRODUCTION

In the field of X-ray technology, the pursuit of creating miniature yet efficient X-ray sources is gaining increasing attention. Such sources have the potential for wide application in medical diagnostics, industrial processes, and scientific research. Despite their widespread use, conventional X-ray sources have limitations, such as high energy consumption, complex construction, and significant weight. As the demand for portable and efficient X-ray systems grows, so does the interest in alternative technologies for their miniaturization.

Trends towards the miniaturization of X-ray devices are evident in the work of numerous research teams focusing on the use of X-rays in miniature systems, such as MEMS (Micro-Electro-Mechanical Systems) and μTAS (micro Total Analysis Systems). Research includes analyzing the effects of radiation on cells and microorganisms, studying heavy metal contamination in water through microchannels, and characterizing chemical reactions in liquid samples [1-4].

Over the years, various miniature X-ray sources utilizing different mechanisms of electron emission and radiation conversion have been developed and created [5-10]. However, producing a complete X-ray device exclusively using MEMS technology has been challenging, primarily due to issues with maintaining high and stable vacuum within the structure.

In this article, we present the first X-ray source entirely manufactured using MEMS technology, encompassing all key components. This innovation allows for potential integration with other experimental systems. Additionally, the source can generate an X-ray beam with specific characteristics (energy and spectrum) thanks to its adaptive polarization conditions and target design.

II. DESIGN

The developed MEMS X-ray source is constructed entirely from silicon and glass wafers. The chip contains silicon electrodes: a cathode with a field emitter, an extraction gate, a focusing electrode, and a target [Fig.1].

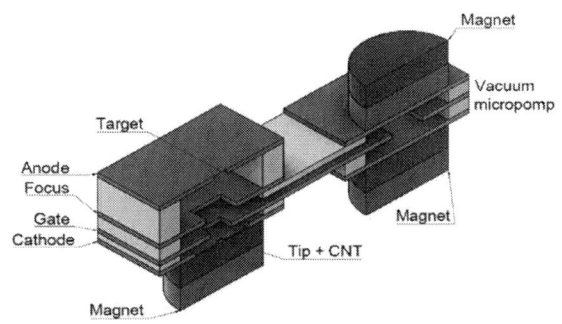

Fig. 1. Construction diagram of a MEMS X-ray source

Borosilicate glass is utilized for spacers between the electrodes. The electron source comprises a composite of carbon nanotubes and PVP (polyvinylidene) deposited onto a silicon substrate using thermo-mechanical method. The extraction electrode, made of silicon, has an etched through-hole and is used to control the emission current and, as a result, the intensity of the X-rays. The focusing electrode enables electrostatic focusing of the electron beam. The target is a silicon chip with a thin membrane of about 15 μm, which can be covered with a thin nanometric metallization layer. The X-ray source is integrated with an ion-sorption pump to maintain high vacuum conditions. The external dimensions of the complete structure are 30×12 mm², with a column height of 6.2 mm. Anodic bonding is employed for connecting silicon and glass elements. The resulting encapsulated MEMS X-ray source [Fig.2a] is sheathed in a polymer layer to prevent electrical breakdowns between the electrodes; as a result, it is a complete self-contained device [Fig.2b].

979-8-3503-7977-8/24 $31.00 © 2024 IEEE

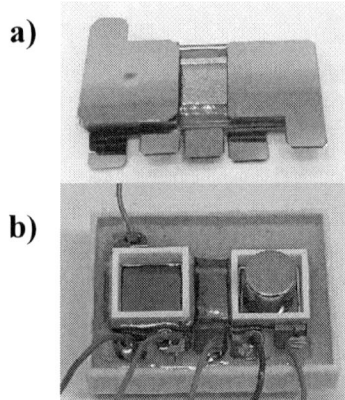

Fig. 2. a) Image of the MEMS X-ray source structure, b) complete, encapsulated MEMS X-ray source with a protective polymer layer

III. MEASUREMENTS AND TESTS

The X-ray source was hermetically sealed, enabling measurements to be conducted outside a vacuum chamber. The ion-sorption pump maintained a stable high vacuum within the structure during device operation. The CNT electron source provided a high emission current, even at low voltages on the extraction electrode, ultimately leading to efficient X-ray beam generation. The experimental results demonstrate the effectiveness of this innovative solution, showcasing excellent performance in capturing precise images of small objects [Fig. 3].

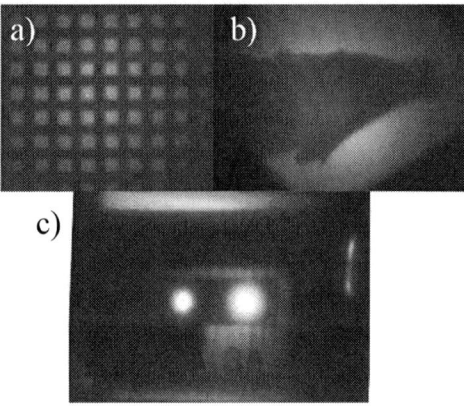

Fig. 3. X-ray image of: a) a silicon mesh (UA = 10kV), b) a leaf (UA = 12kV, c) PCB: light spots – via-holes, black area – metal, green area – laminate (UA = 30kV)

Measurements of the radiation spectra generated using the constructed X-ray source were performed for a distance of 10 cm between the X-ray source and the head of the spectrometer. The measured spectra showed high Bremsstrahlung signal for energies between 4 and 8 keV and only a low intensity of characteristic silicon peaks of 1.46 keV and 1.71 keV [Fig. 4]. This is due to the strong attenuation of low-energy X-ray radiation (<5keV) by air. The design of the source allows for free manipulation of the energy of the emitted radiation by changing the voltage that accelerates the electron beam applied to the target or by changing the target material.

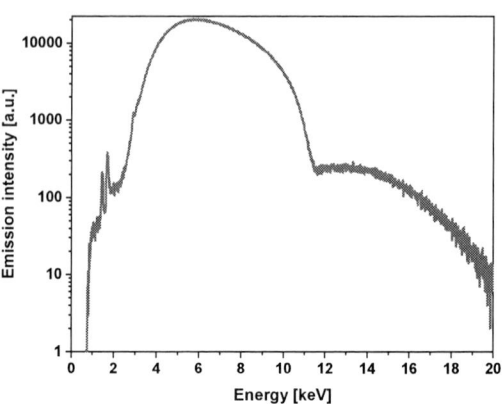

Fig. 4. X-ray spectrum emitted by a MEMS X-ray source for a voltage of 12kV accelerating the electron beam.

IV. CONCLUSION

The article describes a groundbreaking achievement in the field of MEMS and vacuum technology – the development of the first fully integrated X-ray source based on MEMS technology. This X-ray source has been designed to allow precise control of electron emission and beam focusing, resulting in efficient X-ray generation. Integration with an ion-sorption pump ensures a stable high vacuum level within the device. Experiments have shown that the MEMS source can be an alternative to other miniature X-ray sources. Although it does not achieve the same high energy and power as competing solutions, the applied technology opens up new application possibilities – particularly in MEMS systems used for biological, medical, and analytical research on a micro scale.

Acknowledgment

The work was financed by the project UMO 2021/41/B/ST7/01615 of Polish National Science Centre.

V. REFERENCES

[1] Z. Tang, Y. Akiyama, K. Itoga, J. Kobayashi & T. Okano, "Fabrication of microfluidic device on temperature-responsive cell culture surface for studying the shear stress-dependent cell detachment", 2011 Int. Symp. Micro-NanoMechatronics Human Sci. (MHS), IEEE, 2011.

[2] C. Gosse et al., "Development of a fluidic cell to image precipitation reactions by x-ray microscopy", 2017 19th Int. Conf. Solid-State Sensors, Actuators Microsyst. (TRANSDUCERS), Kaohsiung, Taiwan, IEEE, 2017.

[3] Y. T. Yeh, W. J. Khan, T. R. Xu, D. H. Wang i S. Y. Zheng, "Temperature-induced nanochannel array synthesis in microchannels", w 2013 IEEE 13th Int. Conf. Nanotechnol. (IEEE-NANO), IEEE, 2013.

[4] M. Sampietro et al., "Biosensors and Molecular Imaging", IEEE Pulse, vol. 2, no. 3, pp. 35–40, 2011, https://doi.org/10.1109/mpul.2011.941521.

[5] A. Górecka-Drzazga, "Miniature X-Ray sources", Journal of Microelectromechanical Systems, vol. 26, 1 (2017) 295-302.

[6] Technical Data Sheet, accessed on 07.03.2024. [Online]. Available: www.moxtek.com/x-ray-products/

[7] Technical Data Sheet, accessed on 07.03.2024. [Online]. Available: www.oxford-instruments.com/xt

[8] S. D. Kovaleski, A. Benwell, E. Baxter, B. T. Hutsel, T. Wacharasindhu, & J. W. Kwon, "Ultra-compact piezoelectric transformer charged particle acceleration," PowerMEMS Conf., pp. 399–402, 2009.

[9] A. L. Benwell, "A high voltage piezoelectric transformer for active interrogation," Ph.D. dissertation, Faculty Grad. School, 2009.

[10] Technical Data Sheet, accessed on 07.02.2024. [Online]. Available: www.amptek.com/pdf/coolx.pdf

A Gold Needle Tip Array Ultrafast Electron Source with High Beam Quality

Leon Brückner, Constantin Nauk, Philip Dienstbier, Constanze Gerner, Bastian Löhrl, Timo Paschen, Peter Hommelhoff

Department of Physics, Friedrich-Alexander-Universität Erlangen-Nürnberg, 91058 Erlangen, Germany
leon.brueckner@fau.de

Abstract—**Nanometer-sharp needle tips are the electron sources with the highest beam quality. However, for a single needle tip, the total emission current is limited. Combining multiple needles into an array of tips should allow for higher currents while preserving the properties of the individual emitters. Previous studies on tip arrays have shown impressive emission characteristics, but could not demonstrate favourable transversal beam properties such as a well-behaved spot shape or low emittance. Here, we introduce an ultrafast electron source composed of a lithographically fabricated array of nanometer-sharp gold tips illuminated by 25 fs laser pulses. By harnessing the emission of multiple needles, we achieve a high-current electron beam while preserving individual emitter properties. Our source delivers up to 2000 electrons per pulse at moderate laser peak intensities of $10^{11} \mathrm{W/cm^2}$, with a narrow energy width of 0.5 ± 0.05 eV and a well-behaved Gaussian profile. The resulting electron beam exhibits high collimation and a small normalized emittance on the order of nm·rad, making it ideal for applications requiring both high current and spatial resolution, such as free-electron light sources and chip-based particle accelerators.**

I. INTRODUCTION

Ultrashort electron pulses are essential for applications such as ultrafast electron microscopy and diffraction. Sources of such pulses are typically either planar photocathodes or sharp needle emitters. Planar cathodes provide large bunch charges, however they require powerful laser systems and have limited transversal coherence. Needle emitters have superior transversal beam properties, but can only supply small currents. Combining the emission of dozens or even thousands of needles in an array enables large currents while preserving the favorable properties of single emitters [1], [2].

However, previous studies on the transversal properties of emitter arrays [3] showed irregular beam shapes and rather large emittance values. Here, we present an ultrafast electron source consisting of an array of sharp gold needle tips. We investigate the electron emission and transversal beam properties.

II. EXPERIMENTAL SETUP

A. Gold needle tip arrays

The tip arrays were fabricated on a glass substrate via electron beam lithography and subsequent gold deposition, resulting in sharp cone-like tips (Fig. 1). The tips appear highly uniform, with an average radius of curvature of 7.7 ± 1.9 nm, opening half-angle of 13.5 ± 1.9°, and height of 110 ± 10 nm.

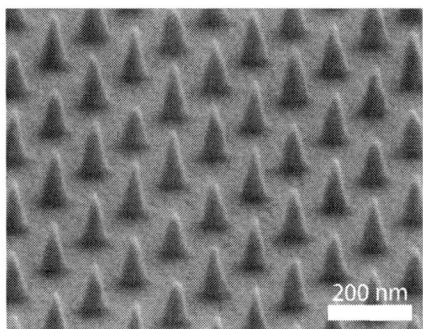

Fig. 1. Scanning electron microscope image of the tip array [4].

B. Measurement setup

The arrays were mounted in an ultra-high-vacuum chamber and biased at -200 V. Photoemission was triggered with 25 fs, 800 nm laser pulses from a Titanium:Sapphire oscillator with a repetition rate of 80 MHz. The transversal electron distribution was observed on a microchannel plate detector and the energy distribution of the electrons was measured with a retarding field spectrometer.

III. ELECTRON EMISSION FROM TIP ARRAYS

The emitted current from an array of 10 x 10 tips is plotted against the incident laser intensity in Fig. 2a). The overall emission is more than 2000 electrons per pulse or 25 nA average current for a moderate incident intensity of $3 \cdot 10^{11} \mathrm{W/cm^2}$ (0.56 nJ pulse energy).

The slope in the double-logarithmic plot is 3.4 for low intensities, indicating multiphoton photoemission. Towards higher intensities, the slope bends towards unity. This behavior can be attributed to strong-field effects [1] as well as the space-charge effect [5]. The measured rate is compared to the Keldysh rate as well as a full time-dependent Schrödinger equation (TDSE) simulation in Fig. 2b). This is assuming a field enhancement factor of 10, which was extracted from a measurement of the electron energy spectrum. The TDSE exhibits a significant bending of the rate due to the first channel closing [6]. However, it considers only single electrons and neglects multi-electron effects. When multiple electrons are emitted during a laser pulse, those emitted first will shield the tip apex from the incoming field, reducing the nonlinearity of the emission. The

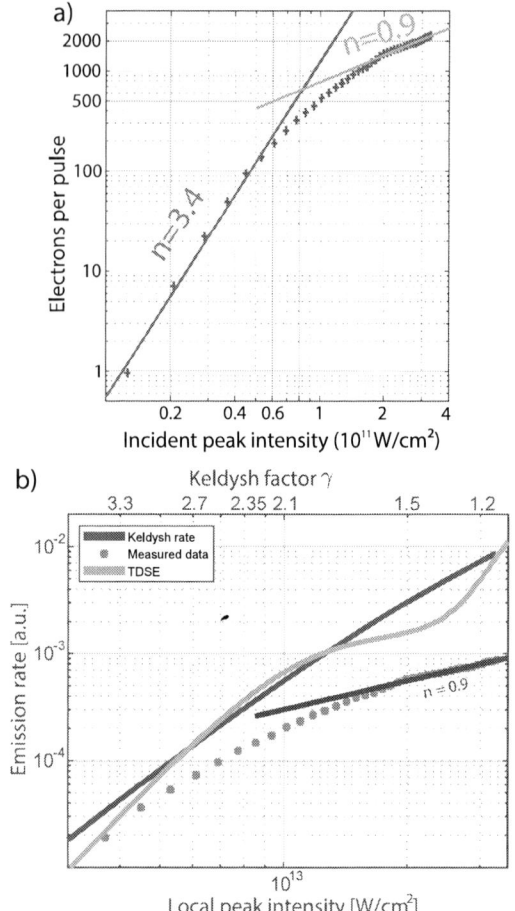

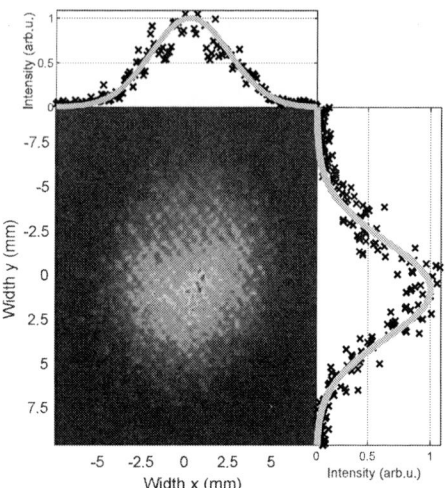

Fig. 3. Transversal distribution of emitted electrons. The spot has a well-behaved Gaussian shape [4].

Fig. 2. a) Log-log plot of the emitted current against the incident laser intensity. b) Comparison of the measured rate (scaled for visibility) to the Keldysh rate and a TDSE simulation. [4].

effective local intensity at the apex is also reduced, making it unlikely for strong-field effects to have a significant influence here. The emission behavior at high incident fields is thus likely dominated by space charge.

IV. TRANSVERSAL BEAM PROPERTIES

Fig. 3 shows the transversal distribution of the emitted electrons. The spot has a well-behaved Gaussian shape, owing to the high uniformity of the individual tips. Calculating the normalized transverse rms emittance via the beam divergence and the rms source size of 0.67 μm yields a low value of 1.1 ± 0.2 nm$\cdot$rad. This is significantly smaller than previously reported array sources [3] and planar photocathodes [7]. The energy width of the source is also low at 0.5 ± 0.05 eV. These properties were measured at low currents where space-charge effects are negligible.

V. OUTLOOK

The gold tip arrays are shown to be a promising ultrafast electron source, providing high charge yields at moderate laser intensities as well as a low emittance and energy width. They are an excellent candidate for applications that need both high currents and good spatial resolution, such as free-electron light sources or attosecond pulse generation schemes. The simple and consistent fabrication process can also easily be adapted to create sharp tips from other materials.

ACKNOWLEDGMENT

We acknowledge funding by the Gordon and Betty Moore Foundation, Deutsche Forschungsgemeinschaft, ERC and BMBF.

REFERENCES

[1] R. G. Hobbs, Y. Yang, A. Fallahi, P. D. Keathley, E. D. Leo, F. X. Kärtner, W. S. Graves, K. K. Berggren, "High-yield, ultrafast, surface plasmon-enhanced, Au nanorod optical field electron emitter arrays", ACS Nano 2014, 8, 11474-11482.

[2] M. E. Swanwick, P. D. Keathley, A. Fallahi, P. R. Krogen, G. Laurent, J. Moses, F. X. Kärtner, L. F. Velásquez-García, "Nanostructured ultrafast silicon-tip optical field-emitter arrays", Nano Letters 2014, 14, 5035-5043.

[3] S. Tsujino, P. D. Kanungo, M. Monshipouri, C. Lee, R. J. Miller, "Measurement of transverse emittance and coherence of double-gate field emitter array cathodes", Nature Communications 2016, 7, 13976.

[4] L. Brückner, C. Nauk, P. Dienstbier, C. Gerner, B. Löhrl, T. Paschen, P. Hommelhoff, "A gold needle tip array ultrafast electron source with high beam quality", Nano Letters 2024, 24, 5018-5023

[5] J .Schötz, L. Seiffert, A. Maliakkal, J. Blöchl, D. Zimin, P. Rosenberger, B. Bergues, P. Hommelhoff, F. Krausz, T. Fennel, M. F. Kling, "Onset of charge interaction in strong-field photoemission from nanometric needle tips", Nanophotonics 2021, 10, 3769-3775

[6] R. Kopold, W. Becker, M. Kleber, G. G. Paulus, "Channel-closing effects in high-order above-threshold ionization and high-order harmonic generation", J. Phys. B: At. Mol. Opt. Phys. 2002, 35 217

[7] H. Ye, S. Trippel, M. D. Fraia, A. Fallahi, O. D. Mücke, F. X. Kärtner, J. Küpper, "Velocity-map imaging for emittance characterization of multiphoton electron emission from a gold surface", Physical Review Applied 2018, 9, 044018.

A MEMS-Type Ionization Vaccum Sensor Based on An On-Chip Thermionic Electron Source

Yanqing Zhao，Zhiwei Li, Dengzhu Guo, Xianlong Wei*

Key Laboratory for the Physics and Chemistry of Nanodevices, Peking University, Beijing, China
*Corresponding atuhor: weixl@pku.edu.cn

Abstract—We present a MEMS-type ionization vacuum sensor based on an on-chip yttrium oxide thermionic source. The sensor has overall dimensions of $12 \times 12 \times 3.3$ mm³, fabricated by anodic bonding technology. Because of the efficient and stable electron emission of an yttrium oxide electron source, the sensor exhibits a working voltage of no more than 210 V and a wide measurement range of 6 orders, covering both high and medium vacuum regimes. The advantages of miniature size, wide measurement range, low working voltage, high stability make our sensors promising for vacuum measurement.

Keywords—ionization vacuum sensor, microelectromechanical system (MEMS), yttrium oxide, thermionic emission.

I. INTRODUCTION

Ionization vacuum sensor (IVS) is the most practical vacuum metrology equipment to high and ultrahigh vacuum regimes below 10^{-1} Pa[1]. With the development of microelectromechanical systems (MEMS) and microfabrication, MEMS-type IVSs have been reported in many literatures[2-5]. However, these MEMS-type IVSs suffer from the instability of the on-chip electron source or electrical discharge processes, resulting in several problems, including a limited measurement range of no more than 4 orders, unstable working performance and high operating voltage.

To solve this problem, We present a MEMS-type IVS based on the efficient and stable electron emission from an on-chip yttrium oxide electron source. The sensor employs a multilayered stacked structure with $12 \times 12 \times 3.3$ mm³, showing a working voltage of no more than 210 V and a wide measurement range of 6 orders from 1×10^{-4} to 100 Pa, covering both high and medium vacuum regimes[6].

II. EXPERMENTAL RESULTS

A. Device Structure

Fig. 1(a) shows the separated components of our MEMS-type IVS, including a silicon on-chip electron source layer, the first glass spacer layer, a silicon electron collector layer, the second glass spacer layer, and a silicon ion collector layer. All the five layers are sequentially stacked and bonded together by anodic bonding process, constructing a step-like semi-closed structure shown in Fig.1(b). The sensor has overall dimensions of $12 \times 12 \times 3.3$ mm³, and weighs 0.6 g. The thermionic emission structure is a Pt-Ir/Y_2O_3 wire, which is bonded to two electrodes on the TSV substrate as an on-chip thermionic emission filament (Fig. 1(c)).

B. Working Principles

Fig. 2 (a) shows the working principle and the corresponding voltage settings of the device. A driving voltage (V_{dri}) is applied to the filament to enable thermionic electron emission. To accelerate emitted electrons, a

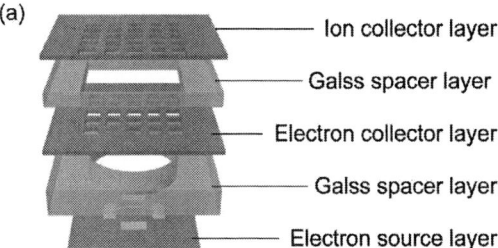

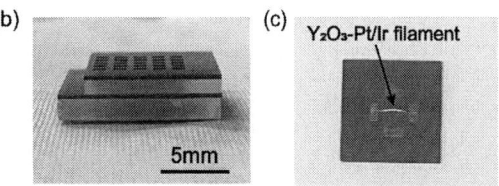

Fig.1. (a) The component structure of the MEMS-type IVS. (b) The side photograph of the MEMS-type IVS. (c)The photograph of the yttrium oxide thermionic electron sources.

positive accelerating voltage (V_{acc}) is applied to the middle silicon grid layer. To collect ions generated by impact ionization, a negative bias voltage (V_{col}) is applied to the top silicon grid layer. Because of the acceleration by V_{acc} and deceleration by V_{col}, electrons emitted from the source will move back and forth around the grid of the electron collector, and finally pour into the electron collector as electron current (I_e). Gaseous molecules are ionized due to impact ionization by energetic electrons. These ions are captured by the ion collector to form an ion current (I_i). The ion current (I_i) is proportional to the pressure (P) and the electron current (I_e) according to the formula for electron impact ionization : $I_i=SPI_e$, where S is the sensitivity of the vacuum sensor[1]. Therefore, the normalized ion current (I_i/I_e) can be used as a parameter to measure the pressure depending on their linear dependence.

C. Measurement Results

To study the measurement range of our vacuum sensors, the device is placed in a vacuum chamber, where the vacuum is obtained by a mechanical pump and a molecular pump to achieve a high vacuum environment. The pressure inside the chamber is controlled through a needle valve, and is measured with a hot cathode ionization gauge less than 10 Pa and a diaphragm gauge in the range of ~10-100 Pa (Fig. 2(a)).

Fig. 2 (b) shows the normalized ion current (I_i/I_e) versus the vacuum pressure measured by commercial vacuum gauges. It can be seen that the normalized ion current of the our sensors show a linear response to the change of vacuum pressure in a range of 5×10^{-4}-50 Pa. By optimizing the

(a)

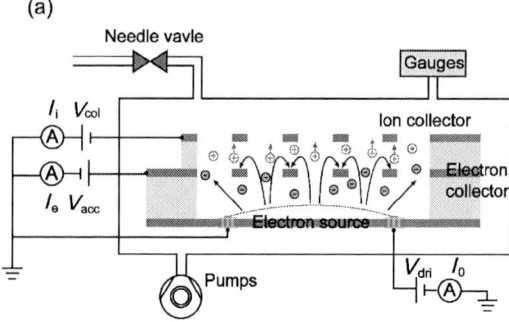

(b)

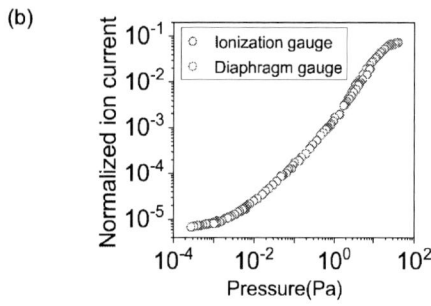

Fig.2. (a) Schematic diagram of working principle and experimental measuring setup of our device. (b) The relationship of normalized ion current (I_i/I_e) and vacuum pressure, at V_{dri}=1.07 V, V_{acc}=100 V, V_{col}=-15 V.

electron accelerating voltage (V_{acc}) and ion collection voltage (V_{col}), the measurement limits of high-vacuum end and high-pressure end reach 1×10^{-4} Pa and 100 Pa[6], respectively, indicating a measurement range of 6 orders for our vacuum sensors.

III. CONCLUSION

In conclusion, we report a MEMS-type ionization vacuum sensor based on yttrium oxide thermionic sources. The overall size of the device is $12 \times 12 \times 3.3$ mm³, presenting a compact structure and a wide measurement range from 5×10^{-4} to 50 Pa. The miniature size, and wide measurement range covering both high and medium vacuum regimes make our sensors promising for vacuum measurement.

REFERENCES

[1] J. F. O'Hanlon and T. A. Gessert, *A users guide to vacuum technology*. John Wiley & Sons, 2023.

[2] C. A. Bower, K. H. Gilchrist, J. R. Piascik, B. R. Stoner, S. Natarajan, C. B. Parker, S. D. Wolter, and J. T. Glass, "On-chip electron-impact ion source using carbon nanotube field emitters," *Applied Physics Letters*, vol. 90, no. 12, p. 124102, 2007.

[3] W. Yang, W. Liu, X. Wang, Z. Li, F. Zhan, G. Zhang, and X. Wei, "A Miniature Ionization Vacuum Sensor With a SiOx-Based Tunneling Electron Source," *IEEE Transactions on Electron Devices*, vol. 68, no. 10, pp. 5127-5132, 2021.

[4] Y. Wang, Y. He, S. Mao, Z. Zhao, W. Yang, P. Liu, and X. Wei, "A Miniaturized Ionization Vacuum Sensor Based on Thermionic Electron Emission From Carbon Nanotubes," *IEEE Transactions on Electron Devices*, vol. 70, no. 6, pp. 2872-2875, 2022.

[5] T. Grzebyk and A. Górecka-Drzazga, "MEMS type ionization vacuum sensor," *Sensors and Actuators A: Physical,* vol. 246, pp. 148-155, 2016.

[6] Y. Zhao, Z. Li, Y. He, S. Mao, D. Guo, and X. Wei, "A MEMS-Type Ionization Vacuum Sensor With a Wide Measurement Range," *IEEE Electron Device Letters,* vol. 45, no. 5, pp. 909-912, 2024.

A Method of Shaping a Carbon Nanotube Pillar Emitter

Zhaoying Xu[1,2], Jun Jiang[1], Yu Zhang[1]*, Yuan Huang[2], Shaozhi Deng[1]

1   State Key Laboratory of Optoelectronic Materials and Technologies, Guangdong Province Key Laboratory of Display Material and Technology, School of Electronics and Information Technology, Sun Yat-sen University, Guangzhou, People's Republic of China
2   School of Microelectronics Science and Technology, Sun Yat-Sen University, Zhuhai, People's Republic of China
*Corresponding author: stszhyu@mail.sysu.edu.cn

Abstract—**Carbon nanotube (CNT) is an excellent kind of field emitter having lots of device application such as the electron source of electron microscope, microwave tube and X ray tube et.al. However, the emission surface on the top is a round area which has a strong edge effect on the circular edge resulting in non-uniform electric field distribution and non-uniform emission. We developed a high-current arcing in-situ treatment to reshape the top morphology of a CNT pillar and obtained an ideal shape for field emission, which can improve the performance in electron transparency and emission uniformity.**

Keywords—*hemispherical CNTs emitter, field emission uniformity, electron transparency*

I. INTRODUCTION

Carbon nanotube (CNT) cold cathode is an important kind of field emitter with many device applications such as the electron source of electron microscope, microwave tube and X ray tube et.al. Edge effect is a common problem of cold cathode and would deteriorate the electron transparency, increase the emission angle and led to emission unevenness. Several ways have proposed to solve this problem[1-3]. For example, A convex carbon nanotubes (CNTs) emitter formed by mechanical polishing treatment is reported to improve the focus and laminar flow by suppressing the edge effect [3].

Here a new method to shape a hemispherical CNTs emitter was proposed. The as-grown CNTs pillar was aged in a destructive field emission process under high voltage. In a cylindrical emitter, the prior emission sites is usually on the circuit edge which has a stronger field intensity and will be destroyed due to Joule heating and atom evaporation [4]. The evaporated atom/molecule further be ionized under the high field and reaccelerate to bombard the emission site. The CNT at the prior emission site was wiped out continually until the remain surface area reaches a uniform electric field intensity. In this way, the surface of the CNT pillar was reshaped. The optimized process parameter was investigated. It is an efficient and low-cost method to obtain uniform emission in a pillar type emitter.

II. EXPERIMENTAL

A. Preparation and Treatment Setup of CNTs

The cylindrical CNT pillar was catalytically grown on silicon substrate using thermal chemical vapor deposition method. As shown in Fig. 1(a), the cylindrical CNT pillar was composed of a bundle of aligned CNTs with the height of 1 mm and diameter of 1 mm. To carry out the arcing treatment, the CNT pillar was fixed on the cathode plate of diode field emission structure with a gap of 1-4 mm, as shown in Fig.1(b). The anode was large aluminum plate. A micrometer on the base was used to adjust the gap between the anode and the cathode.

B. Arcing treatment process

The treatment is carried out in a vacuum chamber with a pressure of 1.0×10^{-5} Pa. High voltage of 4000-7000 V was applied on the CNT pillar surface which was much larger than the threshold voltage of field emission. Then a local glow first appeared at the edge of the pillar (Fig. 1(c)), further a bright sparkle fulfilled the gap (Fig. 1(d)). The CNT pillar undergo a vacuum breakdown and atom evaporation process. In the same time, the emission current suddenly increased and fluctuated; the vacuum pressure also became unstable. After a few seconds, the top of cylindrical CNTs was reshaped to a hemispherical shape, as shown in Fig. 1(e-f).

C. Transparency and Emission Uniformity Measurement

In the same vacuum chamber, the field emission measurement was carried out. A triode field emission structure was used to test its field emission characteristics including electron transparency and emission uniformity as shown in Fig. 2. The CNT pillar emitter was used as the cathode. A stainless-steel plate with an aperture of 2.4 mm was used as the gate electrode. A phosphor screen was used as the anode which can display the uniformity of the emission sites. To illustrate how the shape of emitter influence the characteristics, the field emission characteristics of CNT pillar before and after arcing treatment were both measured and simulated.

III. RESULTS AND DISCUSSION

The field emission characteristics of three CNT pillar sample were measured which are as grown cylindrical CNT pillar, CNT pillar after 7000 V arc treatment and CNT pillar after 4000 V arc treatment. As shown in the IV curve in Fig. 3, applying a cathode voltage of -1400 V, the current of the three samples are 264.9 μA, 162.86 μA and 87.20 μA, and the responding electron transparency is 31.36%, 46.67% and 52.24% (Fig. 3(b)). Obviously, the hemispherical CNT pillar had large current and electron transparency. The reasons are explained as follows. As shown in Fig. 3(f), the cylindrical CNTs emitter has such a sharp edge that the electric field is enhanced at edge. Thus, the current is generated most at the edge. The emission site distribution image in Fig. 3(c) showed a ring shape representing the strong edge effect. The electron trajectory simulation result in Fig.3 (f) and (h) showed that the electric field at the edge is significantly higher than center, which means that the field emission current is concentrated at the edge. The edge current have a large divergence angle. Most of electron emitted from the edge were blocked at the gate electrode. On the contrary, the electric field on the surface of the hemispherical CNT pillar is more uniform as shown in Fig.

979-8-3503-7977-8/24 $31.00 © 2024 IEEE

3(g). As a result, the current distribution from the hemispherical CNT pillar becomes uniform and most electron can pass the gate electrode as shown in Fig. 3(i). In consequence, the electron transparency of the hemispherical CNTs is higher at a lower emission current.

The difference of the two hemispherical CNT pillar treated at 7000 V and 4000 V is explained as follows. As shown in Fig. 2(f), the CNT pillar treated with 7000 V have a thorny surface which may be caused by the uncertainty of the ion bombardment in the case of the higher energy. The rough surface results a larger electric field enhancement factor and consequently a larger emission current. However, it also means less uniformity, which can be seen clearly in the emission site distribution in Fig. 3(d). The CNTs treated at 4000 V has an almost curved surface, as shown in Fig. 2 (e), which shows a better performance in terms of election transparency and emission uniformity (Fig. 3(e)).

IV. CONCLUSION

We developed a new way to fabricate hemispherical CNT pillar field emitter. By applying high voltage, the emitter edge is arced and evaporated, thus the surface of pillar was reshaped according to the local electric field distribution at the emitter surface which can greatly reduce the edge effect, improve the emission uniformity and electron transparency. The simulation and experimental result were agreed with each other confirmed the possibility of this method. As a prospect, it is possible to control the shape of CNTs emitter to achieve an adaptive uniform electric field distribution on the emitter surface. The present results provide a feasible way to fabricate field emitters in a low cost and high efficient way. It also helps to optimize the emitter's performance during its emission process.

ACKNOWLEDGMENT

This work was supported by the National Natural Science Foundation of China (grant no. 62274188, U22A2020), the National Key Basic Research Program of China (grant no. 2019YFA0210201), the Guangdong Basic and Applied Basic Research Foundation (grant no. 2023A1515011876), the Science and Technology Department of Guangdong Province and the Fundamental Research Funds for the Central Universities.

REFERENCES

[1] Sun Y, Song Y, Hoon Shin D, et al, "Fabrication of carbon nanotube emitters on the graphite rod and their high field emission performance," Applied physics letters, 104(4), 2014.

[2] Hyun Jin Kim, Jun Mok Ha, Sung Hwan Heo, and Sung Oh Cho, "Small-sized flat-tip CNT emitters for miniaturized X-ray tubes," Journal of Nanomaterials, p. 3, 2012.

[3] Jiupeng Li, Yu Zhang, Yanlin Ke, Baohong Li, and Shaozhi Deng, "A Cold Cathode Electron Gun Using Convex Carbon Nanotube Emitter," IEEE Transactions on Electron Devices, 69(3), pp. 1457-1460, 2022.

[4] Dean K A, Burgin T P, Chalamala B R, "Evaporation of carbon nanotubes during electron field emission," Applied Physics Letters, 79(12), pp. 1873-1875, 2001.

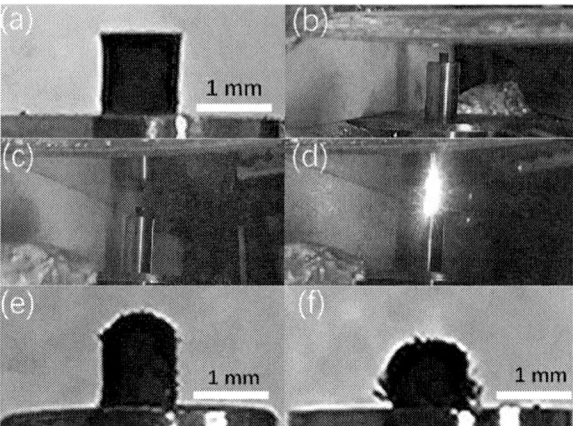

Figure 1 (a) the as-grown cylindrical CNT pillar; (b) the CNT pillar fixing in the diode structure; (c) beginning of arcing: a local glow at the edge of CNT pillar; (d) arcing process: bright sparkling at the gap; (e) the CNT pillar after arc treatment treated at 4000 V and 1 mm gap, and (f) treated at 7000 V and 4 mm gap.

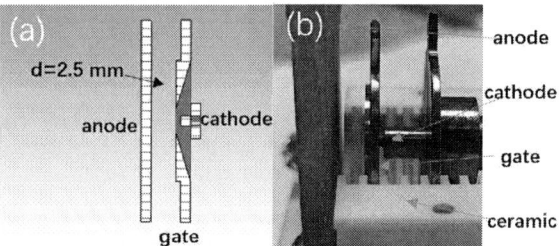

Figure 2 figure of (a) the electron gun for testing and (b)a diagram of the electron gun

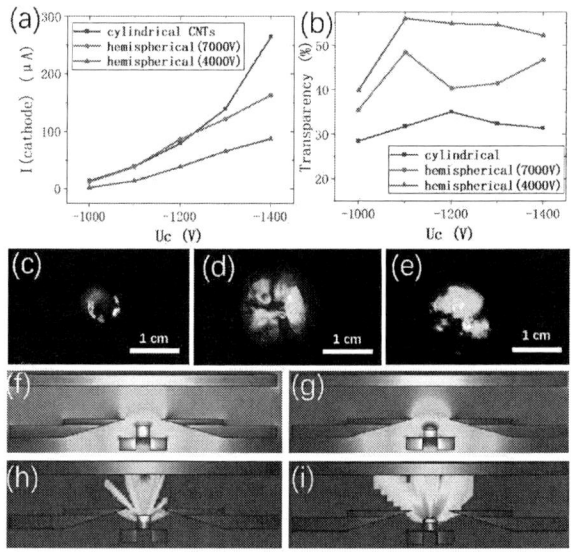

Figure 3 (a) the field emission current and voltage curves and (b) the relationship between electron transparency and cathode voltage of the three kinds of CNT pillars; (c-e) the emission site distribution of the three kinds of CNT pillars; (f) simulation results of electric field intensity distribution map and (h) electron trajectory map of cylindrical CNT pillar; (g) simulation results of electric field distribution map and (i) electron trajectory map hemispherical CNT pillar.

A Nano-Focus X-Ray Source with Nanoneedle Cold Cathode by Simulation

Junhao Zhong, Shuai Tang*, Mingkai Gou, Yan Shen, Yu Zhang, Yifeng Huang, Juncong She, Jun Chen, Ningsheng Xu and Shaozhi Deng

State Key Laboratory of Optoelectronic Materials and Technologies, Guangdong Province Key Laboratory of Display Material and Technology, School of Electronics and Information Technology, Sun Yat-sen University, Guangzhou 510275, People's Republic of China

Corresponding author: tangsh58@mail.sysu.edu.cn

Abstract—An X-ray source based on a nanoneedle cold cathode has been developed. The optimized electron-gun structure and electrical excitation parameters were simulated to obtain a minimum focus spot size (FSS). At the same time, a suppressor electrode structure was introduced in the system to effectively reduce the divergence of the electron beam. The minimum FSS of around 150 nm was obtained which shows nanoscale resolution for X-ray tube. The new X-ray source has the potential to improve the resolution for nanoscale imaging.

Keywords—*LaB₆ nanoneedle, nano-focus X-ray source, electrostatic lens, electron-gun*

I. INTRODUCTION

High spatial resolution X-ray imaging has seen intensively developed in fields such as industry, biology, and medicine due to its excellent penetration depth and non-destructive measurement capabilities. Applications include nanoscale X-ray microscopy for mineralized tissues and soft tissues [1-3]. Although synchrotron radiation sources can achieve X-ray microscopy resolution down to tens of nanometers, their high cost makes them less accessible. Currently, nano-focus electron-beam (E-beam) X-ray source remain a more attainable method for high-resolution X-ray imaging. Consequently, over the past few decades, nano-focus X-ray sources have rapidly advanced. Most nano-focus X-ray sources employ micron or millimeter-scale cathodes and use a combination of electrostatic and magnetic lenses to focus the E-beam onto the anode target. Our group have fabricated high-performance LaB₆ nanoneedle cold cathode [5-6]. The nanoneedle cold cathode can effectively reduce the emission area and results in a high resolution.

Herewith, a nanoneedle cold cathode combined with electrostatic lens was utilized to form an electron gun, aiming to obtain the nano-size electron beam. The electron trajectories of electron gun were simulated using CST software and a FSS of about 150 nm was obtained.

II. METHOD

The cathode is modeled based on the LaB₆ nanoneedle in our previous experiment [5]. The electrical field distribution and electron beam trajectory under different electron gun structure and different electrical excitation parameters are obtained using CST software.

III. RESULTS AND DISCUSSION

The cathode is modeled based on our previous experiment. The LaB₆ nanoneedle was fixed on a tungsten needle. Fig. 1(a) shows the tungsten has a length of 1.35 mm and diameter of 100 μm. Fig. 1(b) shows the length and tip radius of LaB₆

nanoneedle are 3 μm and 10 nm respectively. An initial X-ray source structure was illustrated in Fig. 1(c). The parameters for Fig. 1(c) are configured as follows:

(1) Gate-cathode spacing: 2 mm, gate-thickness: 0.5 mm; diameter of gate-aperture: 1 mm.

(2) Anode-cathode spacing: 9 mm, V_{anode}: 20 kV, foucs lens-cathode spacing: 5 mm, thickness of focus lens: 0.5 mm, diameter of focus lens aperture: 1.5 mm.

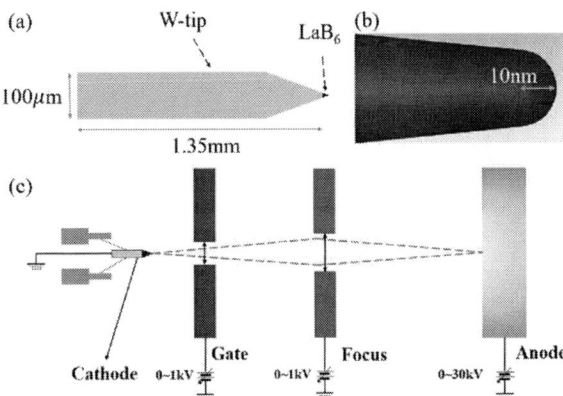

Fig. 1. (a) Schematic of cold cathode LaB₆ nanoneedle modelling; (b) Enlarged view of the nanoneedle, featuring a tip radius of 10 nm; (c) Initial structure of X-ray source.

Firstly, the electric field surrounding the tip and divergence angle at different gate voltages were simulated and shown in Fig. 2.

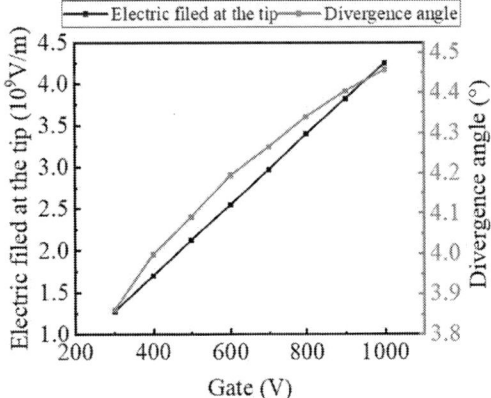

Fig. 2. Effect of gate voltage on electric field and divergence angle.

979-8-3503-7977-8/24 $31.00 © 2024 IEEE

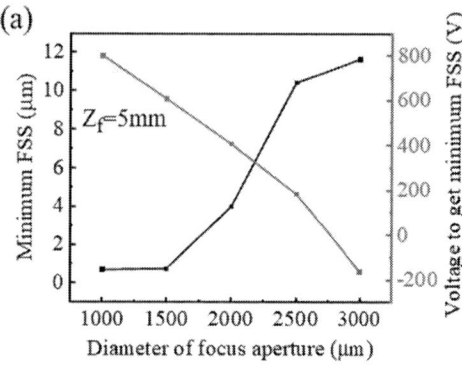

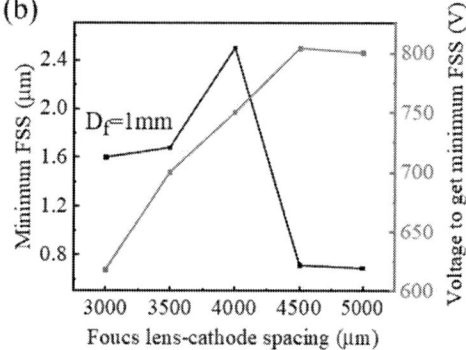

Fig. 3. (a) Effect of diameter of focus aperture on minimum FSS with optimized focus voltage; (b) Effect of focus lens-cathode spacing on minimum FSS with optimized focus voltage.

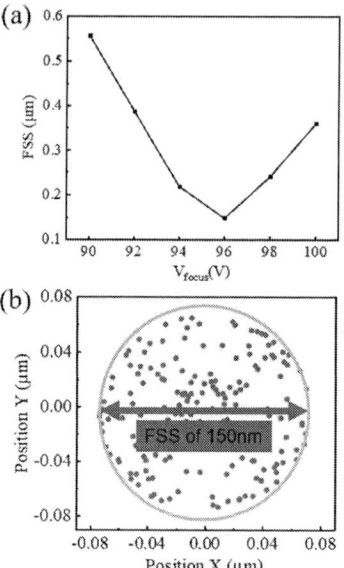

Fig. 4. (a) Relationship between FSS and the focus voltage; (b) Minimum FSS on the anode target.

IV. CONCLUSION

Herewith, a nanoneedle cold cathode X-ray source with electrostatic lens was designed. The electron-gun structure and electrical excitation parameters were simulated and a minimum FSS of 150 nm was obtained under Z_f, D_{f1}, D_{f2} and focusing voltage at 5 mm, 1.54 mm, 2.34 mm, 96 V respectively. Those results show the nanoneedle cold cathode X-ray source has the potential to offer more compact and higher resolution X-ray imaging.

ACKNOWLEDGMENT

This work was supported by the National Key Basic Research Program of China (Grant no. 2023YFF0719004, 2019YFA0210201, 2019YFA0210200), the National Natural Science Foundation of China (Grant no. 62301619), the Key Field Research Program of Guangdong Province (Grant no. 2022B0303030001), the Special Topic on Basic and Applied Basic Research of Guangzhou (Grant no. SL2023A04J01793), the Fundamental Research Funds for the Central Universities, Sun Yat-sen University (Grant no. 23ptpy04) and Start-up Funds for the Central Universities, Sun Yat-sen University.

The electric field and divergence angle both increased with the increasing gate voltage. According to our previous results, the tip electric field was set between 2×10^9 V/m to 2.5×10^9 V/m to keep a stable emission [5-6], so gate voltage was set as 500 V to ensure sufficient electric field strength to generate currents and also reduce the divergence of the E-beam.

Secondly, the optimized diameter of focus electrode aperture (D_f) and focus lens-cathode spacing (Z_f) were studied and shown in Fig. 3. In the simulation, the D_f from 1 mm to 3 mm, and Z_f varied from 3 mm to 5 mm. The focus electrode voltage was regulated to generate smallest beam on the anode, and the corresponding FSS was recorded.

Apparently, When Z_f set as 5 mm and D_f increased, the focus voltage decreased to get minimum FSS in Fig. 3(a). When D_f set as 1 mm and Z_f increased, the V_f increased firstly and then decreased as shown in Fig. 3(b). Finally, the minimum FSS of 0.7 µm was obtained at the optimized parameters of Z_f, D_f and focusing voltage at 1 mm, 5 mm, 800 V respectively.

Based on the above optimized structures, a suppressor is introduced with an aperture of 1 mm to minimize the divergence of the cathode emitted E-beam. Meanwhile, the focus electrode is designed as a conical center electrode to reduce aberration. The cone half-angle between the cone and the axis is 38.66° in which D_{f1} is 1.54 mm and D_{f2} is 2.34 mm. Finally, by adjusting the voltage of the focus electrode, a focal spot of 150 nm was obtained at the focus electrode voltage of 96 V in Fig. 4, which demonstrates the nanoscale resolution for X-ray tube.

REFERENCES

[1] L. De Chiffre, S. Carmignato, J.-P. Kruth, R. Schmitt, and A. Weckenmann, "Industrial applications of computed tomography," CIRP Annals, vol. 63, no. 2, pp. 655–677, 2014.

[2] S. Park et al., "A Fully Closed Nano-Focus X-Ray Source With Carbon Nanotube Field Emitters," IEEE Electron Device Letters, vol. 39, no. 12, pp. 1936–1939, 2018.

[3] M. Shentcis et al., "Tunable free-electron X-ray radiation from van der Waals materials," Nat. Photonics, vol. 14, no. 11, pp. 686–692, 2020.

[4] P. F. Wang, Q. L. Dong, W. P. Li, and J. B. Liu, "High beam-current density of a 10-keV nano-focus X-ray source," Nuclear Instruments and Methods in Physics Research Section A: Accelerators, Spectrometers, Detectors and Associated Equipment, vol. 940, pp. 475–478, 2019.

[5] S. Tang et al., "A stable LaB₆ nanoneedle field-emission point electron source," Nanoscale Adv., vol. 3, no. 10, pp. 2787–2792, 2021.

[6] S. Tang et al., "A stable LaB₆ nanoneedle field-emission electron source for atomic resolution imaging with a transmission electron microscope," Mater. Today, vol. 57, pp. 35-42, 2022.

A Nanoscale Vacuum Transistor Based on Vertical Silicon Dioxide Tunneling Junctions

Yidan He, Zhiwei Li, and Xianlong Wei*

Key Laboratory for the Physics and Chemistry of Nanodevices, School of Electronics, Peking University, Beijing, People's Republic of China
*Corresponding author: weixl@pku.edu.cn

Abstract—An on-chip vacuum transistor relying on electron emission from SiO_x tunneling junctions formed at the side surface of a thin SiO_2 film is proposed, where electron transport in a vacuum channel between the electron emitter and collector electrode is modulated by an underneath gate electrode. The proposed vacuum transistor achieves an ON/OFF current ratio of 10^4 and a subthreshold slope of ~5 V/dec.

Keywords—vacuum transistor, vacuum channel, SiO_x tunneling junctions

I. INTRODUCTION

Vacuum transistors have received extensive attention due to their advantages associated with both conventional vacuum devices and solid-state devices. Compared with solid-state devices, vacuum transistors have the benefits of being free of carrier scattering, immunity to extremely harsh environments [1], higher breakdown voltage [2], etc. Compared with conventional vacuum devices, vacuum transistors have the advantages of much smaller size, ease of integration, lower cost of batch fabrication, etc.

Up to now, nanoscale vacuum transistors have evolved into two major types when considering the mechanisms of electron emission. Firstly, most previously published devices are based on field emission electron sources, which require ultrahigh vacuum conditions of better than 10^{-6} Pa to exhibit long lifetime and stable electron emission. Secondly, vacuum channel transistors relying on thermionic electron emission were also reported[3], which shows obvious disadvantages of high working temperature of more than $1000\,°C$ and thus high power consumption.

Tunneling electron emission sources with typical structures of metal-insulator-metal and metal-oxide-semiconductor tunneling junctions were found to show a stable electron emission even at a rough vacuum pressure of up to 1 Pa[4], providing a possibility to obtain a stable and long lifetime electron emission in an on-chip encapsulated vacuum cavity without active pumping.

Here, we propose a vacuum channel transistor based on electron emission from SiO_x tunneling junctions. An ON/OFF current ratio up to 10^4 and a subthreshold slope (SS) of ~5 V/dec are obtained by experiment.

II. STRUCTURE OF THE DEVICE

Fig. 1(a) illustrates the schematic structure of the device and its measurement setup. A square island with an edge length of 20 μm and the stacked structure of metal (Au/Ti)-SiO_2-doped silicon is employed as the electron emitter. The emitter is surrounded by gate and collector electrodes in three directions. The device is fabricated on a heavily doped Si (N^{++}) substrate with a 300 nm oxide layer. A scanning electron microscopy (SEM) image of an as-fabricated device

is shown in Fig. 1(b). The device shows a distance of 7μm between the electron emitter and the gate electrode and a thickness of 110 nm for the collector electrode.

III. WORKING PRINCIPLES AND RESULTS

To activate an electron emitter, a bias voltage (V_{bias}) is applied between the top metal electrode and the N^{++}-doped

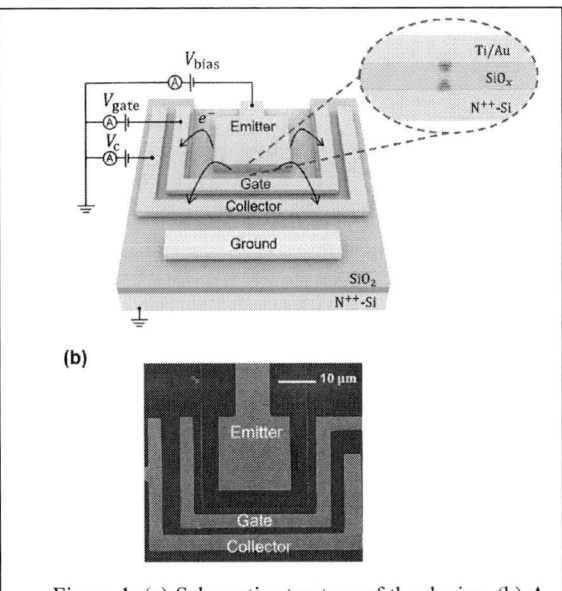

Figure 1. (a) Schematic structure of the device. (b) A scanning electron microscopy (SEM) image of an experimental device.

silicon layer to make the sandwiched SiO_2 layer electroformed. Electrical conduction between the top metal electrode and the bottom silicon layer is attributed to the formation of conducting filaments at the side surface of SiO_2 layer after it is electroformed[5], and reversible connection and rupture of conducting filaments are thought to be responsible for the observed resistive switching behavior[6]. It has been demonstrated in previous papers that electron emission from electroformed SiO_2 can take place when it is in a high-resistance state with conducting filaments ruptured[7]. Electron emission performance of an as-activated electron emitter is shown in Fig. 2(a). An emission current from the emitter becomes measurable at V_{bias}~8 V, where a sudden conduction drop with a resistive switching from a low-resistance state to a high-resistance state is observed in $I_{conduction}$-V_{bias} curve. Then the emission current rises with V_{bias} and hits about 100 nA around V_{bias}=18 V.

979-8-3503-7977-8/24 $31.00 © 2024 IEEE

Fig. 2(b) shows transfer characteristic curves ($I_{collecting}$-V_{gate} curves) of the device under different V_{bias}, when $V_{collecting}$ is set at 30 V. It is clear that $I_{collecting}$ is at a noise level with no gate tunability when V_{bias} is 8 V due to negligible electron emission from the emitter as shown in Fig. 2(a). With the increase of V_{bias}, $I_{collecting}$-V_{gate} curves exhibit obvious gate tunability and the ON/OFF current ratio increases with V_{bias} increase. It can be seen that $I_{collecting}$ increases from ~3 pA to ~100 nA when V_{gate} sweeps from -30 V to 30 V, showing an ON/OFF current ratio of 10^4 and SS of ~5 V/dec, respectively.

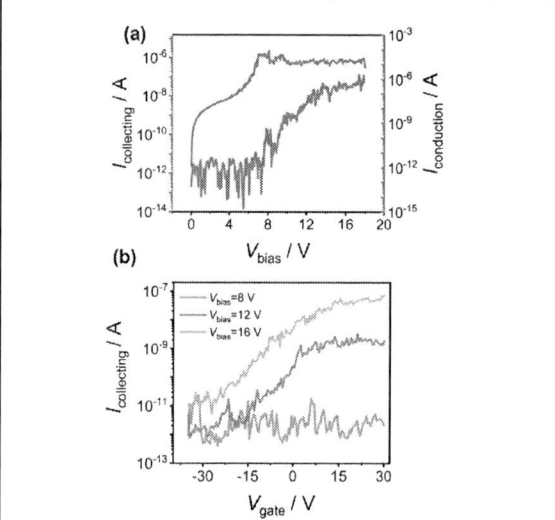

Figure 2. (a) Conduction current ($I_{conduction}$, green curve) and emission current ($I_{collecting}$, red curve) of an activated electron emitter. (b) Transfer characteristic curves ($I_{collecting}$-V_{gate} curves) at different V_{bias}.

IV. SUMMARY

To conclude, an on-chip vacuum transistor is proposed based on electron emission from SiO_x tunneling junctions formed at the side surface of a thin SiO_2 film. Based on microfabrication technologies, the devices with a surrounding structure are demonstrated in experiments, and an ON/OFF current ratio of 10^4 and SS of ~5 V/dec are obtained. By optimizing device structures, device performances are expected to be significantly improved in experiments. Good tolerance to the low vacuum of SiO_x tunneling electron emitter makes our vacuum transistor able to work in an on-chip encapsulated vacuum cavity without active pumping.

REFERENCES

[1] H. K. Kim, "Vacuum transistors for space travel", *Nature Electronics*, vol. 2, pp. 374-375, September 2019.

[2] R. S. Symons, "Tubes: still vital after all these years", *IEEE Spectrum*, vol. 35, pp. 52-63, April 1998..

[3] G. Wu, X. Wei, Z. Zhang, et al., "A Graphene-Based Vacuum Transistor with a High ON/OFF Current Ratio", *Advanced Functional Materials*, vol. 25, pp. 5972-5978, August 2015.

[4] K. Yokoo, H. Tanaka, S. Sato, et al., "Emission characteristics of metal–oxide–semiconductor electron tunneling cathode", *Journal of Vacuum Science & Technology B: Microelectronics and Nanometer Structures Processing, Measurement, and Phenomena*, vol. 11, pp. 429-432, March 1993.

[5] J. Yao, Z. Sun, L. Zhong, et al., "Resistive Switches and Memories from Silicon Oxide", *Nano Letters*, vol. 10, pp. 4105-4110, October 2010.

[6] Y. Yang, P. Gao, S. Gaba, et al., "Observation of conducting filament growth in nanoscale resistive memories", *Nature Communications*, vol. 3, pp. 732, March 2012.

[7] G. Wu, Z. Li, Z. Tang, et al., "Silicon Oxide Electron-Emitting Nanodiodes", *Advanced Electronic Materials*, vol. 4, pp. 1800136, June 2018.

A Photo-electric Co-excited Self-focusing Nano-cold-cathode Electron Gun Device

Pengbin Xu, Yan Shen*, Dong Han, Xiaoyu Qin, Yu Zhang, Ningsheng Xu and Shaozhi Deng*

State Key Laboratory of Optoelectronic Materials and Technologies, Guangdong Provincial Key Laboratory of Display Materials and Technologies, School of Electronics and Information Technology, Sun Yat-sen University, Guangzhou 510275, China

*Corresponding authors: shenyan7@mail.sysu.edu.cn; stsdsz@mail.sysu.edu.cn

Abstract—**A photo-electric co-excited self-focusing electron gun device based on carbon nanotubes cold cathode has been meticulously designed and experimentally implemented, that utilizes a synergistically mixed field comprising an ultrafast laser field and an electrostatic field, to generate electron beam with picosecond pulse-width. The architecture of the electron gun combines the cathode and focusing electrode into an integrated design, thereby facilitating superior focusing performance.**

Keywords—*field emission, carbon nanotubes, nano cold cathode, electron gun*

I. INTRODUCTION

Nano-cold-cathode possesses advantages such as fast response, room-temperature operation, and high emission current density and brightness [1]. In recent years, various vacuum microwave devices based on nano-cold-cathodes have shown significant application potential [2,3]. Specifically for the terahertz (THz) radiation sources are concerned, people have demonstrated that ultrafast laser-induced optical field emission (OFE) can directly generate THz electromagnetic waves from a carbon nanotubes (CNTs) nano-cold-cathode [4]. In this process, the emitted electron beam is highly controlled by the laser excitation, enabling the generation of electron pulses with similar spatio-temporal characteristics. This makes the relevant electron gun core devices appear particularly important. This study presents a photo-electric co-excited self-focusing nano-cold-cathode electron gun for THz radiation sources and investigates the impact of device structure on the focusing and transmitting properties of the device, to achieve better output performance and efficiency.

II. DEVICE STRUCTURE DESIGN

The structure diagram of the photo-electric co-excited nano-cold-cathode electron gun is illustrated in Fig. 1a, which consists of cathode, focusing, grid, and collector electrodes. The cathode is adopted by vertically aligned CNTs bunch with individual diameter of 2 mm and height of 500 μm, prepared *via* chemical vapor deposition method. The focusing electrode, depicted in Fig. 1b, integrates with the cathode and features a wedge-shaped inclined surface with a tilt angle (θ) and a depth (H) to generate lateral focusing electric field at the cathode edge. The grid electrode is equipped with a square aperture molybdenum grid (Fig. 1c) to induce the cathode into a critical field emission state by applying voltages. Within the encapsulated electron gun, a picosecond laser is focused onto the cathode surface through a convex lens, generating a polarized optical field component perpendicular to the cathode surface, inducing the ultrafast field-induced electron emission and achieving a picosecond pulsed electron beam synchronized with the laser pulses. Subsequently, under the influence of the electric field

between the cathode and the collector, the pulsed electron beam drifts through the grid and emits into free space. Fig. 1d displays a photograph of the assembled electron gun device.

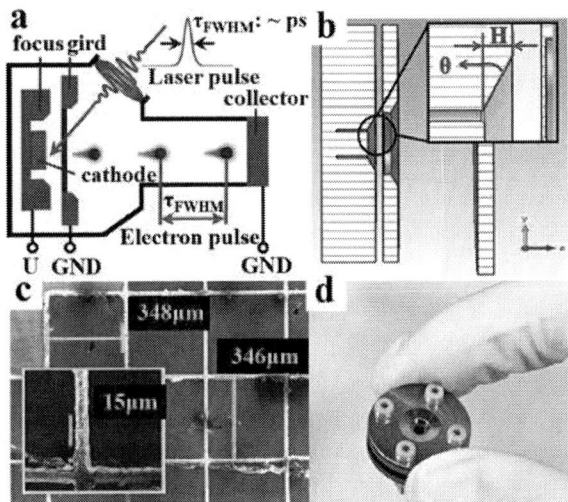

Fig. 1. (a) Structure diagram of the electron gun; (b) Structure of the wedge-shaped self-focusing electrode; (c) SEM images of the molybdenum grid; (d) Photograph of the assembled electron gun device.

According to our design, this electron gun features an integrated cathode and focusing electrode, enabling effective electron beam self-focusing under the application of the same driving voltage and laser excitation. Moreover, the grid electrode is grounded, requiring only a single negative driving voltage source, reducing the device complexity and eliminating the need for magnetic focusing devices, to facilitate the device miniaturization and integration. The molybdenum grid boasts over 92% aperture ratio, providing excellent coupling between the incident laser and the cathode, that contributes to a final transmittance exceeding 91.5% of the total emission current and enhances the output performance and energy conversion efficiency of the picosecond pulsed nano-cold-cathode electron gun.

III. RESULTS AND DISCUSSION

Before actual device measurement, the electron gun device was modeled and optimized using the CST software, with the voltages of the cathode (with focusing electrode), grid electrode, and collector electrode being set to -5000 V, 0 V, and 0 V, respectively. In the absence of a focusing magnetic field, to minimize the divergence of the electron beam propagating to a distant location (such as ~20 mm) under the influence of the focusing electrode, the tilt angle (θ)

979-8-3503-7977-8/24 $31.00 © 2024 IEEE

of the proposed wedge-shaped focusing electrode was carefully designed. The root mean square (RMS) of distance between the electron distribution and the device axis at this position was taken as a reference, resulting in the curve shown in Fig. 2a. Optimal focusing was thereafter achieved when the tilt angle (θ) of the focusing electrode was set to 25°. Fig. 2b depicts the electron beam trajectory and current density spatial distribution under this structural parameter, indicating minimal divergence of the electron beam after drifting 20 mm along the axial direction, with over 95% of the emission current concentrated within a circular area equivalent to the cathode radius (~1 mm).

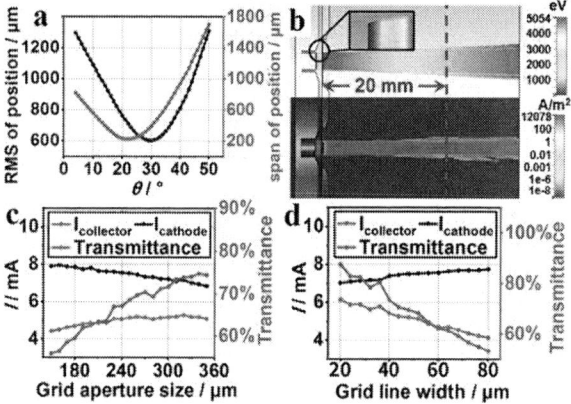

Fig.2. (a) Influence of the tilt angle (θ) on the focusing performance; (b) Trajectory and current density distribution of the electron beam when $\theta = 25°$; (c, d) Influences of the aperture size and line width of the molybdenum grid on the performance of the electron gun.

Fig. 2c and 2d further illustrate the influence of the structure parameters of grid electrode on the emission performance of the electron gun. The square aperture size and the line width on the grid affect the distribution of the localized electric field on the cathode surface and the drift of the generated electron beam, thereby impacting the emission current and electron transmittance. Reducing the width of the grid lines and increasing the size of the apertures gradually decrease the cathode current while significantly increase the transmittance, resulting in an increase in the collector current (final beam current). To maximize the beam current and transmittance as much as possible, a grid with a line width of 15 μm and aperture size of 350 μm was selected under permissible process conditions, ultimately achieving a grid transmittance of up to 91.5%.

The overall device performance of the fabricated electron gun has been comprehensively measured and demonstrated. Specifically, the electron gun device was measured in a vacuum environment (~10^{-5} Pa). The grid electrode and collector electrode were grounded, while the CNTs cold cathode was subjected to a DC bias and co-excited by a picosecond ultrafast laser with continuous wavelengths ranging from 430 nm to 2400 nm, pulse width of 100 ps, repetition frequency of 1 MHz, and oblique incidence angle of 45°.

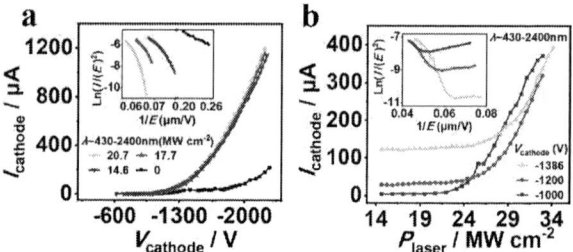

Fig. 3. (a) $I_{cathode}$ *versus* $V_{cathode}$ curves under different incident laser power densities (P_{laser}). Inset: the corresponding F-N plots in which E consists of the fixed E_{laser} and varying $E_{cathode}$; (b) $I_{cathode}$ *versus* P_{laser} curves under different voltages of cathode ($V_{cathode}$). Inset: the corresponding F-N plots in which E consists of the fixed $E_{cathode}$ and varying E_{laser}.

Fig. 3a and its inset show the $I_{cathode}$ *versus* $V_{cathode}$ curves under different incident P_{laser} values, as well as the corresponding field emission Fowler-Nordheim (F-N) plots. One can observe that the laser excitation effectively enhances the electron emission performance of the CNTs cold cathode. With a peak laser power density of 20.7 MW/cm², the turn-on voltage of the electron gun was reduced from -1211 V to -1014 V, and the maximum $I_{cathode}$ increased by up to 891% under the same DC bias conditions. Fig. 3b and its inset illustrate the variation of $I_{cathode}$ with P_{laser} and the corresponding F-N plots at different cathode voltages ($V_{cathode}$). One can see when $V_{cathode}$ is held constant, $I_{cathode}$ increases exponentially with P_{laser}, although the rate of increase slows as $|V_{cathode}|$ decreases. In addition, when the incident P_{laser} is strong, $I_{cathode}$ of the electron gun device follows the classical Fowler-Nordheim law, which indicates the existence of a photo-assisted field emission (PFE) mechanism; however, when P_{laser} is weak, $I_{cathode}$ exhibits an inverse trend, suggesting that the emission mechanism at lower P_{laser} range may involve a multiphoton emission (MPP) mechanism (see the insets in Fig. 3). The detailed device pulse output characteristics and electron transmittance measurements will be presented at the IVNC 2024.

ACKNOWLEDGMENT

This work was supported by National Key Basic Research Program of China (2019YFA0210201, 2019YFA0210200), National Natural Science Foundation of China (52072416, 51702372), and the Science and Technology Planning Project of Guangdong Province (2023B1212060025).

REFERENCES

[1] N. S. Xu and S. E. Huq, "Novel cold cathode materials and applications". Materials Science and Engineering: R: Reports, vol. 48, pp. 47-189, Jan. 2005.

[2] X. Yuan, et al. "A fully-sealed carbon-nanotube cold-cathode terahertz gyrotron". Scientific Reports, vol. 6, pp.32936, Sept. 2016.

[3] Y. Xing, et al. "A Cold-Cathode Microwave and Terahertz Radiation Source: Experimental Realization @10's GHz and Computational Design @THz," in IEEE Electron Device Letters, vol. 40, no. 9, pp. 1534-1537, Sept. 2019.

[4] Y. Shen, Z. Song, N. Xu, H. Chen and S. Deng, "Generation of Submillimeter Terahertz Radiation by Ultrafast Optical-Field-Emission," in IEEE Electron Device Letters, vol. 44, no. 7, pp. 1204-1207, July 2023.

A Self-consistent Combined Laplace-Schroedinger Solution of Field Emission from the Hemisphere on a Post Model

G.C. Kokkorakis, D.Karaoulanis, J.A. Roumeliotis, J.P. Xanthakis[*]

ECE department, National Technical University of Athens, Athens 15700, Greece
[*]Corresponding Author: jxanthak@central.ntua.gr

Abstract—It is customary to calculate the field emission (FE) from nanometric curvature surfaces by relying only on the solution of the Laplace equation for the form of the barrier while the transmission coefficient and the supply function are obtained from 1-dimensional expressions valid for planar surfaces. In this work we examine the validity of such a hypothesis for nanometric curvature surfaces by performing a self-consistent Schroedinger-Laplace calculation of field emission from the most commonly used model of FE, that of the hemisphere on a cylindrical post model (HCP).

Keywords—field emission, tunneling, supply function, hemisphere on a post model.

I. INTRODUCTION

The simulation of vacuum field emission (FE) devices with even nanometric curvature relies exclusively on the solution of the Laplace equation while the quantum mechanical transmission coefficient T and the supply function is obtained from one-dimensional (1D) expressions valid for planar surfaces. The implicit assumption is that the emission from a curved surface can be obtained by applying such 1D expressions along the radial lines of the emitting surface. However such an approach necessitates radial symmetry for both the external potential V and the wavefunctions Ψ, a questionable assumption for nanometric surfaces. In this work we examine the validity of such a hypothesis

II. METHOD

The hemisphere on a cylindrical post=model emitter is shown in Figure 1 together with the relevant dimensions and the external potential $V(r,\theta)$ where (r,θ) are the spherical distance and azimuthal angle respectively of a spherical coordinate system centred on the centre of the hemisphere. We solve the FE problem in two stages: first we set V=infinite everywhere outside the emitter and we calculate the eigenfunctions inside it and hence the incident current density onto the hemispherical surface; then we relax the condition on V and apply an external V such that FE can occur from that surface and we then calculate the transmission coefficient and the emitted current density. In each part of the model- (hemisphere, cylinder)-a different form of the wavefunction Ψ will hold. By separation of variables of the Schroedinger equation $H\Psi=E\,\Psi$ we obtain:

A) For the cylindrical post in cylindrical coordinates:

$$\Psi 1(E)=\sum_{n=1}^{\infty} A_n J_0(k_n*rt)\sin(k_z^n(z+h)) \quad (1)$$

where rt=is the distance from the z axis, J_0 is the Bessel function of order zero, $k_n=\rho_n/R$ with ρ_n the roots of J_0 and $k_z^n=\sqrt{E-k_n^{\wedge}2}$. Note that we have taken Ψ to be rotationally invariant so as to resemble more the plane waves commonly used in FE.

A) For the hemisphere on the post in spherical coordinates

$$\Psi 2(E)=\sum_{n,p}^{\infty} B_{n,p}j_n(k_{n,p}*r)*P_n(\cos\vartheta) \quad (2)$$

where j_n is the nth order spherical Bessel Function, P_n is the corresponding Legendre polynomial, $k_{n,p}=\rho_{n,p}/R$ with $\rho_{n,p}$ the roots of the spherical Bessel function. The A_n and B_n are calculated by demanding that the wave function and its derivative at the common interface (i.e. the z=0 plane) are continuous. This leads to a set of linear homogeneous equations for the A, B coefficients which is solved for each E. One should note that the Ψ appearing in (1) and (2) are standing waves and to calculate the transmitted current one should extract the incident part of each standing wave Ψ. For each such wave we can write $\Psi=\Psi^{inc}+\Psi^{refl}$.

A few remarks are necessary at this point. In both equations (1) and (2) the wavefunctions are not states of definite parallel and normal energies but linear combinations of them. It is only the total energy which characterizes these states. This is most evident in equation (2). Hence the convenient Cartesian separation into normal and parallel energies does not hold here. The above peculiarities are a direct result of matching wavefunctions with different symmetries in the various parts of the emitter. However this is what actually happens at the apex of a nanometric emitter. It is not artificial. Furthermore, only the normally incident part of each wave will be transmitted. Its corresponding current density is

$$j_r^{inc}(E,\theta)=\frac{2e}{h}*Im(\Psi^{inc*}\nabla_r\Psi^{inc}) \quad (3)$$

We now allow the HCP to emit. The potential barrier in this case outside the hemispherical surface of the emitter can be described by the relation

$$V(r,\theta)=E_{vac}-V_{im}^{sph}(r)-V_{els}(r,\theta) \quad (4)$$

where E_{vac} is the vacuum energy, $V_{im}^{sph}(r)$ is the spherical image potential and $V_{els}(r,\theta)$ is the electrostatic potential, i.e. the solution to the Laplace equation. The electrostatic potential is not separable and hence one cannot obtain an 1D Schroedinger equation along the normal direction of emission. This- together with the above remarks on the wave functions makes it impossible to use the traditional Cartesian expression for the tunneling current density. However, given that expressions for the transmission coefficient $T(E,\theta)$ suitable for our case are given in the work of Das and Mahanty [1] as corrected by A. Patterson [2], we can write for the normally transmitted current density

$$j_r^{trans}(\theta) = \sum_E j_r^{inc}(E,\theta) * T(E,\theta) \qquad (5)$$

III. RESULTS

Our first set of results is shown in Figure 2 where we portray the radially incident current density $j_r^{inc}(E=EF=10eV)$ at various distances r from the centre of an HCP with R=1nm and h=20nm as a function of the angle θ. If $j_r^{inc}(E)$ were spherically symmetric the graph would have been a straight horizontal line. Instead the graphs have a peak at $\theta=0$. This contrasts sharply with the uniform incidence of FN theory. Fig. 3 shows the radially incident density $j_r^{inc}(\theta,E=10eV)$ exactly at the hemispherical surface for 3 HCP with a constant h=20R and R=2.5, 3.75, 5nm. We observe that for all HCP with R>1nm additional features are present compared to the simpler case of R=1nm.

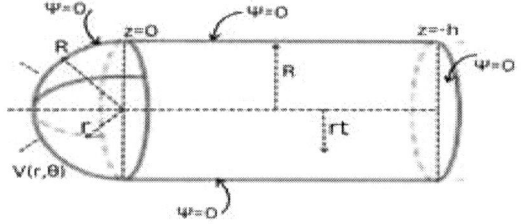

Fig. 1 An HCP model of FE with characteristic lengths shown. A V(r, θ) is applied onto the HCP from the left once the $\Psi=0$ condition is relaxed.

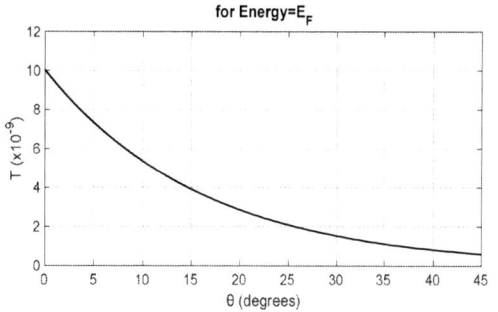

Fig. 4 Transmission coefficient at the Fermi level as a function of angle θ when an electric field=5V/nm is applied at the apex.

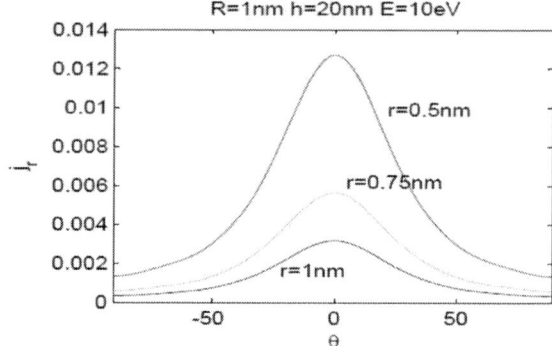

Fig. 2 Radially incident current density j_r^{inc} (E=10eV) of an HCP with R=1nm at various distances from the centre of the hemisphere.

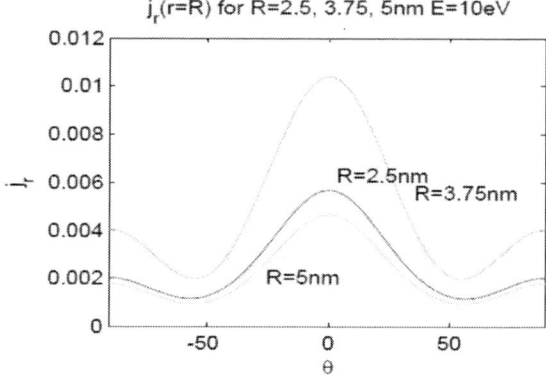

Fig. 3 Normally incident current density j_r^{inc} (E=10eV) at the emitting surface r=R of three HCP with R=2.5,3.75,5.0 nm .

We now turn our attention to the $T(E,\theta)$. Its variation with angle θ is shown in Figure 4. It portrays the expected behavior seen in previous publications. The rapid decrease with θ means that the additional features seen in Figure 3 at higher angles will be wiped out. However the features near $\theta=0$ will survive. The variation with E (below E_F) is more interesting but space does not allow for a discussion here.

IV. CONCLUSIONS

A Schrodinger-Laplace calculation for FE has shown that the usual 1D formulae valid for large R are not applicable for nanometric curvature surfaces

REFERENCES

[1] B. Das, and J. Mahanty, "Spatial distribution of tunnel current and application to scanning-tunneling microscopy: A semiclassical treatment" Phys. Rev. B, vol. 36, pp. 898-903, 1987.

[2] A. A. Patterson, and A.I. Akinwande, "Semiclassical theory of cold field electron emission from nanoscale emitter tips" J. Vac. Sci. Tecnol. B, vol. 38, 023206 2020.

[3] J. P. Xanthakis, and G. C. Kokkorakis, "Theoretical calculation of spatial variation of the transmission coefficient of closed carbon nanotubes" Surf. Inteface Anal., vol. 36, pp. 391-394, 2004..

Advancing Debris-free EUV Light Source Technology with Reduced Visible Light Effect by Cold Cathode Electron Beam (C-Beam) Irradiation

Umesh Balaso Apugade[1], Bishwa Chandra Adhikari[1], Iksu Kim[1], Kyu Chang Park[1*]

[1] *Kyung Hee University, Information display, Seoul, 02447, Republic of Korea,*
Email: kyupark@khu.ac.kr (Prof. Kyu Chang Park)

Abstract— **The purpose of this study to develop debris free extreme Ultraviolet (EUV) light source technology for producing next-generation semiconductor devices, offering enhanced resolution and precision. Key feature of this study will be to create efficient EUV radiation with reduced visible light effects and debris free source. This study introduces a novel solution using cold cathode electron beam (C-beam) irradiation to tackle this issue as our approach includes the utilization of mirrors for optimal EUV light reflection, and filters for selective attenuation of visible and UV-A/B/C wavelength and use debris block for debris free EUV source. The authors believe our sophisticated portable EUV source will pave next generation EUV inspection technology**

Keywords— *Extreme ultraviolet (EUV), Cold Cathode electron beam (C-beam), Vertically Aligned Carbon Nano Tube (VACNT), Plasma enhanced chemical vapor deposition (PECVD), Photodiode (PD) (SXUV100, OPTODIODE).*

I. INTRODUCTION

Extreme Ultraviolet (EUV) lithography has emerged as a pivotal technology for fabricating next-generation semiconductor devices, offering finer resolutions and higher precision[1]. It facilitates the production of semiconductor devices with higher precision and finer resolutions compared to traditional lithographic techniques. However, the efficient generation of EUV radiation while minimizing visible light effects and debris remains a significant challenge[2]. Here, we utilized CNT-based cold cathode electron beams, or C-beams, to EUV illumination by using them as FE electron sources[3]. Our technique is based on the EUV light generated when electrons directly bombard Sn (anode)[4]. The vertically aligned carbon nanotubes perform uniform field emission with the high stability of the electron beam. CNTs have attracted a lot of attention in the field of vacuum nanoelectronics applications, particularly in the area of cold cathode-based vacuum devices. The attractive material with superior structural, chemical, and electrical properties makes them ideal for cold cathode electrode[5][6] . Thus, the generation of the EUV light depends on the electron emission of the field emission mechanism of the CNT emitters. C-beam with a triode structure for producing field emission that created UV and EUV lighting[7]. In this report we have discussed about our proposed portable EUV light source integrates mirrors and filters focusing on achieving debris-free EUV light sources with minimal visible light interference. By employing C-Beam irradiation we achieve precise control over EUV radiation, ensuring enhanced efficiency.

II. EXPERIMENT DETAILS

Fabrication of vertically aligned CNT emitters

In order to establish portable EUV light source and to investigate debris free operation and reduced visible light effect I used vertically aligned carbon nanotube (CNT). Vertically aligned carbon nanotubes (VACNTs) are grown on a patterned silicon wafer using a triode direct current plasma-enhanced chemical vapor deposition (PECVD) technique[4][8]. Scanning electron microscopy (SEM) is used to view the morphological structure of CNTs, which have height lengths of 40 μm, 15 μm, and dot diameters of 3 μm, respectively. An SEM image of the 88 arrays of vertically aligned CNT emitters is displayed (Fig. 1).

III. METHODOLOGY

A. C-Beam based EUV Irradiation Technique

The C-beam technique involves the use of a cold cathode electron beam to irradiate materials, using Tin (Sn) as a anode to generate EUV light (fig. 2-a). The C-beam module contains the cathode and gate electrode, which is insulated by the ceramic body. The CNT emitters are located on the cathode, and gate is used to control and switch the electrons emitted from the CNT emitters. The cold cathode setup ensures minimal thermal effects, reducing the likelihood of debris formation. The electron beam parameters are carefully controlled to optimize EUV radiation output.

B. Integration of Filters

To address the issue of debris as well as visible light contamination and enhance the quality of EUV light, the EUV light source incorporates mirrors and filters specifically designed in between EUV source to photodetector (measurement scheme) to reflect EUV light while attenuating visible and UV-A/B/C wavelengths. Zirconium and molybdenum filters are used; these are the candidate materials in the EUV range due to their absorption characteristics. The filters are composed of materials that selectively absorb unwanted wavelengths, ensuring that only EUV light reaches the target.

979-8-3503-7977-8/24 $31.00 © 2024 IEEE

C. Experimental Setup and Verification

The experimental setup includes white condition which is without any block, dark condition with block . Particularly block is used to cover light source. The EUV light generated is then characterized using IV measurements and photo-current measurements with a photodiode (PD) (SXUV100, OPTODIODE) at different condition accordingly. These measurements provide data on the transmittance, efficiency, and purity of the generated EUV light.

IV. RESULTS AND DISCUSSION

Efficiency of EUV Generation

The experimental results indicate a high efficiency of EUV generation using the C-beam irradiation technique. The IV characteristics show a transmittance of approximately 80 %, When the gate voltage applied, the anode current is measured to be 3.86 mA which confirms the emission (fig. 2-b). The transmittance of the electron beam is defined as the ratio of the anode current to the cathode current, which is found 80 %. Eexperiment suggesting effective electron beam interaction with the Tin (Sn) targets. The photo-current measurements further confirm the generation of EUV light with minimal visible light contamination which is after adaption of block, we achieved better EUV efficiency, reducing visible light effects significantly and gets EUV efficiency I_{ph} about 5.27 E-07 A, which is significantly lower from white condition 1.02 E-06 A. Total 52% reduction measured. The results show high level of selective attenuation of non-EUV wavelengths, confirming reducing losses and improving the intensity of the EUV radiation. The integration of block, filters and by controlling the thermal effects at anode that can ensures that only the desired EUV wavelengths are utilized so that visible light effect get reduced and further enhancing the cleanliness of the operation.

V. CONCLUSION

The This study presents a novel approach to generating debris-free EUV light with reduced visible light effects using a C-beam irradiation technique. The integration of VACNTs, block and filters results in a highly efficient and clean EUV light source suitable for next-generation semiconductor fabrication. We achieved total 52% reduction of visible light effect. The experimental results demonstrate the effectiveness of the proposed system in achieving high transmittance, minimal debris, and high-quality EUV light. The quality of EUV light evaluated with multi-layer-mirror and EUV filters. More detail of the EUV lighting properties with various filtering and mirror effect would be discussed.

ACKNOWLEDGMENT

This work was supported by the Technology Innovation Program (No. 20013595, Extreme ultraviolet light source using nano electron beam) funded by the Ministry of Trade, Industry & Energy (MOTIE, Korea).

REFERENCES

[1] N. Fu, Y. Liu, X. Ma, and Z. Chen, "EUV lithography: State-of-the-art review," *J. Microelectron. Manuf*, vol. 2, no. 2, pp. 1–6, 2019.

[2] A. Egbert *et al.*, "Compact electron-based extreme ultraviolet source at 13.5 nm," *J. Micro/Nanolithography, MEMS MOEMS*, vol. 2, no. 2, pp. 136–139, 2003.

[3] J. S. Kang, J. H. Hong, M. T. Chung, and K. C. Park, "Highly stable carbon nanotube cathode for electron beam application," *J. Vac. Sci. \& Technol. B*, vol. 34, no. 2, 2016.

[4] S. T. Yoo and K. C. Park, "Extreme ultraviolet lighting using carbon nanotube-based cold cathode electron beam," *Nanomaterials*, vol. 12, no. 23, p. 4134, 2022.

[5] K. C. Park and J. Lim, "Direct grown vertically full aligned carbon nanotube electron emitters for X-ray and UV devices," in *Nanostructured Carbon Electron Emitters and Their Applications*, Jenny Stanford Publishing, 2022, pp. 245–267.

[6] K. C. Park and S. T. Yoo, "Flat panel deep Ultraviolet (UV) light sources with carbon nanotube cold cathode," in *25th International Display Workshops, IDW 2018*, 2018, pp. 1324–1326.

[7] Y. Y. Yu and K. C. Park, "Gate offset and emitter design effects of triode cold cathode electron beams on focal spot sizes for x-ray imaging techniques," *J. Vac. Sci. \& Technol. B*, vol. 40, no. 2, 2022.

[8] S. T. Yoo, J. S. Kang, and K. C. Park, "High performance carbon nanotube cold cathode for x-ray and UV devices," in *2018 31st International Vacuum Nanoelectronics Conference (IVNC)*, 2018, pp. 1–2.

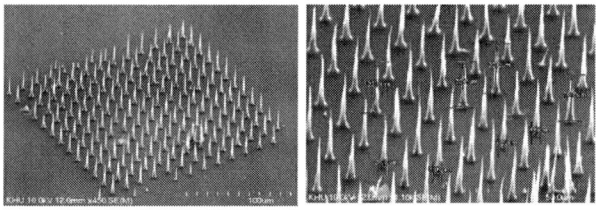

Fig. 1. SEM images of vertically aligned carbon nanotube (VACNT), with dot diameter 2.5-3.5 μm, height 40 - 45 μm

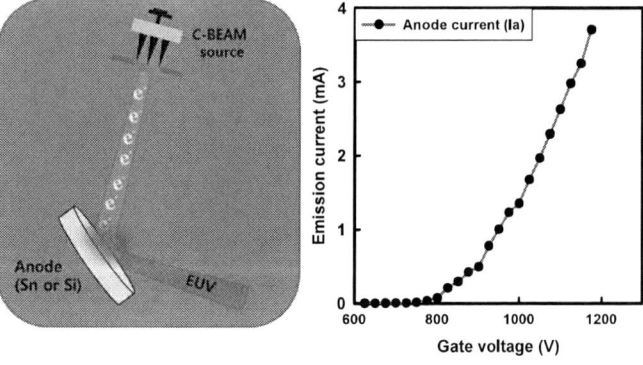

Fig. 2. (a) Beam irradiation technique and the EUV irradiation layout of EUV beamline and (b) IV characteristics of vertically aligned CNT emitters in triode structure.

Alternative Definition of the Apex Field Enhancement Factor for a Regular Array of Electrostatically Interacting Post Emitters

Thiago A. de Assis[1,2,*], Richard G. Forbes[3] and Fernando F. Dall'Agnol[4]

[1]*Institute of Physics, Federal University of Bahia, Salvador, Bahia, Brazil 40170-115*
[2]*Institute of Physics, Fluminense Federal University, Niterói, Rio de Janeiro, Brazil 24210-340*
[3]*School of Mathematics and Physics Faculty of Engineering and Physical Sciences*
University of Surrey Guildford, Surrey, GU2 7XH, UK
[4]*Department of Exact Sciences and Education, Federal University of Santa Catarina,*
Blumenau, Santa Catarina, Brazil 89036-004
Corresponding author: thiagoaa@ufba.br

Abstract—**In cold field electron emission (FE), the so-called "applied electrostatic field" is an important parameter used to model an ideal FE device. Normally it is used to define the local field enhancement factor (FEF), arguably the most analyzed parameter used to assess the performance of an emitter. However, a consistent definition of the applied field is not straightforward when emitters are close to the counter-electrode, since the local field distribution is then not uniform in the FE device, due to electrostatic interactions between emitters and with the counter-electrode. In this work, we report consistent new definitions of the applied electrostatic field (and of the related FEF), suitable for the "close-proximity" situation. These new definitions are based on an "effective distance of field penetration" measured from the counter-electrode. Advantages and disadvantages of these new definitions are pointed out.**

I. INTRODUCTION

Interesting progress has been made in exploring the electrostatics of conductive post-like emitters, including counter-electrode proximity effects [1,2]. In the last two decades, technological interest in large-area high-current-density electron sources has stimulated extensive theoretical electrostatics (ES) work on numerous emitting materials and for many configurations of large-area field electron emitters (LAFEs) (e.g., [3–6]). LAFEs utilizing carbon nanotubes (CNTs) have emerged as highly promising [3]. Thus, significant focus has been on simple ES models that involve arrays of posts on a plane—often (for initial simplicity) regular arrays of identical posts. For CNTs, the "hemisphere-on-a-cylindrical-post" (HCP) shape model is convenient and much used.

Typically, posts are modeled as standing on one of a pair of well-separated parallel planar plates that have a lateral extent significantly larger than the plate separation d_{sep}. This arrangement is known as "parallel planar plate (PPP) geometry".

A relevant parameter in a field electron emission (FE) device is the "applied field". At present, two main types are used in PPP geometry: inter-plate fields (E_{P}) and gap fields (E_{G}) [7]. Here the symbol E denotes classical ES field (negative

for FE). For a given inter-plate voltage, V_{P}, E_{P} is defined as $-V_{\text{P}}/d_{\text{sep}}$ [see Fig. 1]. In this case, the apex plate-field enhancement factor (plate-FEF) is defined by $\gamma_{\text{Pa}} \equiv E_{\text{a}}/E_{\text{P}}$, where E_{a} is the local ES field at the post apex. E_{G} is defined as the average ES field between the post apex and the adjacent counter-electrode, when a voltage V_{G} is applied between them, i.e. $E_{\text{G}} = -V_{\text{G}}/d_{\text{gap}}$, where d_{gap} is the gap length [see Fig.1]. In the electronically ideal systems [8] considered in this work, $V_{\text{G}} = V_{\text{P}}$. Gap fields are useful in system geometries that involve close proximity of counter-electrodes to emitters [1,9–20]. The apex gap-FEF is defined by $\gamma_{\text{Ga}} \equiv E_{\text{a}}/E_{\text{G}}$.

In this work, we report consistent definitions of a new type of applied ES field and of the related FEF. These are based on a distance d_{eff} (see Fig. 1) that recognises the limited field penetration into the space between the emitters in a relatively close-packed regular array. This distance can then be used to define a *penetration distance* d_{pen} that better characterises the field penetration into the array. d_{eff} can also be used to define a new "effective applied field" $E_{\text{e}} \equiv -V_{\text{P}}/d_{\text{eff}}$ and the related "effective FEF" $\gamma_{\text{ea}} \equiv E_{\text{a}}/E_{\text{e}}$.

II. SIMULATION RESULTS AND DISCUSSION

For a regular square array of identical posts, we use h to denote the total post height and c to denote the nearest-neighbour distance between emitter axes. Figure 2 shows that, as the ratio c/h decreases from a large value, the ratio d_{pen}/h decreases from unity to zero. Between these limiting cases, d_{eff} varies from d_{sep} to a value close to d_{gap}. These results show that, as the array gets increasingly close-packed, it is increasingly able to screen, from the interior of the array, the field associated with the counter-electrode.

When discussing proximity effects, there are a minimum of four independent system parameters that need to be considered, and multiple ways of combining them into dimensionless ratios. As a way in to the electrostatics, we present results for a regular square array of HCP-model emitters that each have a total-height $h = 5~\mu\text{m}$ and apex radius $r_a = 50$ nm, and

---

979-8-3503-7977-8/24 $31.00 © 2024 IEEE

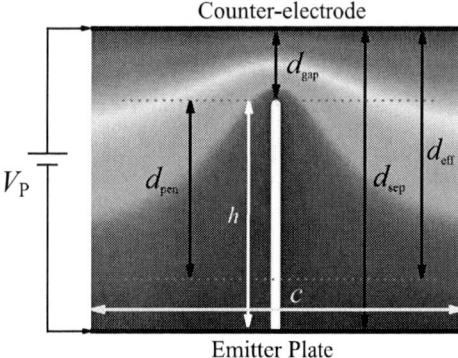

Fig. 1. Definitions of the variables in our ideal FE device. The HCP emitter is shown in a unit cell. The color map indicates the electrostatic potential distribution numerically calculated.

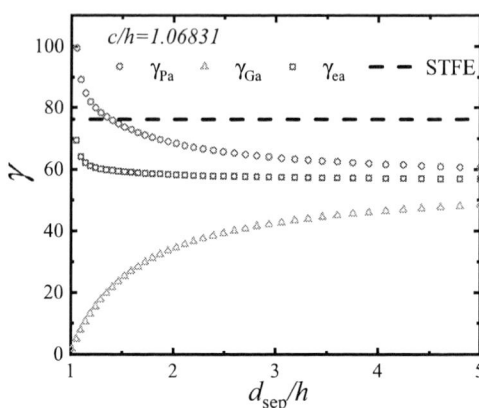

Fig. 3. γ_{Pa}, γ_{Ga}, and γ_{ea} as a function of d_{sep}/h for $c/h = 1.06831$. The horizontal dashed line indicates γ_{Pa} (= 76.28276) numerically calculated for a single-tip HCP emitter with $\sigma_{\mathrm{a}} = 100$ using formulae (13) of Ref. [21].

thus have an *apex sharpness ratio* $\sigma_{\mathrm{a}} \equiv h/r_{\mathrm{a}} = 100$. Results are shown for the array characterization ratio $c/h = 1.06831$. These results are drawn from a larger set in which c/h was increased in exponential steps. Hence the non-simple value of c/h for the results shown here.

Figure 3 shows relevant FEF-values derived using finite-element simulation methods [7]. The FEF-values are shown as functions of d_{sep}/h. Significant variations in γ_{Pa} and γ_{Ga} are seen even when $d_{\mathrm{sep}}/h > 2$, suggesting that experimental variations in the counter-electrode position could in some cases lead to significant (unwanted) variations in plate or gap FEFs.

In the limiting case of sufficiently large d_{sep}/h the FEFs γ_{Pa} and γ_{Ga} tend to a common well defined value, about 56. (For very large c/h this limiting value would be near 76.)

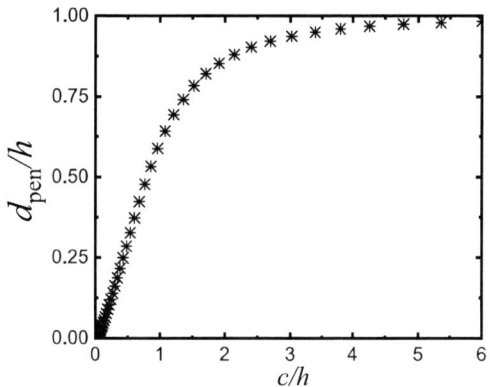

Fig. 2. d_{pen}/h versus c/h for $d_{\mathrm{gap}}/h = 1.5$.

For intermediate values of d_{sep}/h, it could be useful to have a FEF-type parameter that is nearly constant as a function of d_{sep}/h. Clearly, our new FEF γ_{ea} has this property. However, for this FEF to be useful there would have to exist tables or fitted formulae specifying how d_{eff} could be derived from the system geometry. In fact, it is arguable that (in the close-proximity situation) it might be simpler to analyse experimen-

tal current-voltage data by using local conversion lengths. This alternative will be examined elsewhere.

Acknowledgment

This work was supported by the Conselho Nacional de Desenvolvimento Científico e Tecnológico (CNPq), Grant No. 305688/2023-5 (TAdA).

References

[1] D. Biswas and R. Rudra, *J. Vac. Sci. Technol. B* **38**, 023207 (2020).

[2] C. Franey C, B. M. Al-Ameri, G. I. Acosta and M. Ghashami, *Phys. Rev. Appl.* **21**, 024004 (2024).

[3] Y. Saito *Carbon Nanotube and Related Field Emitters* (Wiley-VCH, Weinheim) (2010).

[4] N. S. Xu and S. E. Huq, *Mater. Sci. Eng. R Rep* **48**, 47 (2005).

[5] T. Zhai, L. Li, Y. Ma, M. Liao, X. Wang, X. Fang, J. Yao, Y. Bando and D. Golberg, *Chem. Soc. Rev.* **40**, 2986 (2011).

[6] N. Dwivedi, C. Dhand C, J. D. Carey, E. C. Anderson, R. Kumar, A. K. Srivastava, H. K. Malik, M. S. M. Saifullah, S. Kumar, R. Lakshminarayanan, S. Ramakrishna, C. S. Bhatia and A. Danner, *J. Mater. Chem. C* **9**, 2620 (2021).

[7] T. A. de Assis, F. F. Dall'Agnol and R. G. Forbes, *Journal of Physics: Condensed Matter* **34**, 493001 (2022).

[8] R. G. Forbes, *J. Vac. Sci. Technol. B*, **41**, 028501 (2023).

[9] H. C. Miller, *J. Appl. Phys.* **38**, 4501 (1967).

[10] V. A. Nevrovskii and V. N. Yarolavskii, *Sov. Phys. Tech. Phys.* **27**, 180 (1982).

[11] H. C. Miller, *J. Appl. Phys.* **55**, 158 (1984).

[12] J. M. Bonard, K. A. Dean, B. F. Coll and C. Klinke, *Phys. Rev. Lett.* **89**, 197602 (2002).

[13] X. Q. Wang, M. Wang, P. M. He, Y. B. Xu and Z. H. Li, *J. Appl. Phys.* **96**, 6752 (1982).

[14] R. C. Smith, J. D. Carey, R. D. Forrest and S. R. P. Silva, *J. Vac. Sci. Technol. B* **23**, 632 (2005).

[15] R. C. Smith, D. C. Cox and S. R. P. Silva, *Appl. Phys. Lett.* **87**, 103112 (2005).

[16] K. F. Hii, R. R. Vallance, S. B. Chikkamaranahalli, M. P. Mengüç and A. P. Rao, *J. Vac. Sci. Technol. B* **24**, 1081 (2006).

[17] Z. Xu, D. Bai and E. G. Wang, *Appl. Phys. Lett.* **88**, 133107 (2006).

[18] A. J. le Fèbre, L. Abelmann and J. C. Lodder, *J. Vac. Sci. Technol. B* **26**, 724 (2008).

[19] W. Zeng, G. J. Fang, N. S. Liu, L. Y. Yuan, X. X. Yang, S. S. Guo, D. J. Wang, Z. Q. Liu and X. Z. Zhao, *Diam. Relat. Mater.* **18**, 1381 (2009).

[20] S. G. Sarka, R. Kar, J. Mondal, L. Mishra, N. Maiti, R. Tripathi and D. Biswas, *Carbon Trends* **2**, 100008 (2021).

[21] F. F. Dall'Agnol, S. V. Filippov, E. O. Popov, A. G. Kolosko and T. A. de Assis, *J. Vac. Sci. Technol. B* **39**, 032801 (2021).

An In-depth Analysis of the Impact of Vacuum Conditions on the Field Emission Characteristics of a Cold Cathode C-beam.

Ketan R. Bhotkar[1], Jaydip Sawant[1], Ravindra Patil[1], Kyu Chang Park[1*]

[1] Kyung Hee University, Information display, Seoul, 02447, Republic of Korea,
*Email: kyupark@khu.ac.kr (Prof. Kyu Chang Park)

Abstract—The present study explores the impact of the vacuum conditions on the field emission characteristics of a cold cathode C-beam on important x-ray beam characteristics, with particular attention on the line phantom, x-ray dose rate and focal spot size. The study examines how changes in vacuum level affect the accuracy of the focal spot and the rate at which x-ray dose rate are given by delving into the dynamic link between the vacuum conditions and these crucial parameters. The goal of the research is to shed light on the complex interactions between the vacuum conditions and basic beam properties through methodical experimentation and analysis. Understanding these influences is essential for ensuring accurate and reliable diagnostic imaging outcomes while minimizing radiation exposure. The findings of this study contribute to the ongoing advancement of x-ray technology, with implications for improving healthcare diagnostics and patient care.

Keywords—field emission, line phantom, x – ray dose, c – beam.

I. INTRODUCTION

The performance of x-ray imaging systems is significantly influenced by the vacuum conditions within the x-ray tube and the x-ray chamber. The vacuum environment is essential for the production and transmission of x-rays, which directly affects the quality of the image, the amount of radiation, and the size of the focal spot [1][2][3][4][5][6]. Comprehending the influence of vacuum conditions on these crucial parameters is vital for maximizing the efficiency of x-ray imaging systems and guaranteeing the safety and effectiveness of diagnostic and therapeutic operations.

The x-ray tube is the central component of an x-ray imaging system, and its functioning relies significantly on the vacuum conditions inside the tube. An exceptional vacuum is essential to avoid the ionization of gas molecules and preserve the integrity of the electron beam, which is responsible for generating x-rays. When the electron beam is exposed to low-vacuum circumstances, it can be dispersed or redirected, resulting in a decline in the production of x-rays and a drop in the precision of the focus point. Additionally, the existence of leftover gas molecules in the x-ray tube might result in the creation of x-ray absorption artifacts, which may undermine the quality of the obtained images.

The x-ray chamber necessitates a controlled vacuum environment to guarantee best performance. It accommodates both the x-ray tube and the patient or object being scanned. The vacuum conditions within the x-ray chamber have a significant impact on both the x-ray dosage rate and the size of the focus point. These characteristics are essential in x-ray imaging. In conditions of high vacuum, the pace at which x-rays are delivered is usually greater, as the x-rays may pass through the chamber with low absorption and scattering. In contrast, when the vacuum level is low, the pace at which x-rays are delivered may decrease because the x-rays are more likely to interact with the gas molecules that remain in the chamber. The size of the focus spot, which indicates how the x-ray beam is spread out in space, is also affected by the vacuum conditions in the x-ray chamber. With the enhancement of vacuum conditions, the size of the focal spot often decreases, resulting in increased spatial resolution and enhanced image quality. Ultimately, the level of vacuum present in both the x-ray tube and the x-ray chamber greatly influences the effectiveness of x-ray imaging systems.

II. RESULTS AND DISSCUSION

The schematic in Fig. 1. depicts x – ray chamber and the x -ray tube. We evaluated the I-V characteristics of the C-beam in the triode mode, prior to acquiring the x-ray images. The gate voltage was varied from 400 to 2,000 V in increments of 50 V, with the anode voltage fixed at 5 kV. In the chamber, the I-V measurement displayed a maximum anode current of 1.0 mA at a gate voltage of 1,750 V. The analysis of the anode current of our C-beam in both the x – ray chamber and the x – ray tube showed that the threshold voltage of the tube was increased by 250 V in comparison to the x – ray chamber as shown in Fig 2. To evaluate the x-ray imaging, we conducted an examination of the focal spot size (FSS) and line phantom, and we quantified the x-ray dose. We did a performance comparison of our C-beam under x-ray chamber conditions and tested its integration into an x-ray tube. FSS is defined by two components: a horizontal FSS with a short axis of the beam (HFSS) and a vertical FSS with a long axis of the beam (VFSS). The EN-12543-5 standard and tungsten (W) wire were used to test the blurriness at the air-high absorption interface. X-ray image of line phantom and tungsten cross wire is shown in Fig. 3 and Fig 4 respectively. The measured focal spot size (FSS) for the

979-8-3503-7977-8/24 $31.00 © 2024 IEEE

tungsten cross wire in the x-ray chamber and x-ray tube was 0.33 × 0.49 mm and 0.40 × 0.60 mm, respectively. Similarly, the line phantom was measured for both systems using a chamber, and the images obtained were substantially clearer compared to those obtained using an x-ray tube. The inconsistencies in the x-ray imaging are attributed to the deteriorated pressure of the x-ray tube. Finally, dose measurement was done at 100 cm. it revealed that the vacuum condition has small effect on the dose rate with 172 μGy/s in chamber and 160 in x – ray tube for the same anode current. Additional information regarding the results and experiment will be published in the near future. The summary of x – ray chamber and x – ray tube is depicted in Table I.

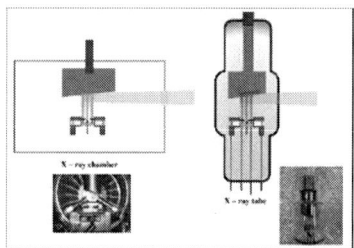

Fig. 1.   Schematic of x – ray chamber and x -ray tube

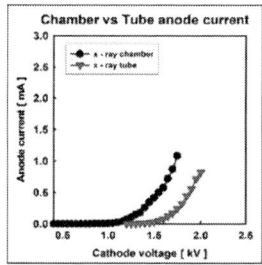

Fig. 2.   Comparison of anode current (chamber and tube).

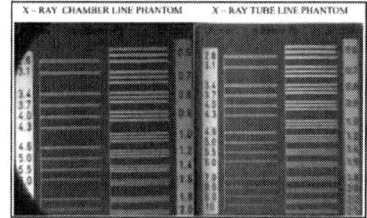

Fig. 3.   Phantom line (chamber and tube).

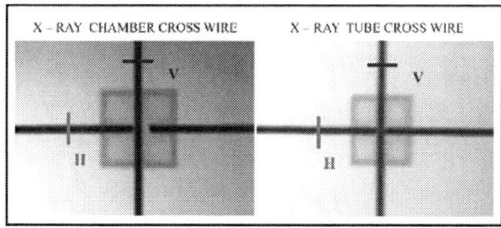

Fig. 4.   Tungsten crosswire (chamber and tube).

TABLE I.   SUMMARY OF X – RAY CHAMBER VS/ TUBE

Parameter	Chamber	X-ray tube
Anode current (mA)	1.08 @ 1750 V	0.825 @ 2000 V
Focal spot size (mm) (H × V)	0.33 × 0.49	0.40 × 0.60
Dose rate (μGy/s)	172 @ 100 cm	160 @ 100 cm

CONCLUSION

This work yielded significant insights into the performance of the C-beam in x-ray imaging. During the experiment, the chamber was subjected to I-V measurements. These measurements revealed that the maximum current flowing through the anode was 1.08 mA when the gate voltage was set at 1,750 V. It is worth noting that the threshold voltage in the x-ray tube was 250 V higher than the threshold voltage observed in the chamber. The focal spot size (FSS) and line phantom images were assessed, revealing an FSS of 0.33 × 0.49 mm in the chamber and 0.40 × 0.60 mm in the tube. The clarity of line phantom images was enhanced in the chamber, presumably as a result of improved pressure conditions. Measurements taken at a distance of 100 cm showed that the vacuum conditions had a slight impact on the dose rate. The dose rate in the chamber was 172 μGy/s, but in the tube, it was 160 μGy/s, despite having the same anode current. These findings demonstrate that the quality of images and the rate at which radiation is delivered are influenced by the pressure conditions of the x-ray tube. This impact is particularly noticeable in terms of the size of the focal point and the clarity of the resulting images. Additional information will be released in the near future

ACKNOWLEDGMENT

This work was supported by the Technology Innovation Program (No. 20013595, Extreme ultraviolet light source using nano electron beam) funded by the Ministry of Trade, Industry & Energy (MOTIE, Korea).

REFERENCES

[1]   R. Behling, "Medical X-ray sources now and for the future," *Nucl. Instruments Methods Phys. Res. Sect. A Accel. Spectrometers, Detect. Assoc. Equip.*, vol. 873, pp. 43–50, 2017.

[2]   I. V. Plotnikova, N. V Chicherina, S. S. Bays, R. G. Bildanov, and O. Stary, "The selection criteria elements of X-ray optics system," in *IOP Conference Series: Materials Science and Engineering*, 2018, p. 12029.

[3]   E. Ammann and W. Kutschera, "X-ray tubes—continuous innovative technology," *Br. J. Radiol.*, vol. 70, no. Special-Issue-1, pp. S1--S9, 1997.

[4]   Y. Y. Yu and K. C. Park, "Fabrication of high quality X-ray source by gated vertically aligned carbon nanotube field emitters," *J. Vac. Sci. \& Technol. B*, vol. 41, no. 2, 2023.

[5]   R. Behling, "Latest high performance X-ray tubes for medical imaging," in *2014 Tenth International Vacuum Electron Sources Conference (IVESC)*, 2014, p. 1.

[6]   T. Kariyawasam, "Field emission of carbon nanotubes," *Online Doc. 2005 June 28], Available HTTP http//www.physics.uc.edu/\~jarrellCOURSES Electrodyn. jects/tharanga/review. pdf*, 2005.

Area Scaling of Dense Si Field Emitter Arrays

Shabnam Ghotbi, Saeed Mohammadi
School of electrical and computer engineering
Purdue University
West Lafayette, IN, USA
saeedm@purdue.edu

Abstract—High density silicon field emitter arrays (FEAs) with high aspect ratio tips are fabricated and characterized. These dense FEAs are patterned with electron beam lithography and formed using inductively coupled plasma (ICP) etching. Device characterization shows a turn-on voltage of 110 V, and a maximum current of 2.7 mA for the large array of 640K emitter tips. Characterization of devices with different emission area reveals the current is scalable with device area for large array sizes (> 160 K emitters). It is presumed that the current in larger area FEAs is dominated by the field emitters in the entire area rather than that from emitters on the four sides of the array.

Keywords—Area scalability, electron beam lithography, Fowler-Nordheim, silicon field emitter arrays, vacuum electronics.

I. INTRODUCTION

Field emitter arrays (FEAs) find applications in scientific and research equipments such as X-ray sources, electron microscopy, and microwave and high powere vacuum devices [1]-[4]. The current density of FEAs nearly obeys Fowler-Nordheim equation and depends on the value of electric field at the tip of the emitters (emission sites), emitter geometry and work function [5], [6]. Operating Anode-Cathode voltage can be reduced by sharpening the emitter tips, bringing anode plate into close proximity of the emitter tips and employing low-work function emitter material.

II. FABRICATION

Sharpening emitter tips, not only reduces the turn on voltage and operating voltage of FEAs but also allows the emitter tips to be packed closer to each other to achieve high current densities [7-8]. In this work, we report fabrication and characterization of field emitter arrays using electron-beam lithography process to achieve dense arrays. FEA fabrication process begins with the selection of highly N-doped 400 µm thick Si substrate. The substrate has an epitaxial layer on top with 400 Ω.cm^{-1} resistivity. Employing a high throughput electron-beam lithography system (JBX-8100FS), positive tone ZEP 520A resist is patterned and developed in ZED N-50. Nickel (Ni) with a thickness of 90 is deposited at a rate of 1 Å/sec. Next, the substrate is immersed in Remover PG for lift-off process. ICP etching is done using Ni as mask inside a Panasonic E620 ICP etcher. The CF$_4$ etching was accomplished in 20 minutes under a pressure of 1 Pa, DC power of 50 W, RF power of 700 W, and gas flow rate of 40 sccm. After the dry etching, Ni is removed with commercial TFB Ni etchant. The scanning electron micrograph picture of a dense FEA is shown in Fig. 1. The cathode and anode substrates are then bonded together. The anode substrate is formed by depositing and patterning 4 µm of low-temperature SiO$_2$ on a low resistivity Si substrate. The patterned SiO$_2$ layer acts as a spacer layer and its opening is larger than the active areas of the FEAs to achieve 2 µm anode-emitter tip distance after bonding. The details of the process are provided in [9]. Scanning electron microscope image (Fig. 1) shows the field emitters are very uniform with emitter yield of nearly 100%.

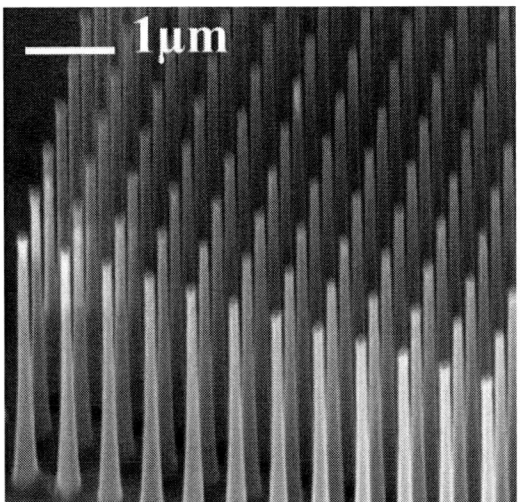

Fig. 1. High aspect ratio Si FEAs (length ~2.5 µm, pitch 0.5 µm, tip diameter 100 nm) carved out from Si substrate using e-beam lithography.

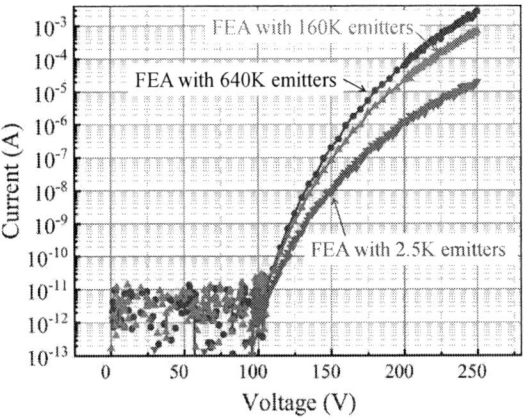

Fig. 2. Log-linear IV characteristics of the three FEAs studied here. Extracted turn-on voltages are about 110 V for all three FEAs.

III. CHARACTERIZATION AND INTERPRETATION OF THE RESULTS

This work investigates device reliability and scalability for arrays of 50×50, 400×400, and 800×800 emitters with very dense emitter spacing of 500 nm. Electrical characterizations of the FEAs were done with a source meter (Keithley 2410C) at a vacuum setting of 10^{-6} Torr. Maximum anode-cathode voltage V_{AK} = 250 V was used, which is lower than the FEA breakdown voltage. No emission current was observed under reversed-bias conditions. Log-linear IV characteristics of the three FEAs shown in Fig. 2 indicates turn-on voltages of about 110 V for all three devices. Fig. 3 shows IV characteristics dominated by tunneling emission mechanism with an observed maximum current of 2.7 mA for the device with 640K emitters at V_{AK} = 250 V. All FEAs show excellent

979-8-3503-7977-8/24 $31.00 © 2024 IEEE

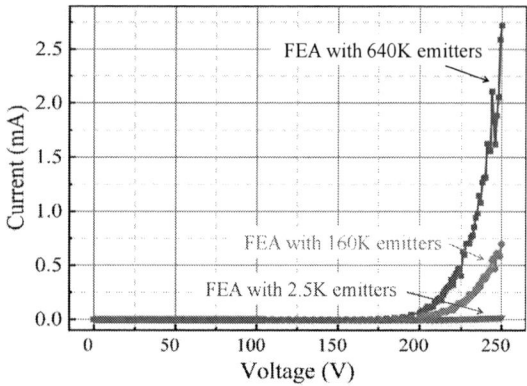

Fig. 3. IV characteristics of the three FEAs studied in this work.

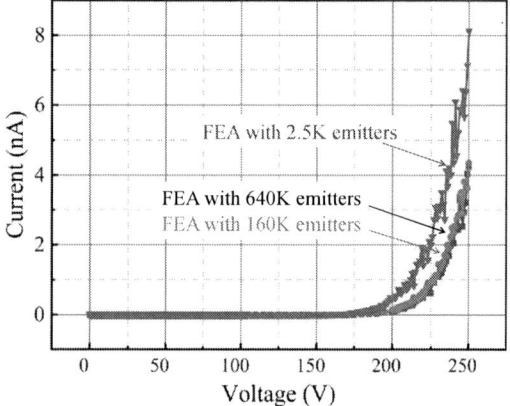

Fig. 4. Average emitter current-voltage characteristics of the three FEAs studied in this work.

reliability characteristics with no significant current degradation under an applied anode-cathode voltage V_{AK} of 150 V in 24 hours.

Fig.4 shows the average tip current for the three FEAs as functions of applied anode-cathode voltage. This value of the average current for the two large arrays is identical at every applied anode-cathode voltage, indicating the current mechanism in these two devices is dominated by emission from the entire device area. The maximum average tip current reaches 4.4 nA at the highest applied anode-cathode voltage of $V_{AK} = 250$ V. The average tip current for the smallest FEA (2.5K emitters) is higher than that of the other two devices and reaches 8.2 nA/emitter at $V_{AK} = 250$ V. The higher average tip current is observed over the entire range of V_{AK}. As emitter tips on the four sides of FEA experience less screening effect, their current could be much higher than the ones in the middle of the array. Therefore, a significant portion of the array current may stem from emitters on the sides of the array which are under less screening effect. As the size of array increases, the ratio of the emitter tips on the four sides to the total number of emitters decreases, hence the significance of the emission from the edges of the FEA decreases. Therefore, the higher average tip current in the small FEA (2.5K emitters) is due to the fact that 8% of the emitter tips ($4\times50/50\times50$) are on the four sides of the array as opposed to 1% and 0.5% for the other

two devices. One can calculate the average current of the emitter tips on the sides of the array to be ~52 nA at 250 V vs. ~4.2 nA for the ones in the middle of the array. Even though the emitters on the sides emit almost an order of magnitude higher electron current than those in the middle, the devices still show excellent reliability characteristics. This is presumably due to the fact that their current is regulated by the low emitter doping in these devices ($\sim10^{13}$ cm-3) that sets a bottleneck on the number of available electrons for emission.

Overall, the study demonstrates the potential for producing scalable and reliable Si-based FEAs with identical turn on voltages and average currents when large arrays are used. These dense FEAs were optimized to suppress screening effect by following the methods suggested in [7]. Further geometry and doping optimization can lead to even less screening effect and improved device area scalability for even small arrays.

IV. ACKNOWLEDGEMENT

The authors would like to thank helpful discussions with Profs. Enrico Bellotti and Dimitris Pavlidis and the support provided by US Airforce office of scientific research (AFOSR) through Contract No. FA9550-19-1-0349.

REFERENCES

[1] S. Itoh, M. Tanaka, and T. Tonegawa, "Development of field emission displays," *Journal of Vac. Sci. Technol. B Microelectron. Nanom. Struct.*, vol. 22, no. 3, p. 1362, 2004, doi: 10.1116/1.1691409.

[2] J. W. Jeong, J. W. Kim, J. T. Kang, S. Choi, S. Ahn, and Y. H. Song, "A vacuum-sealed compact x-ray tube based on focused carbon nanotube field-emission electrons," *Nanotechnology*, vol. 24, V. 8, March 2013, doi: 10.1088/0957-4484/24/8/085201.

[3] S. Ghotbi and S. Mohammadi, "Effect of Substrate Conductivity on Si Self-Assembled Field Emission Arrays," *2021 34th Int. Vac. Nanoelectron. Conf. IVNC 2021*, doi: 10.1109/IVNC52431.2021.9600721.

[4] E. J. Radauscher *et al.*, "Improved Performance of Field Emission Vacuum Microelectronic Devices for Integrated Circuits," *IEEE Trans. Electron Devices*, vol. 63, no. 9, pp. 3753–3760, Sep. 2016, doi: 10.1109/TED.2016.2593905.

[5] R. G. Forbes, "The Murphy-Good plot: A better method of analysing field emission data," *R. Soc. Open Sci.*, vol. 6, no. 12, Dec. 2019, doi: 10.1098/RSOS.190912.

[6] E. L. Murphy and R. H. Good, "Thermionic Emission, Field Emission, and the Transition Region," *Phys. Rev.*, vol. 102, no. 6, p. 1464, Jun. 1956, doi: 10.1103/PhysRev.102.1464.

[7] S. Ghotbi and S. Mohammadi, "Reducing Screening Effect in Close-Packed Si Field Emitter Arrays," in *IEEE Transactions on Electron Devices*, vol. 70, no. 8, pp. 4387-4393, Aug. 2023, doi: 10.1109/TED.2023.3281296.

[8] S. Ghotbi and S. Mohammadi, "Close-packed silicon field emitter arrays with integrated anode fabricated by electron-beam lithography," *Journal of Vac. Sci. Technol. B*, vol. 41, no. 1, p. 013202, Jan. 2023, doi: 10.1116/6.0002295.

[9] S. Ghotbi and S. Mohammadi, "Low-Power Field Emission Arrays with Sharp Emitters," *24th International Vacuum Electronics Conference (IVEC)*, Chengdu, China, 2023, doi: 10.1109/IVEC56627.2023.10157710.

Beam Divergence and Interference Characteristics of Field Emission Beam: Effects of Transverse Momentum and Source Spatial Coherence

Soheil Hajibaba[1,2*], Soichiro Tsujino[1]
[1] *Paul Scherrer Institut, 5232 Villigen PSI, Switzerland*
[2] *Institute of Physics, University of Zurich, 8057 Zurich, Switzerland*
*Email: soheil.hajibaba@psi.ch

Abstract—**In this study, we explore the intricate interplay between the electronic properties within the solid and field emission beams characteristics. By examining the fundamental principles governing field emission, we investigate how transverse momentum impact the propagation of the emitted electron beams. We demonstrate that transverse momentum and the spatial coherence of the emission source significantly affect beam properties, leading to notable changes in interference patterns.**

Keywords—*field emission, transverse momentum, transverse coherence*

I. Introduction

It has been well established that the energy spectrum of field emission electron beam is affected by the electron distribution and electronic states in solid. These also governs the beam properties such as the beam divergence and the spatial coherence of the field emission beam. Although these are important properties for the application of field emitters for microscopy and accelerators, in-depth study is still needed to further exploit the field emitters fabricated not only from conventional materials but also from novel systems such as two-dimensional materials, topological insulators, etc. Recently, we reported that the average transverse momentum of field emission electrons governs the intrinsic normalized emittance [1], and also determines the transverse structure of the wave function, together with the minimal source size as determined by the uncertainty relationship [2,3].

In the present work, we wish to expand our study by considering the influence of the Fermi energy of the solid, that has a direct influence on the beam divergence of the 'point-source' field emitter. We further consider the case when the phase coherence of electrons at the cathode surface is finite, as opposed to the fully incoherence cathode case, that has been traditionally assumed, therefore the spatial coherence of the beam is completely determined by the effective source size of the cathode as the van Cittert-Zernike theorem assumes. The temperature-dependent interference experiment of a tungsten etched-wire needle-shaped cathode [4] and the analysis of the transmission low-energy electron diffraction experiment using a double-gated single Mo nanotip field emitter [5] indicated the finite spatial coherence of metal field emitters at the source surface, even though field emitters are normally assumed to be fully incoherent, as in the case of thermionic emission and photoemission.

II. Methods

To study the beam property of a finite source size but finite source spatial coherence cathode, we applied an optical theory of partially coherent beam and numerically evaluated the influence of the source spatial coherence on the interference pattern.

III. Results

The average transverse momentum of the electrons dictates the minimal source size and the beam divergence. The balance between the field emission tunneling parameter and the Fermi energy further influences those (Fig. 1).

When the source spatial coherence is finite, high-contrast interference is maintained for high scattering angles (Fig. 2). This is advantageous for electron microscopy, in particular for the lens-less method such as the ptychographic imaging [6].

Acknowledgment

We would like to thank S. Purcell and A. Ayari (Univ. Lyon), K. Leifer (Univ Uppsala), A. Kyritsakis (Univ. Tartu), and T. Latychevskaia (U. Zürich) for discussions on the beam coherence and experimental possibilities. This work is partially supported by Swiss National Science Foundation Nr. 200021_212748.

References

[1] TSUJINO, S. et al. Measurement of transverse emittance and coherence of double-gate field emitter array cathodes. *Nature communications*. 2016. Vol. 7, no. 1, p. 13976.

[2] TSUJINO, S. Transverse structure of the wave function of field emission electron beam determined by intrinsic transverse energy. *Journal of Applied Physics*. 2018. Vol. 124, no. 4.

[3] TSUJINO, S. On the brightness, transverse emittance, and transverse coherence of field emission beam. *Journal of Vacuum Science & Technology B*. 2022. Vol. 40, no. 3.

[4] CHO, B. et al. Quantitative evaluation of spatial coherence of the electron beam from low temperature field emitters. *Physical review letters*. 2004. Vol. 92, no. 24, p. 246103.

[5] LEE, C. et al. Transmission low-energy electron diffraction using double-gated single nanotip field emitter. *Applied Physics Letters*. 2018. Vol. 113, no. 1.

[6] RODENBURG, J and MAIDEN, A. Ptychography. In: HAWKES, P W and SPENCE, J C H (eds.), *Springer Handbook of Microscopy*, 2019. p. 819–904.

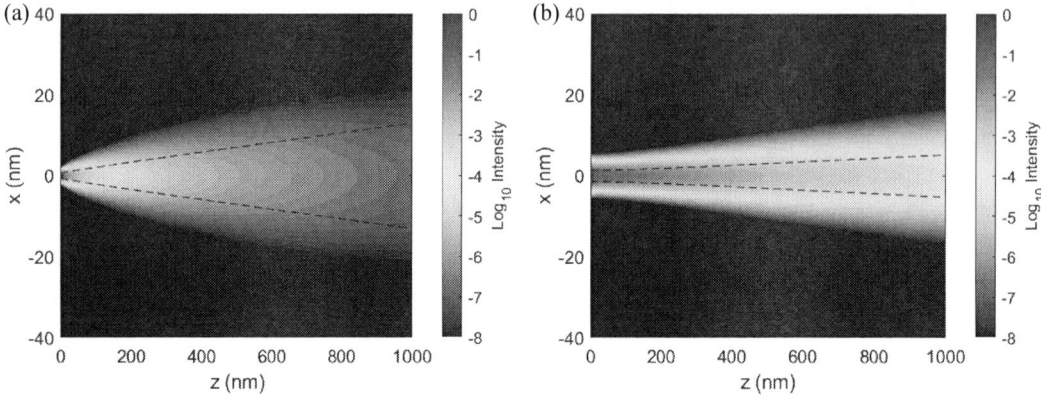

Fig. 1. The square of the wave function (normalized by the maximum value) for the case when the electric field is rapidly decreasing from F = 3 GV/m from the field emitter surface with the increase in the distance, W = 4.5 eV, and d_F = 0.13 eV. (a) E_F = 5 eV. (b) E_F = 50 meV.

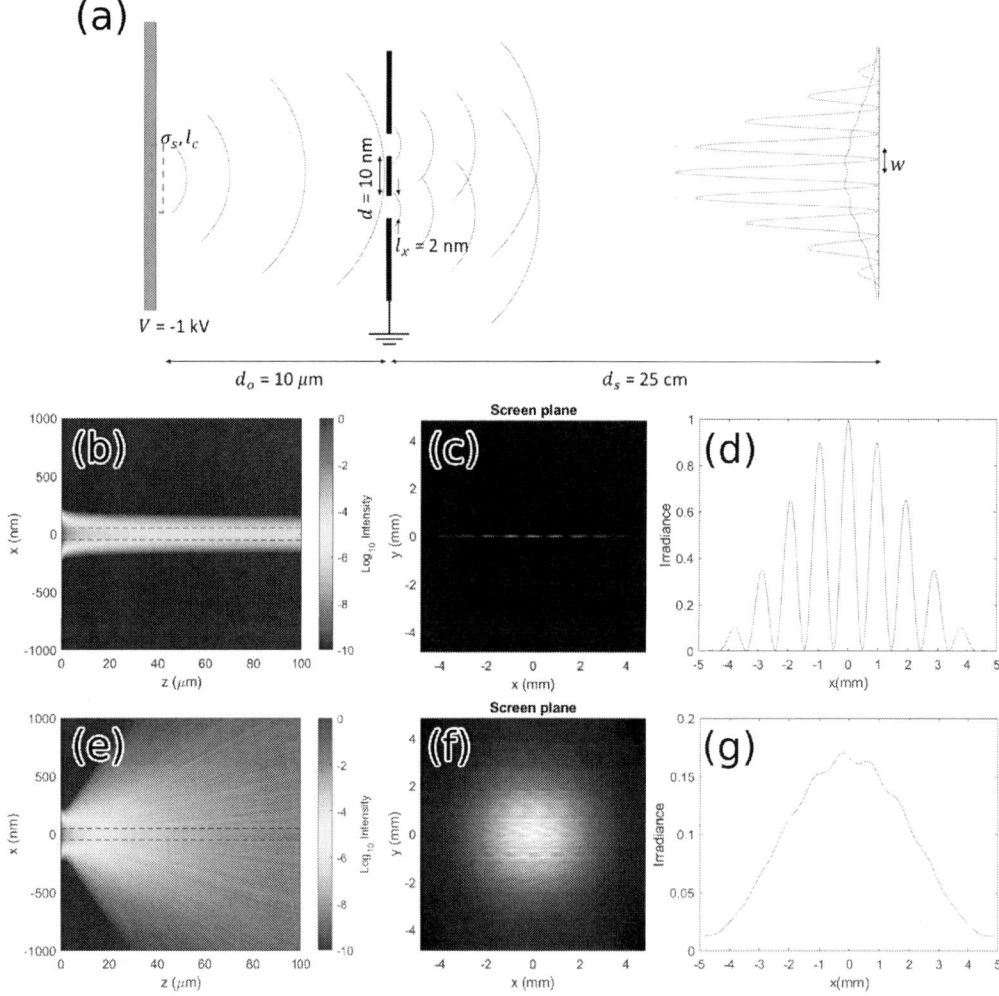

Fig. 2. (a) Schematic of the simulation geometry and parameters. The electric field is rapidly decreasing from F = 3 GV/m from the field emitter surface with the increase in the distance, W = 4.5 eV, d_F = 0.13 eV, E_F = 5 eV, and σ_S = 50 nm. (b,e) The square of the wave function (normalized by the maximum value). (c,f) The diffraction pattern on the screen. (d,g) The diffraction profile along the x axis. (b-d) fully coherent case. (e-g) incoherent case.

979-8-3503-7977-8/24 $31.00 © 2024 IEEE

Coherent Particle Acceleration on a Nanophotonic Chip

Leon Brückner[1], Tomas Chlouba[1], Roy Shiloh[1,2], Stefanie Kraus[1], Julian Litzel[1], Peter Hommelhoff[1]

[1]*Department of Physics, Friedrich-Alexander-Universität Erlangen-Nürnberg, 91058 Erlangen, Germany*
[2]*Institute of Applied Physics, Hebrew University of Jerusalem (HUJI), Jerusalem, Israel*
leon.brueckner@fau.de

Abstract—**Dielectric Laser Acceleration (DLA) offers a pathway to miniaturized particle accelerators capable of generating high-energy, ultrashort electron bunches. A typical DLA nanostructure comprises two rows of silicon pillars illuminated by an infrared laser pulse, creating an oscillating nearfield mode. Electrons that are phase-matched to this mode gain energy as they pass through the channel between the pillars. Transverse particle confinement is achieved by exploiting the transverse forces exerted by the nearfield, facilitating beam transport over considerable distances. By combining transverse confinement and a tapering of the structure period for continual phase-matching, acceleration over long distances is made possible. Experimental results demonstrate successful capture, confinement, and acceleration of electron bunches, with a notable energy gain of 12.3 keV, or 43% increase compared to the starting energy, in a 500 μm long acceleration structure. Further refinement and extension of interaction length hold promise for enabling applications such as high-energy electron imaging and diffraction as well as biomedical applications. Exploration of diverse dielectric materials and excitation wavelengths may yield even higher acceleration gradients.**

I. INTRODUCTION

Dielectric laser acceleration (DLA) is a scheme that combines the high peak fields of femtosecond laser systems with the high damage threshold of dielectric materials to enable miniaturized particle accelerators (Fig. 1) with high acceleration gradients [1]. Typically, a periodic dielectric structure is illuminated with an infrared laser pulse. Electrons whose velocity is matched to the resulting near-field's phase velocity are accelerated as they pass through the structure. By leaving field-free gaps in the structure, it is possible to alternately inject the electron into different phases of the light field, which results in alternating between transversally focusing and defocusing forces and thus a net guiding effect [2].

Here, we present a structure that combines the aforementioned guiding with a continuous tapering of the structure periodicity, which allows for phase-synchronous acceleration over long distances, leading to significant energy gains.

II. STRUCTURE DESIGN AND WORKING PRINCIPLE

The working principle of the accelerating structure is illustrated in Fig. 2. Electrons (green) travel in z-direction through the channel of the structure (grey). The Lorentz forces experienced by an electron that is phase-synchronous to the light field are drawn above and below the structure. Initially,

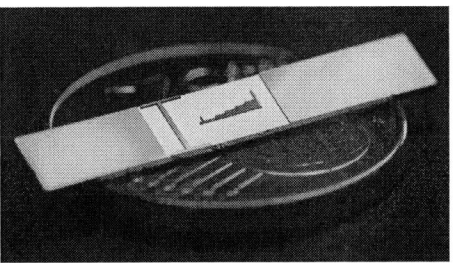

Fig. 1. Photograph of a chip containing dozens of microscopic accelerators. [3].

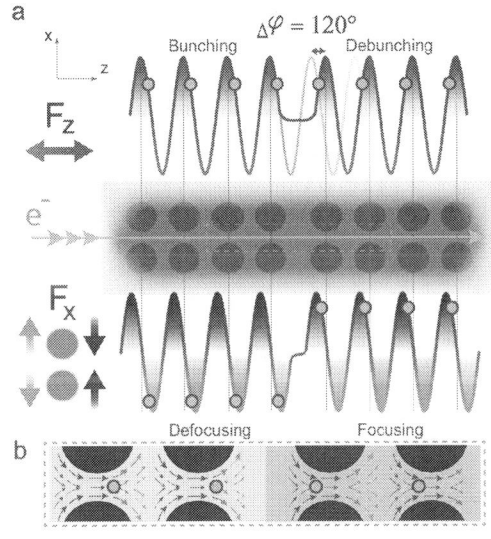

Fig. 2. a) Sketch of a part of the structure (grey) together with the forces F_z and F_x experienced by the electrons (green). See text for details. b) Total force at one instant in time acting on an electron. [4].

979-8-3503-7977-8/24 $31.00 © 2024 IEEE

the electron feels an accelerating force F_z in z-direction and a defocusing force F_x in x-direction, which repeats in every cycle of the light field. After passing through the field-free gap, the phase of the light field that the electron sees has shifted by $120°$. F_z is still accelerating, but F_x is now transversally focusing. This is later reversed by a $240°$ phase jump to return to transversal defocusing.

The structure design uses 26 of such focusing and defocusing elements over a total length of 500 μm. The structure period is continuously increased to guarantee coherent, phase-synchronous interaction as the electron gains energy.

III. EXPERIMENTAL SETUP

The accelerator structures are fabricated from silicon via electron beam lithography and reactive ion etching. An ultrafast scanning electron microscope delivers 700 fs long, 28.4 keV electron bunches that are injected into the structure channel. The structure is illuminated with 250 fs long, 1.93 μm laser pulses. Their pulse front is tilted by $71°$ to guarantee interaction along the whole structure length. The local incident peak field strength at the structure is 600 MV/m. The electron energy is measured with a magnetic spectrometer.

IV. COHERENT PARTICLE ACCELERATION RESULTS

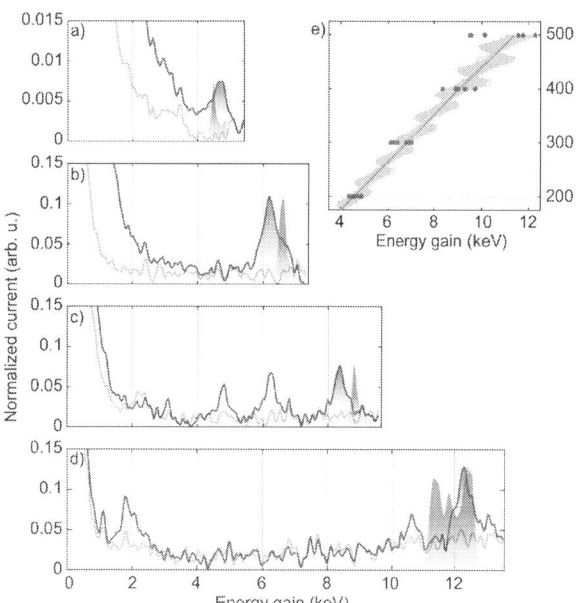

Fig. 3. a)-d) Electron energy spectra from 200, 300, 400 and 500 μm long accelerator structures. The signal is plotted in red, the background (without laser) in grey, and the simulated acceleration peak in blue. e) shows the energy gain (red dots) from several structures for each length together with the simulated gain (blue). [4]

The experimental results are plotted in Fig. 3. a)-d) show electron energy spectra from 200, 300, 400 and 500 μm long accelerator structures. The spectra show a noticeable peak,

which corresponds to the electron bunches that get guided and accelerated through the whole structure length. We observe an energy gain of 12.3 keV, or 43% for the longest structure. The acceleration gradient is 22.7 MeV/m. Similar results have been achieved in [5].

V. MID-INFRARED DLA

Practical applications of DLAs will require higher electron currents as well as higher gradients than what has been demonstrated here. One possible way to achieve this is to change the excitation wavelength to mid-infrared [6], which will enhance current throughput due to the larger structure as well as enable higher gradients due to the higher damage threshold.

In a demonstration experiment using a silicon DLA illuminated with 10 μm wavelength laser pulses, we observed 1.4 keV energy gain over 15 μm interaction length, resulting in a 93 MeV/m gradient for an incident peak field of 1 GV/m. The damage threshold of the structure pillars exceeds 5.2 ± 0.9 GV/m, the highest field strength currently available in our labs. [7]

VI. OUTLOOK

The on-chip nanophotonic accelerator shown here is the light-wave, miniaturized analogue to the well-known radiofrequency accelerators. Further development will necessitate the extension of the alternate phase focusing scheme to three dimensions to fully confine the electron bunch [8]. Higher currents and gradients can be reached by exploring different combinations of materials and wavelengths. A complete, miniaturized, high-gradient accelerator would enable novel applications in the fields of compact light sources and radiotherapy.

ACKNOWLEDGMENT

We acknowledge funding by the Gordon and Betty Moore Foundation, Deutsche Forschungsgemeinschaft, ERC and BMBF.

REFERENCES

[1] R. Shiloh, et al., "Miniature light-driven nanophotonic electron acceleration and control," Adv. Opt. Photon. 2022, 14, 862-932

[2] R. Shiloh, J. Illmer, T. Chlouba, P. Yousefi, N. Schönenberger, U. Niedermayer, A. Mittelbach, P. Hommelhoff, "Electron phase-space control in photonic chip-based particle acceleration", Nature 2021, 597, 498-502

[3] Picture courtesy of S. Kraus and J. Litzel.

[4] T. Chlouba, R. Shiloh, S. Kraus, L. Brückner, J. Litzel, P. Hommelhoff, "Coherent nanophotonic electron accelerator", Nature 2023, 622, 476-480

[5] P. Broaddus, T. Egenolf, D. S. Black, M. Murillo, C. Woodahl, Y. Miao, U. Niedermayer, R. L. Byer, K. J. Leedle, O. Solgaard, "Subrelativistic alternating phase focusing dielectric laser accelerators", Phys. Rev. Lett. 2024, 132, 085001

[6] X. Mei, R. Zha, S. Pan, S. Wang, B. Sun, C. Lei, C. Ke, Z. Zhao, D. Wang, "Dielectric laser accelerators driven by ultrashort, ultraintense long-wave infrared lasers", Ultrafast Sci. 2023, 3;0050

[7] L. Brückner, T. Chlouba, Y. Morimoto, N. Schönenberger, T. Shibuya, T. Siefke, U. D. Zeitner, P. Hommelhoff, "Mid-infrared dielectric laser acceleration in a silicon dual pillar structure", unpublished

[8] U. Niedermayer, T. Egenolf, O. Boine-Frankenheim, "Three dimensional alternating-phase focusing for dielectric-laser electron accelerators", Phys. Rev. Lett. 2020, 125, 164801

Cold Cathode Coherent-Structure Flat Panel X-Ray Source Using Microarray Metal Target

Guicai Qi[1], Qi Liu[1], Song Kang[1], Zhuoran Ou[1], Wangjiang Wu[2], Yuan Xu[2], Linghong Zhou[2],
Juncong She[1], Shaozhi Deng[1], Ningsheng Xu[1], Jun Chen[1,*]

[1] *State Key Laboratory of Optoelectronic Materials and Technologies, Guangdong Province Key Laboratory of Display Material and Technology, School of Electronics and Information Technology, Sun Yat-sen University, Guangzhou 510275, Guangdong Province, People's Republic of China.*

[2] *School of Biomedical Engineering, Southern Medical University, Guangzhou 510515, Guangdong Province, People's Republic of China.*

*E-mail: stscjun@mail.sysu.edu.cn

Abstract—**Traditional X-ray imaging has difficulty in differentiating between soft biological tissues due to low absorption contrast. However, the Talbot-Lau grating interferometer (GI) could be a promising solution to this problem by enabling X-ray phase contrast imaging (XPCI). Unfortunately, the G0 grating has some limitations regarding fabrication and efficiency. To overcome this issue, we have developed a new cold cathode coherent-structure flat panel X-ray source (FPXS) that eliminates the need for G0. This innovation has the potential to significantly improve the capabilities of Talbot-Lau GI for phase contrast imaging in the future.**

I. INTRODUCTION

Since Roentgen's discovery in 1895, X-ray imaging has rapidly advanced and become widely adopted. Traditional X-ray imaging depends on X-ray absorption properties to create images. However, for low-Z specimens such as soft biological tissues, the variation of absorption coefficients within the tissue is minimal. This makes absorption imaging ineffective for producing clear images. Instead, X-rays passing through low-Z samples induce significant phase changes. Consequently, XPCI shows greater sensitivity when compared to absorption imaging [1], with potential applications in scientific research and disease diagnosis. One of the most advanced systems of XPCI is the Talbot-Lau GI, which is preferred for its compatibility with existing low-coherence laboratory X-ray sources and imaging setups [2].

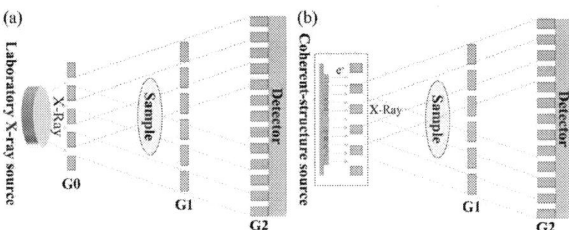

Fig. 1. (a) Traditional and (b) proposed cold cathode coherent-structure FPXS Talbot-Lau GI.

As shown in Fig.1 (a), a typical Talbot-Lau GI setup comprises a traditional laboratory X-ray source with a thermionic cathode, a source grating G0, a phase grating G1, an absorption grating G2, and a position-sensitive detector. The grating G0 is designed to supply individually coherent but mutually incoherent line sources, ensuring the essential coherent-structure X-ray for Talbot-Lau GI. When X-rays pass through G1, self-imaging of G1 occurs at specific fractional distances because of the Talbot effect. An absorption grating is placed at the Talbot image position for image analysis. The periodic structure of G2 and the periodicity of the G1 self-imaging combine to produce moiré fringes beyond G2, which are captured by the detector. Placing an imaging object in front of G1 results in the object's refractive effect on the X-rays, distorting the incident X-ray wave-plane on the phase grating. The gradient of the X-ray phase shift caused by the object reflects its structure information at various positions. However, the source grating G0 poses significant challenges in terms of fabrication and efficiency limitations.

To address these issues, a new cold cathode coherent-structure FPXS that utilizes a microarray metal anode target was proposed [4,5]. By using a large-area nanomaterial cold cathode, the large-area flat-panel coherent-structure X-ray source can be made. This innovative coherent-structure X-ray source, depicted in Fig.1 (b), eliminates the need for G0. In this study, a coherent-structure FPXS was fabricated and its characteristics were studied.

II. DEVICE STRUCTURE AND FABRICATION

Figure 2(a) depicts the schematic diagram of the coherent-structure X-ray source. Arrays of patterned ZnO nanowires on the ITO glass substrate act as the cathode. The anode is metal target in strip-shape microarrays located on an ITO glass substrate. Each metal bar is 10 μm wide, and there is a 20 μm spacing between adjacent metal bars.

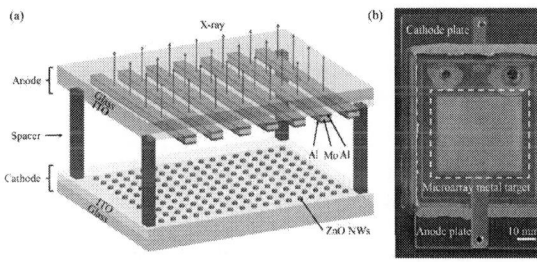

Fig. 2. (a) Schematic diagrams of cold cathode coherent-structure FPXS and (b) the vacuum-encapsulated coherent-structure FPXS.

The ZnO nanowire arrays were prepared using the process described in our previous study [3]. The microarray metal target was formed by deposition of Mo target film and followed by a wet-etching process. The vacuum-encapsulated coherent-structure X-ray sources were fabricated using a standard vacuum-sealing process.

Fig.2 (b) shows the fabricated fully vacuum-encapsulated device. Fig.3 (a) presents SEM pictures of the ZnO nanowire arrays created on the cathode panel, with the nanowires having lengths of approximately 2-5 μm. Fig.3 (b) displays SEM images of the microarray metal target, with the stripe period matching the design specifications.

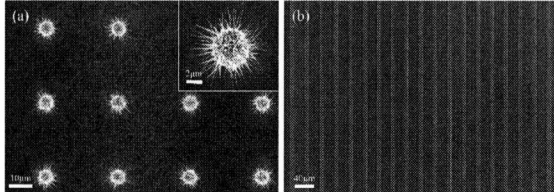

Fig. 3. The SEM pictures of (a) the ZnO nanowire arrays and (b) the microarray metal target.

III. SIMULATION AND EXPERIMENTAL RESULTS

A. Simulation of coherent-structure FPXS

As coherent-structure X-ray illumination cannot be directly measured by a real detector, the Monte Carlo simulation using EGSnrc was conducted to acquire the spatial distribution of X-ray beneath the metal target of the coherent-structure X-ray sources. The number of simulated electrons and the acceleration voltage were set to be 2×10^9 and 28 kV, respectively. A local area of the coherent-structure X-ray source was simulated with an area size of 0.6 mm × 0.6 mm. Other parameters are chosen according to the fabricated coherent-structure X-ray source described above. Fig.4 (a) shows the coherent-structure X-ray emission obtained at the surface under the Mo strips, due to the narrow width of each metal target bar, the X-rays emitted by each bar metal target are regarded as spatially coherent. The simulation verified the feasibility of realizing G0 using the structured microarray line sources as required by Talbot-Lau GI.

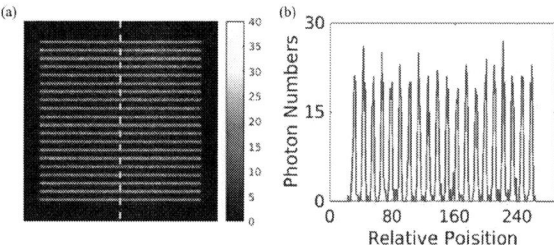

Fig. 4. (a) X-ray spatial distribution at the plane under Mo target of FPXS, and (b) corresponding line profile along the yellow line in (a).

B. Experimental Results

High voltage was applied to the anode to test the device, and the X-ray emission characteristics were analyzed. Fig.5 (a) shows the I-V characteristics and the corresponding F-N plot obtained from one of the coherent-structure X-ray sources. The maximum voltage and current recorded were 26 kV and 1.18 mA, respectively. Fig.5 (b) shows the X-ray energy spectrum measured 20 cm in front of the coherent-structure X-ray source, obtained at an anode voltage of 26 kV.

Since the GI system utilizing the coherent-structure X-ray source is still under construction. Here we present the results of attenuation projection imaging using the fabricated coherent-structure X-ray sources. Fig.6 shows the images of starfish specimens and PCBs, demonstrating that the X-ray source is capable of producing clear projection images.

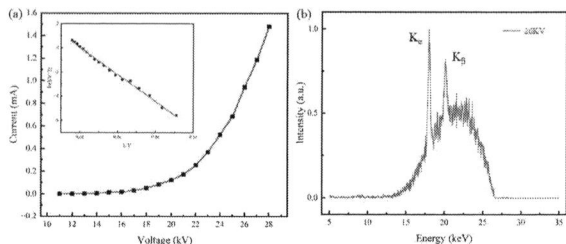

Fig. 5. (a) I-V characteristics (inset: F–N plot) and (b) X-ray spectrum of the coherent-structure FPXS.

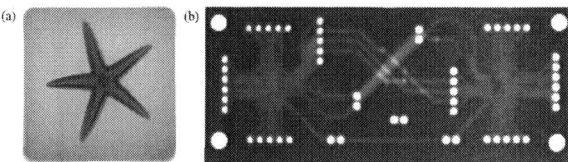

Fig. 6. X-ray images of (a) starfish specimens and (b) PCB.

IV. SUMMARY

Our work verified the feasibility of fabricating the coherent-structure FPXS used for Talbot-Lau GI. This innovative X-ray source eliminates the necessity for G0 and have potential application in X-ray phase contrast imaging.

ACKNOWLEDGMENTS

Thanks for the financial support from the National Key R&D Program of China (Grant No. 2022YFA1204200), and the S&T Department of Guangdong (Grant No. 2023B1212060025).

REFERENCES

[1] A. Momose, W. Yashiro, Y. Takeda, Y. Suzuki, and T. Hattori, "Phase Tomography by X-ray Talbot Interferometry for Biological Imaging, "Jpn. J. Appl. Phys., vol. 45, no. 6A, pp. 5254-5262, 2006.

[2] F. Pfeiffer, T. Weitkamp, O. Bunk, and C. David, "Phase retrieval and differential phase-contrast imaging with low-brilliance X-ray sources, " Nature Phys., vol. 2, no. 4, pp. 258-261, 2006.

[3] L. Wang et al., "Fabrication of large-area ZnO nanowire field emitter arrays by thermal oxidation for high-current application, " Applied Surface Science, vol. 484, pp. 966-974, 2019.

[4] W. Wu et al., "Simulation study of a novel ZnO nanowire cold cathode flat-panel x-ray source using EGSnrc for Talbot-Lau type grating interferometry, " in Medical Imaging 2023: Physics of Medical Imaging, R. Fahrig, J. M. Sabol, and L. Yu, Eds., San Diego, United States: SPIE, Apr. 2023, p. 9.

[5] W. Wu et al., "Feasibility Study of a Cold-cathode Flat-panel X-ray Source with Micro-array Anode Target for Grating Interferometer Computed Tomography, " IEEE Trans. Nucl. Sci., vol. 70, no. 12, pp. 2553-2560, 2023.

Compact and Low-cost Scanning Electron Microscope

Casimir Kuzyk*, Alexander Dimitrakopoulos*, R. Fabian Pease†, Alireza Nojeh*‡

*Department of Electrical and Computer Engineering, and Quantum Matter Institute
University of British Columbia, Vancouver BC, Canada
†Department of Electrical Engineering
Stanford University, Stanford CA, USA
‡Corresponding author: alireza.nojeh@ubc.ca

Abstract—The scanning electron microscope (SEM) yields lifelike images with inherently better resolution and depth of focus than a light microscope, but can be prohibitively expensive. About a decade ago we started a project to create a compact SEM with extremely low cost [1-3]. Here we present a working prototype for the first time.

Index Terms—scanning electron microscope, compact, low-cost, carbon nanotube forest, electron source, broad access

I. INTRODUCTION

The scanning electron microscope is used for high-resolution imaging in a broad range of research and application areas. However, even the least expensive commercial instruments are beyond the reach of many potential users.

Several efforts have been made over the years to create simple SEMs, dating back to the early days of the development of the instrument [4, 5]. A notable contemporary example is the systematic work of researchers at the Wrocław University of Science and Technology, who have reported on various aspects of their progress including electron sources, scanning, and signal detection toward making electron microscopes based on micro-electromechanical systems technologies [6].

We have taken a different approach, capitalizing on a unique electron source, to create a compact and low-cost SEM without use of specialized technologies, with the goal of significantly broadening access to electron microscopy.

II. METHODOLOGY AND RESULTS

Our approach has centred on a thermionic electron source featuring localized optical heating of a carbon nanotube array [7, 8]. This source uses very low power, does not require thermal management, emits from an optically-controlled spot, and is rugged and tolerant of poor vacuum. The rest of the column uses simple, off-the-shelf components, and widely available prototyping technologies. An example micrograph of the head of an insect obtained using our instrument is presented in Fig. 1, exhibiting the familiar real-life quality of scanning electron microscope images.

Funding: Natural Sciences and Engineering Research Council of Canada, Canada Foundation for Innovation, British Columbia Knowledge Development Fund, Canada First Research Excellence Fund (Quantum Materials and Future Technologies Program)

Fig. 1. Image of the head of an insect obtained using our compact and low-cost scanning electron microscope prototype.

REFERENCES

[1] M. Chang, K. Dridi, A. Nojeh, and R. F. Pease, "Carbon nanotube thermionic emitter for a compact SEM gun," 2016 29th International Vacuum Nanoelectronics Conference (IVNC), Vancouver BC, 11-15 July 2016.

[2] C. Kuzyk, M. Chang, C. Aiello, K. Jessen, R. F. Pease, and A. Nojeh, "Opto-thermionic cathodes for SEM," 2018 62nd International Conference on Electron, Ion, and Photon Beam Technology and Nanofabrication (EIPBN), Puerto Rico, 29 May-1 June 2018.

[3] C. Kuzyk, E. Blankenburg, G. Robinson-Leith, A. Dimitrakopoulos, H. Li, M. Chang, M. Cen, B. Ye, G. Hu, K. Jessen, R. F. Pease, and A. Nojeh, "Removing barriers to innovation with a tabletop, low-cost SEM," 2019 63rd International Conference on Electron, Ion, and Photon Beam Technology and Nanofabrication (EIPBN), Minneapolis MN, 28-31 May 2019.

[4] P. J. Spreadbury, "Investigations relating to the design of a simple scanning electron microscope," MSc Dissertation, Cambridge University, 1958.

[5] A. Khursheed, "Recent developments in scanning electron microscope design," Adv. Imag. Elec. Phys., P. W. Hawkes, Ed., Elsevier, 2001, vol. 115, pp. 197–285.

[6] M. Białas, T. Grzebyk, M. Krysztof, and A. Górecka-Drzazga, "Signal detection and imaging methods for MEMS electron microscope," Ultramicroscopy, vol. 244, p. 113653, February 2023.

[7] P. Yaghoobi, M. Vahdani Moghaddam, and A. Nojeh, ""Heat trap": Unusual light-induced-heat localization in carbon nanotube arrays," Solid State Comm., vol. 151, pp. 1105–1108, September 2011.

[8] A. Nojeh, "Carbon nanotube photo-thermionics: toward laser-pointer-driven cathodes for simple free-electron devices and systems," MRS Bullet., vol. 42, pp. 500–504, July 2017.

979-8-3503-7977-8/24 $31.00 © 2024 IEEE

Compact Modeling Approach of Field Emitter Arrays

Youngjin Shin, Nedeljko Karaulac, Winston Chern, Akintunde I. Akinwande

Electrical Engineering and Computer Science
Massachusetts Institute of Technology
Cambridge, MA 02139, USA
ys18@mit.edu

Abstract—Silicon field emitter arrays (FEAs) are cold electron sources for devices such as x-ray sources, ion sources, high-power microwave amplifiers or multi-beam electron lithography. For each device, it is imperative to optimize the anode structure and device package to obtain reliable performance. It is generally assumed that a parallel configuration of the FEA cathode-anode structure is the most ideal due to the compactness and symmetry. However, little work has been done to study the behavior of the FEA devices with this structure and consequently there is no compact model for the device. In this work, we report an unexpected yet repeatable negative differential resistance (NDR) in the device output characteristics, suggesting a need for an optimal FEA cathode-anode configuration. A compact model for the FEA cathode-anode parallel-plate configuration is proposed for anode-to-emitter distances ≤100μm.

Index Terms—Field Emitter Array, Negative Differential Resistance

I. INTRODUCTION

Cold-cathodes based on silicon field emitter arrays (FEAs) have shown promise for devices that operate in harsh environments due to the temperature independence of tunneling current and the property of ballistic transport [1-2]. Examples include x-ray sources and high power, high frequency sources. In any such application, it is ideal to have a compact model that predicts the device current-voltage characteristics. In doing so, the devices must be packaged such that the FEA cathode and anode have a well-defined and robust vacuum channel length to predict reasonably accurate device behaviors. In general, a parallel-plate configuration between the FEA cathode-anode assembly is ideal as the emitter-to-anode distances are well defined and the electron trajectory between the emitter and anode is approximated by 1D transport perpendicular to the anode plane. Under these circumstances, the widely accepted 1D Child-Langmuir space charge limited current should apply. Additionally, from a packaging standpoint, a parallel-plate configuration allows for compactness and scalability. In this work, we study and model the FEA characteristics utilizing an on-chip flat silicon anode with well-defined stand-offs which determine the emitter-to-anode distances of $L_{AE} \leq 100\mu$m.

II. RESULTS AND DISCUSSION

The characteristics of the FEAs were obtained by using an on-chip flat anode with well-defined stand-off heights. The anode was fabricated from silicon with the bottom of the stand-offs having a 2μm layer of oxide for electrical isolation and a backside deposition of TiAu, as shown in Fig. 1(a). The experimental schematic is displayed in Fig. 1(b), where the anode distance is determined by the etch depth of the anode stand-offs. The measurements were conducted using pulsed I-V through a B2902A power supply.

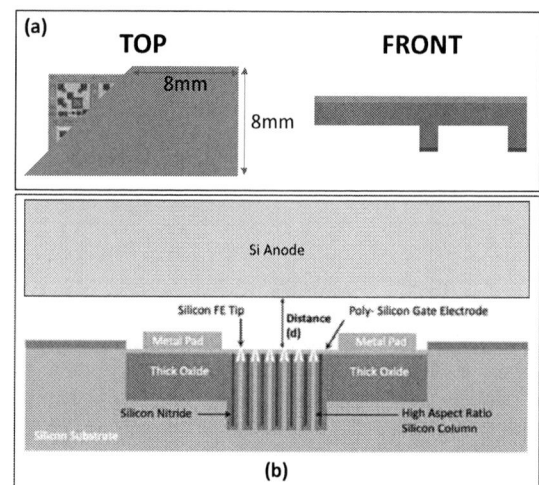

Fig. 1. (a) Final anode design of three layers consisting of silicon oxide (purple), silicon (grey), and TiAu metal contact (gold). The relative size of the flat anode is compared to a sample device chip. (b) Schematic of the flat anode experimental configuration relative to device. The anode is large relative to a single FEA device; hence, the stand-offs are not pictured.

The output and transfer characteristics are shown in Fig. 2 and Fig. 3, respectively. The output characteristics reveal the device behavior for the parallel-plate FEA cathode-anode assembly. We observe an unexpected yet repeatable NDR region within the space charge limited regime of the device. This results in a double saturation behavior with the second saturation occurring when the anode-to-emitter voltage is greater than the gate-to-emitter voltage, $V_{AE} \geq V_{GE}$. Initially, this phenomena was hypothesized to be a result of semiconductor physics of the silicon anode, but even with the metallization of the anode with nickel silicide, there was no observed difference in the device characteristics. The electrostatic simulation reveals

979-8-3503-7977-8/24 $31.00 © 2024 IEEE

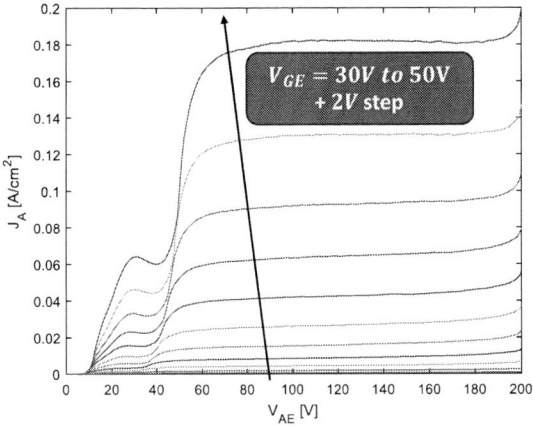

Fig. 2. Output characteristics of a 1000x1000 FEA with an emission area of $1mm^2$. V_{GE} is swept from 30V to 50V at 2V increments. There is an NDR region that is reflected within the space charge limited regime which is repeated in each curve.

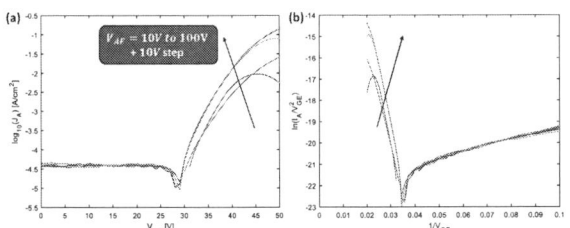

Fig. 3. (a) Transfer characteristics of a 1000x1000 FEA with an emission area of $1mm^2$. V_{AE} is swept from 10V to 100V at 10V increments. Space charge effects are more easily observable at low V_{AE}. (b) Fowler-Nordheim curves confirm the current is field emission.

that when $V_{AE} < V_{GE}$, the plate geometry of the gate and anode introduces a 1D deceleration region in the vacuum space between the FEA cathode and anode. This explains the double saturation behavior observed in the output characteristics as the deceleration causes the emitted electrons coming out of the emitter to be reflected away from the anode unless the emission angle of the electron is near vertical. The NDR region was recorded when $L_{AE} \leq 100\mu$m. This result would also suggest that there is 2D electron transport between the FEA and anode plates that cannot be approximated to 1D formulations at close anode-to-emitter distances.

To model the device behavior in a compact model, we calculated the final velocity of an electron at the anode based on the electric potential and electric field the electron experiences when exiting the gate aperture. The final velocity calculations show that at low V_{AE}, only the electrons exiting from the center of the gate aperture has sufficient velocity to reach the anode under the decelerating field and saturates at higher V_{AE} when all electrons regardless of their exit position at the aperture are collected. By approximating the current density distribution at the gate aperture to resemble that of a Gaussian distribution [3], we can determine the proportion of current that is collected at the anode given a value for V_{AE}. Hence, we report a semi-empirical compact model optimized

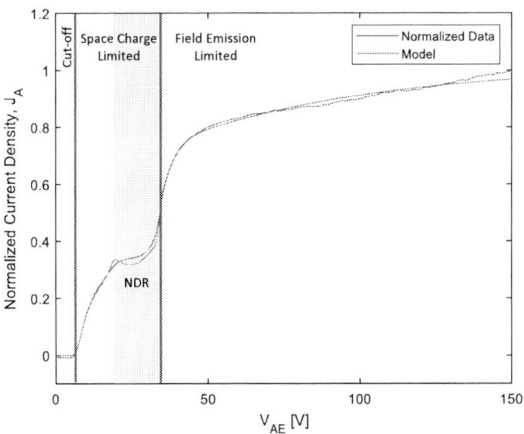

Fig. 4. Optimized model function overlayed on a normalized output characteristic for a 1000x1000 FEA for V_{GE}=34V and L_{AE}=30μm.

to the device output characteristics of gated FEAs, as shown in Fig. 4. The behavior predicted by the compact model is solely determined by four physical parameters consisting of the anode-to-emitter voltage, V_{AE}, gate-to-emitter voltage, V_{GE}, anode work function, ϕ_A, and anode-to-emitter distance, L_{AE}. The model demonstrates good accuracy within the typical operating ranges of the device (V_{AE}=0-200V, V_{GE}=30-50V, ϕ_A=7V, $L_{AE} \leq 100\mu$m).

III. CONCLUSION

We report the behavior of FEA devices for a parallel configuration of the FEA cathode-anode at close anode-to-emitter distances of $L_{AE} \leq 100\mu$m. Due to the electrostatics of the configuration, the device characteristics reveal a NDR region in the space charge limited regime of the output characteristics. The simulation show that there exists a deceleration channel in the vacuum region in between the gate and anode that reflect the electrons, preventing its collection when $V_{AE} < V_{GE}$. The compact model was developed by using the simulation and experimental data to determine the proportion of current density distribution that is collected given a value for V_{AE} and V_{GE}.

ACKNOWLEDGMENT

We acknowledge sponsorship by the Air Force Office of Scientific Research (AFOSR) under Grant No. FA9550-18-1-0436. This work was carried out in part using the MIT.nano facility.

REFERENCES

[1] D. Temple, "Recent progress in field emitter array development for high performance applications," Materials Science and Engineering: R: Reports. 1999. Vol. 24, no. 5, p. 185–239.

[2] R. Bhattacharya, N. Karaulac, W. Chern, A. I. Akinwande, J. Browning, "Temperature effects on gated silicon field emission array performance," Journal of Vacuum Science & amp; Technology B, 2021, Vol. 39, no. 2.

[3] L. Dvorson, M. Ding, A. I. Akinwande, "Analytical electrostatic model of silicon conical field emitters I," IEEE Transactions on Electron Devices. 2001. Vol. 48, no. 1, p. 134–143.

Conductively Coated 3D Printed Emitters for Electron Devices

Daniel Burda[1,2,*], Mario Kandra[3], Mohammad Allaham[1,4], Alexandr Knápek[1], Milan Matějka[1,3,*]

[1]*Electron and Plasma Technologies, Institute of Scientific Instruments of the Czech Academy of Sciences, Brno, Czech Republic*
[2]*Department of Physics, Faculty of Electrical Engineering and Communication, Brno University of Technology, Brno, Czech Republic*
[3]*IQS nano s.r.o., Husinec, Czech Republic*
[4]*Central European Institute of Technology, Brno University of Technology, Brno, Czech Republic*
*Corresponding authors: burda@isibrno.cz, mmatejka@isibrno.cz

Abstract—**The additive manufacturing method is used to produce electron emitters using a two-photon laser printer (2FL). Structures are manufactured on a silicon wafer from photosensitive resins, which form a base shape of the structures on top of which a conductive layer is deposited afterward. These structures can take on various geometries, including pyramidal, conical, or ellipsoidal shapes. Good results in terms of mechanical stability were achieved for structures of square pyramids, 50 μm in height. The structures on chips are measured in a field emission experiment chamber operated under an ultrahigh vacuum (~10 $^{-7}$ Pa).**

Keywords—*cold field emission, two-photon lithography, 2PL, FEM, electron device*

I. INTRODUCTION

Two-photon lithography (2PL) using two-photon polymerization (2PP) enables interesting results in the development of new electron sources, as recently presented [1,2]. The print-in-place approach to fabrication, fast printing speeds, and high resolution of the current generation of 2PL printers make 2PL printing viable for large-area field emission electron sources. The study focuses on optimization of the geometry and physical parameters of structures and exploration of the capabilities of 3D printed arrays as electron emitters.

II. FABRICATION

The additive manufacturing approach was used to manufacture electron emitters using a two-photon lithography (2FL) printer (IQnano3D, IQS nano). Arrays of structures were manufactured on <111> silicon wafers from photosensitive resins, various compositions of resins were investigated. The exposure of a resin results in crosslinking and it remains attached to the substrate. After the development, the exposed resin was further UV cured.

The printed structures were designed from a bitmap model using IQWriter (IQS nano) software. The structures were grown layer by layer. Different geometries such as pillars, cones, and pyramids were investigated. We found that in the case of pillars with high aspect ratios there were issues with adhesion to the substrate and the pillars also tended to bend during the handling. The final structures were made from an array of square pyramids; the height of the pyramids is 50 μm. The spacing between individual pyramids is 60 μm. The 5x5 array of square pyramids is illustrated in Fig. 1. The optimized printing strategy by one volumetric element pixel (voxel) at the very last layer resulted in tip diameter in range from 200 to 300 nm.

The subsequent step was a mild cleaning process of the wafer with structures in reactive plasma etcher (PlasmaPro, Oxford Instruments) for 10 s. After that, the wafer was cleaved into single chips and coated with a 40 nm thick Au layer in a sputter coater (Q150T ES, Quorum), the thickness of the deposited layer was measured in situ using a microbalance quartz method. The tips of coated pyramids had a diameter of 280 to 350 nm, as seen in Fig. 2.

III. PROCEDURE AND MEASUREMENTS

The chips are mounted in a field emission experiment chamber (FEM) operated in an ultra-high vacuum (~10 ^{7} Pa) details on the chamber can be found in [3]. The experimental setup utilizes a planar sample holder, in Fig. 3-A, which can incorporate samples of 8 x 8 mm. The chip with fabricated structure and conductive layer is connected in diode configuration. An 80 μm thick polyimide sheet spacer separates the chip from the counter electrode. The distance between the tips of the array and the counter electrode is 25 to 30 μm.

The fabricated chip is attached to the conductive pad using a tiny amount of conductive silver paste, which is applied from the backside of the silicon substrate and also from all four sides, ensuring an electrical connection between the top Au layer and the surface of the conductive pad. The pad is electrically connected to the high-voltage source, which uses an overcurrent protection set to 50 μA.

The counter electrode consists of a Ce:YAG scintillating monocrystal coated with a 30 nm conductive Al layer. The counter electrode has two functions, it collects the part of emitted current, which is measured using an external pico-ammeter (9103, RBD Instruments), and part of the emission current is converted to photons, which are visible through a vacuum window and captured using a camera. Fig. 3-B shows the emission pattern visible during the measurement of the 5x5 array of square pyramids; it is nonhomogeneous, indicating that only a fraction of tips forming the array contribute to the emission.

IV. RESULTS

The current-voltage (I-V) characteristics obtained in the FEM showed rapid fluctuations in current at a threshold voltage of 1300 V. Further increasing the voltage up to 2000 V did not result in stabilization of current, which hints at the switching-on of only several tips of the 5x5 array. There may be small differences in the diameter and height of each structure, which greatly affects the emission throughout the array [4]. The field emission orthodoxy test [5] applied to the measured I-V characteristics yielded a non-orthodox behavior. Thus, it appears that there may be occurrence of

979-8-3503-7977-8/24 $31.00 © 2024 IEEE

changes to the tip geometry as well during the observed fluctuations.

V. CONCLUSION

Optimization of the shape of resin-based tips yielded chips with a conveniently small diameter of 300 nm or less. This increased to 350 nm after coating with an Au layer. The fabricated chips were functional in FEM, showing electron emission. It was observed that not all tips in the 5x5 array were operating properly, resulting in current fluctuations.

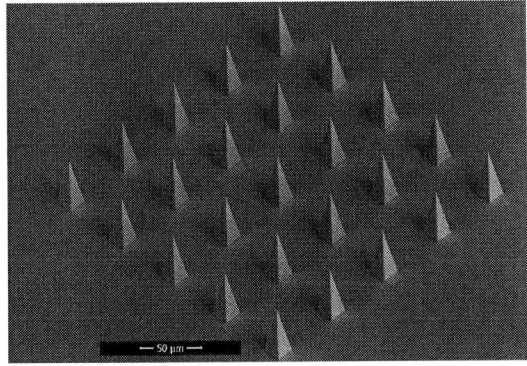

Fig. 1   Micrograph of 5 x 5 array of square pyramids on Si <111> surface, after sputter coating with 40 nm of Au layer. Sample tilted by 45°.

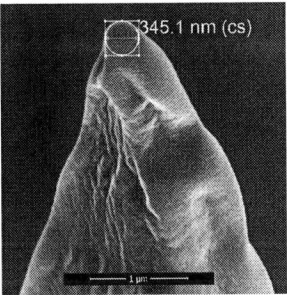

Fig. 2   Detail of tip shape of square pyramid coated with 40 nm of Au layer. The tip has diameter of 345 nm (cs - cross-section correction), tilted by 45°.

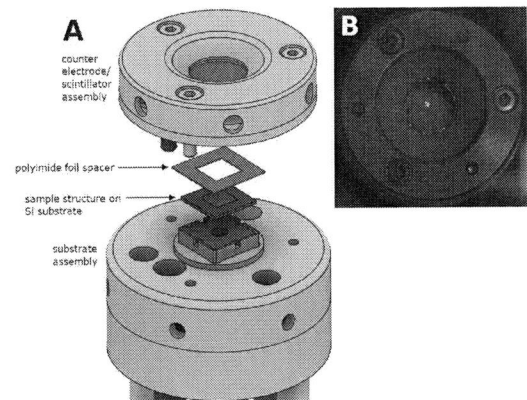

Fig. 3   A) Schematic of sample placed in the internal FEM assembly, B) An emission pattern on the backside of scintillator visible during FEM measurement through a chamber mounted window.

ACKNOWLEDGMENT

We acknowledge support by the Czech Academy of Sciences (RVO:68081731) and CzechNanoLab Research Infrastructure supported by The Ministry of Education, Youth and Sports of the Czech Republic (LM2018110).

REFERENCES

[1]   A. Kachkine, C. E. Owens, A. J. Hart, and L. F. Velásquez-García, "3D-Printed, Non-Planar Electron Sources For Next-Generation Electron Projection Lithography", *in 2023 IEEE 36th International Vacuum Nanoelectronics Conference (IVNC)*, 2023, pp. 128-130.

[2]   Z. Bigelow and L. F. Velásquez-García, "Fully 3D-Printed Miniature Langmuire Multi-Probe Sensor for Cubesat Ionospheric Plasma Diagnostics", *in 2023 IEEE 36th International Vacuum Nanoelectronics Conference (IVNC)*, 2023, pp. 103-105.

[3]   A. Knapek, M. Horacek, F. Hruby, J. Sikula, T. Kuparowitz, and D. Sobola, "Noise behaviour of field emission cathode based on lead pencil graphite", in *2017 30th International Vacuum Nanoelectronics Conference (IVNC)*, 2017, pp. 274-275.

[4]   M. M. Allaham, P. Buchner, R. Schreiner, and A. Knapek, "Testing the performance of Murphy-Good plots when applied to current-voltage characteristics of Si field electron emission tips", in *2021 34th International Vacuum Nanoelectronics Conference (IVNC)*, 2021, pp. 1-2.

[5]   M. M. Allaham, R. G. Forbes, A. Knápek, D. Sobola, D. Burda, P. Sedlák, and M. S. Mousa, "Interpretation of field emission current–voltage data: Background theory and detailed simulation testing of a user-friendly webtool", *Materials Today Communications*, vol. 31, 2022.

Contact Interface of Graphene Sensors

Patrik Staroň[1], Robert Macků[1], Petr Sedlák[1], Nikola Papež[1], Ramazanov Shihkgasan[1], Farid Orudzhev[1], Mohammed A. Al-Anber[2], Dinara Sobola[1]*

[1]Department of Physics, Faculty of Electrical Engineering and Communication, Brno University of Technology, Brno, Czech Republic
[2]Laboratory of Inorganic Materials and Polymers, Department of Chemistry, Faculty of Sciences, Mutah University, Al-Karak, Jordan

*Corresponding author: sobola@vut.cz

Abstract—**Gas sensors come in various types, including chemiresistive, infrared, photoionization, and semiconductor sensors. Each type leverages a distinct physical phenomenon. For carbon-based sensors, they can function under two primary categories. Graphene-metal contacts are crucial because they significantly impact the performance of carbon-based sensors. The geometry of these contacts plays a vital role in determining the sensor's effectiveness. The contact geometry affects how charge carriers are transferred between the graphene and the metal, which in turn influences the sensor's sensitivity, response time, and overall efficiency. Optimal contact geometry can minimize resistance and maximize the active sensing area, thereby enhancing the detection capabilities and reliability of the sensors.**

Keywords— graphene, contact, geometry, metal, interface

I. INTRODUCTION

The performance of graphene-based gas sensors is heavily influenced by the characteristics of the metal contacts. The geometry, material, thickness, and cleanliness of these contacts are all critical factors that determine the sensor's sensitivity, response time, and overall efficiency. Optimizing these parameters is essential for enhancing the performance and reliability of graphene-based sensors, making them more suitable for a wide range of applications [1, 2].

The geometry of metal contacts is vital for sensor performance. Optimal contact geometry ensures efficient charge carrier transfer between the graphene and the metal, which is crucial for high sensitivity and rapid response times. Inadequate contact geometry can lead to increased resistance and reduced sensor efficiency. Different geometrical configurations, such as edge contacts or surface contacts, provide various benefits in terms of conductivity and signal integrity.

Choosing the right material for the metal contacts is equally important. Different metals offer varying levels of conductivity, stability, and compatibility with graphene. Common materials include gold (Au), platinum (Pt), and palladium (Pd). Gold is often chosen for its excellent conductivity and resistance to oxidation. Platinum is known for its high stability and durability, making it suitable for harsh environments. Palladium is recognized for its good hydrogen sensitivity, which can be advantageous in specific sensing applications. The contact material impacts contact resistance and overall sensor performance, making it a critical consideration in sensor design.

The thickness of metal contacts also plays a significant role in the performance of graphene-based sensors. Thicker contacts can provide better durability and mechanical stability, essential for maintaining consistent performance over time. However, excessively thick contacts can increase resistance at the interface, negatively affecting the sensor's sensitivity. Conversely, thinner contacts may reduce resistance but be more susceptible to wear and damage. Therefore, finding the right balance in the thickness of the contacts is crucial for optimizing sensor performance [3-6].

Cleanliness of the metal contacts is another crucial factor that significantly affects sensor efficiency. Any contamination at the contact interface can increase resistance, noise, and reduce sensitivity. Ensuring that contacts are free from impurities and residues during fabrication is essential for maintaining sensor integrity and performance. Techniques such as thorough cleaning of the substrate and using cleanroom environments during assembly help achieve the necessary level of cleanliness.

The methods used to integrate and fabricate contacts with the graphene layer are also significant. Techniques such as electron beam lithography, chemical vapor deposition, and thermal evaporation are commonly used to create high-quality contacts. Each method has its advantages and limitations regarding precision, scalability, and cost. High-quality fabrication processes can help mitigate contact resistance issues and improve sensor reliability.

II. EXPERIMENTAL

We explored the capabilities of pristine graphene sensors, as depicted in figures 1 and 2. To test these sensors effectively, we developed a method in which the sensor is affixed to an FR4 PCB and connected to the circuit traces. Figures 3 illustrates the specific trace layout required for sensor interfacing. Additionally, to facilitate various test configurations, we designed a matrix connection PCB. This design allows for the flexible connection of any connector pin to any sensor pad, enabling the simultaneous measurement of two sensors. This configuration is not only portable but also ensures the sensor is securely mounted on the PCB. Consequently, connecting and disconnecting the sensors for testing in an environment with reduced noise is straightforward. The circuit traces were gold-plated to minimize material transition effects from the pads to the amplifier. However, we are aware that the intermetallic phase of CU-NI-AU might introduce more noise compared to a direct AU to CU transition. Therefore, this setup may not be optimal for experiments requiring ultra-high sensitivity.

979-8-3503-7977-8/24 $31.00 © 2024 IEEE

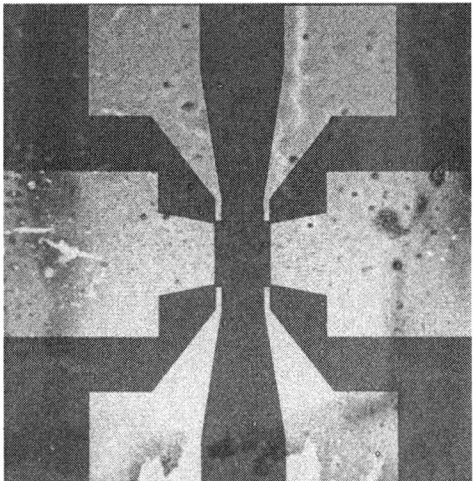

Fig. 1. Star electrode arrangement

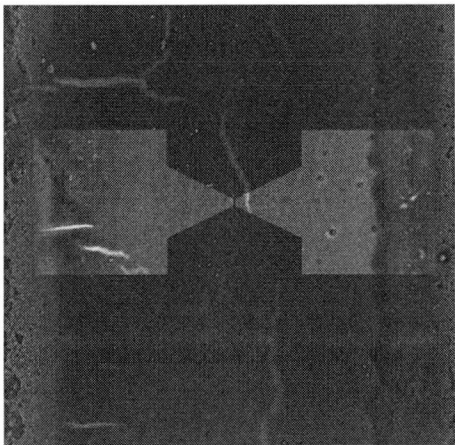

Fig. 2. Two-electrode arrangement

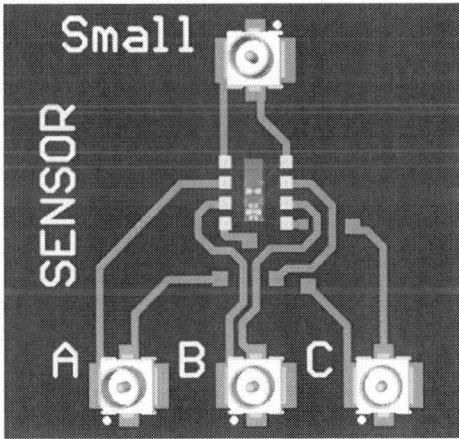

Fig. 3. PCB for the sensor interface with direct connections.

The actual schematic proposed for measuring the graphene sensor in spectral domain is shown in Figure 1. This is a setup that allows to push current through the sensor and a resistor, while measuring the voltage on the resistor. The resistor contributes to the noise, but the tweaking the voltage from the

battery should decrease the influence of the resistive noise so that the dominant noise is the one from the sensor.

III. CONCLUSIONS

The performance of graphene-based gas sensors is significantly influenced by the metal contacts used. The geometry, material, thickness, and cleanliness of these contacts are crucial factors in determining the sensor's overall efficiency and effectiveness.

ACKNOWLEDGMENT

The research described in the paper was financially supported by the Internal Grant Agency of the Brno University of Technology, grant No. FEKT-S-20-6352 and the GACR 23-07384S.

REFERENCES

[1] A. Knápek, J. Sýkora, J. Chlumská, and D. Sobola, "Programmable set-up for electrochemical preparation of STM tips and ultra-sharp field emission cathodes," Microelectron Eng, vol. 173, pp. 42–47, Apr. 2017, doi: 10.1016/J.MEE.2017.04.002.

[2] A. Knápek, D. Sobola, P. Tománek, Z. Pokorná, and M. Urbánek, "Field emission from the surface of highly ordered pyrolytic graphite," Appl Surf Sci, vol. 395, pp. 157–161, Feb. 2017, doi: 10.1016/J.APSUSC.2016.05.002.

[3] D. Sobola, N. Papež, R. Dallaev, S. Ramazanov, D. Hemzal, and V. Holcman, "Characterization of nanoblisters on HOPG surface," Journal of Electrical Engineering, vol. 70, no. 7, pp. 132–136, Dec. 2019, doi: 10.2478/JEE-2019-0055.

[4] A. AlSoud et al., "Electrical properties of epoxy/graphite flakes microcomposite at the percolation threshold concentration," Phys Scr, vol. 99, no. 5, p. 055955, Apr. 2024, doi: 10.1088/1402-4896/AD3B50.

[5] A. Knápek, M. M. Allaham, D. Burda, D. Sobola, M. Drozd, and M. Horáček, "Explanation of the quasi-harmonic field emission behaviour observed on epoxy-coated polymer graphite cathodes," Mater Today Commun, vol. 34, p. 105270, Mar. 2023, doi: 10.1016/J.MTCOMM.2022.105270.

[6] K. Ronoh, S. H. Fawaeer, V. Holcman, A. Knápek, and D. Sobola, "Comprehensive characterization of different metallic thin films on highly oriented pyrolytic graphite substrate," Vacuum, vol. 215, p. 112345, Sep. 2023, doi: 10.1016/J.VACUUM.2023.112345.

Controlled Synthesis of Tungsten Oxide Nanowires Prepared by Thermal Oxidation for Application in Cold Cathode Flat Panel X-ray Source

Qi Liu, Zufang Lin, Song Kang, Guofu Zhang, Shaozhi Deng, Ningsheng Xu and Jun Chen *

State Key Laboratory of Optoelectronic Materials and Technologies, Guangdong Province Key Laboratory of Display Material and Technology, School of Electronics and Information Technology, Sun Yat-sen University, Guangzhou 510275, People's Republic of China

* E-mail: stscjun@mail.sysu.edu.cn

Abstract—Cold cathode flat panel X-ray source has significant uses in medical imaging and industrial non-destructive testing. Tungsten oxide (WO_{3-x}) nanowires have exceptional field emission characteristics, involving low turn-on field, high current density, and excellent field emission stability. However, the controllable preparation of tungsten oxide nanowires on a large area is still an unsolved problem. In this study, the synthesis of WO_{3-x} nanostructures was explored employing a catalyst-free thermal oxidation method by varying the oxidation atmosphere. Controllable growth of large area high density WO_{3-x} nanowires and nanosheets were realized on glass substrate. Field emission measurement results demonstrate that lower turn-on field and higher emission current can be obtained from nanowires than those from the nanosheets. A diode-structure flat panel X-ray source utilizing WO_{3-x} nanowires was investigated. Uniform X-ray emission and projection imaging were achieved.

Keywords—Catalyst-free thermal oxidation, WO_{3-x} nanowires, FPXS

I. INTRODUCTION

Cold cathode flat panel X-ray source (FPXS) has significant employs in medical imaging, industrial non-destructive testing, and other fields [1-2]. Tungsten oxide (WO_{3-x}) nanowires have exceptional field emission capabilities, such as low turn-on field, high current density, and high stability and are considered as a suitable material for large area and high current field emitter [3, 4]. Recently, FPXS using WO_{3-x} nanowires cold cathode prepared by thermal oxidation has been reported [5-6]. However, large area controllable growth is still an issue to be solved for their application in FPXS devices.

In this paper, we studied the growth of controllably synthesized WO_{3-x} nanostructures on glass substrate by thermal oxidation method without using catalyst. The effects of oxidation atmosphere on the morphology of the products were studied. The field emission performance of WO_{3-x} nanostructures was characterized. A diode-structure FPXS device was fabricated using WO_{3-x} nanowires as the cathode.

II. EXPERIMENT DETAILS

WO_{3-x} nanostructures were grown directly from tungsten thin film on glass substrate by catalyst-free thermal oxidation method under different atmosphere conditions. The indium tin oxide (ITO) film and tungsten (W) films were sequentially sputtered on the glass. The W film had a thickness of 500 nm. The samples were heated to 540°C in 90 minutes after placing them in the middle of a quartz tube stove. During the heating-up process, nitrogen (N_2) gas flow was supplied into the quartz tube. A flow rate of 3 standard liters per minute (slm) was applied. The gas flow was stopped when the tube temperature was 540°C. The nanostructures were prepared by adjusting the thermal oxidation atmosphere. Two conditions were used in this study. The nanostructures were prepared under (1) ambient atmosphere or (2) oxygen (O_2) at a flow rate of 1 slm while keeping at 540°C for 90 minutes. The prepared nanostructures were morphologically characterized using SEM, and the crystalline quality was characterized using XRD and TEM. The composition of the nanostructures were characterized by a Raman spectrometer.

The field emission properties of prepared WO_{3-x} nanostructures were measured in a dynamic ultravacuum chamber using a diode structure. The variation characteristics of the field emission current with the applied field voltage were recorded, and field emission image was obtained using ITO-coated glass as the anode.

A molybdenum transmission anode target was assembled with WO_{3-x} nanowires cold cathode to form a FPXS. The cathode panel and anode panel were placed relatively parallel to each other using a glass border with a distance of 5 millimeters as an isolator. A current meter was connected in series with the cathode electrode to record real-time data. The device was measured in an ultravacuum chamber under (P=6.6×10^{-6} Pa). The X-ray emission uniformity, energy spectra, and imaging properties were measured.

III. RESULT AND DISCUSSION

Fig. 1 shows SEM and TEM images of WO_{3-x} nanostructures prepared under two different conditions. The SEM results show that high-density nanowires were obtained under ambient atmosphere (Fig. 1(a)) and high-density nanosheets were obtained (Fig. 1 (b)) when the sample was oxidized under 1 slm O_2. The length of the nanowires is up to 30 μm while the height for the nanosheets is below 10 μm. Single crystalline structures were obtained for the both conditions as shown in Fig. 1(c)-(d) the high-magnification TEM images along with their SAED patterns. The SAED in Fig. 1(c) displays that the diffraction spots extend in a straight line, indicating the presence of defects parallel to the direction of growth [7]. On the contrary, the diffraction spots shown in Fig. 1(d) are circular dots, indicating that introducing oxygen flow during the growth process can reduce defects and enhance the crystallinity of nanostructures. The introduce of oxygen during the growth process favor the nucleation of nanosheets at the grain boundaries.

979-8-3503-7977-8/24 $31.00 © 2024 IEEE

Field emission measurement showed that the turn-on field (the corresponding current density is 10 μA/cm²) of WO_{3-x} nanowires is about 4.85 MV/m and the current density of 9.75 mA/cm² was acquired under electric field of 10.4 MV/m. At an electric field of 8.6 MV/m, the turn-on field of the WO_{3-x} nanosheets was about 5.1 MV/m and the current density was 3.33 mA/cm². The relatively higher current obtained from WO_{3-x} nanowires is due to higher aspect ratio of nanowires compared with the nanosheets, which is consistent with the morphology.

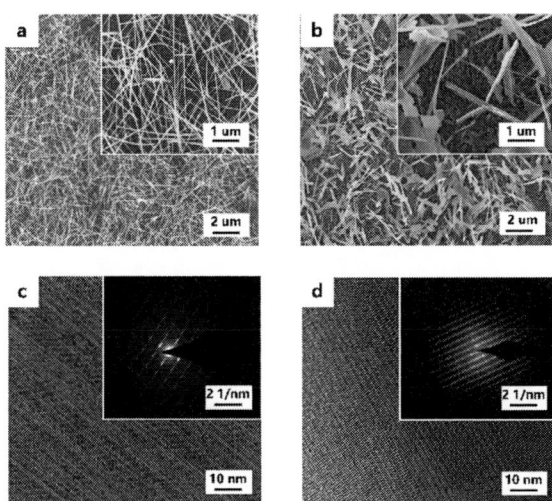

Fig.1 SEM and TEM images of prepared WO_{3-x} nanowires. (a) (c) prepared under ambient atmosphere; (d) (b) prepared under O_2 atmosphere. (Insets in (c) and (d) show the corresponding SAED patterns)

Fig. 2(a) shows the schematic of the diode FPXS device and the testing set-up. The device has an effective emitting area of 5.2 × 5.2 cm². The anode target produced X-rays when a high voltage was supplied to the anode. The curve of anode current versus applied anode voltage is shown in Fig. 2(b). An anode voltage of 29 kV resulted in an emission current of approximately 57 μA. Meanwhile, the inset of Fig. 2(b) displays the uniform emission image observed from the FPXS.

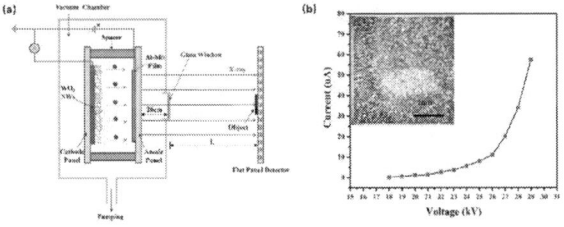

Fig.2 (a) Schematic diagram of the testing set-up for the FPXS device; (b) Current-voltage (I-V) curve measured from the FPXS (inset shows the field emission image)

Fig. 3(a) shows the X-ray energy spectrum obtained from the device at an anode voltage of 27 kV. The energy spectrum of the emitted X-ray exhibited a continuous distribution in the energy range of 0-27 keV, with two distinct molybdenum peaks at 17.89 keV and 20.07 keV. Subsequently, static projection X-ray imaging tests were

carried out using a line-pair card and a CMOS flat panel X-ray detector (Xineos-1515, Thousand Oaks). The line-pair card was attached to the detector surface, which is positioned 60 cm away from the X-ray source. The imaging result is shown in Fig. 3(b). The line pairs in the 2 lp/mm region can be clearly distinguished.

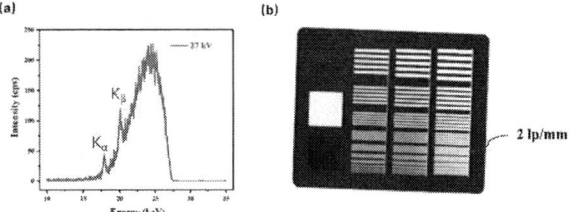

Fig.3 (a) Energy spectrum of WO_{3-x} nanowires cold cathode FPXS; (b) X-ray imaging results of line pair card of the object to be measured.

IV. CONCLUSION

In summary, large-area WO_{3-x} nanowires were prepared by controlled catalyst-free thermal oxidation method directly from tungsten thin film on glass structure. The results of field emission tests indicate that the prepared WO_{3-x} nanowires achieved large-area uniform emission. Diode-structure FPXS employing WO_{3-x} nanowires was fabricated and tested. Uniform X-ray emission and clear projection imaging were achieved. The results are important for the development of large-area FPXS device for X-ray imaging.

ACKNOWLEDGMENTS

The authors gratefully acknowledge the financial support from the National Key R&D Program of China (Grant No. 2022YFA1204200), Key Research and Development Program of Guangdong Province (Grant No. 2023B0101200013), National Natural Science Foundation of China (Grant Nos. 82272131 & 62001527), the Science & Technology Department of Guangdong Province (Grant No. 2023B1212060025), the Science & Technology Program of Guangzhou (Grant No. 2023A04J1666), Fundamental Research Funds for the Central Universities.

REFERENCES

[1] P. Serbun, V. Porshyn, G. Müller, et al "Advanced field emission measurement techniques for research on modern cold cathode materials and their applications for transmission-type x-ray sources", *Review of Scientific Instruments.*, vol. 91, p. 083906, 2020.

[2] Y.F. Chen, S.Z. Deng, N.S. Xu, et al "Recent progress on ZnO nanowires cold cathode and its applications", *Nanomaterials.*, vol. 11, p. 2150, 2021.

[3] K. Huang, Q.T. Pan, F. Yang, et al "The catalyst-free synthesis of large-area tungsten oxide nanowire arrays on ITO substrate and field emission properties", *Materials Research Bulletin.*, vol. 43, pp. 919-925, 2008.

[4] Z.F. Lin, H.J. Chen, J.C. She, et al "WO$_3$ nanowire field emission point electron source with high brightness and current stability", *Vacuum.*, vol. 195, pp. 110660, 2022.

[5] Z.F. Lin, P.B. Xie, R.Z. Zhan, et al "Defect-enhanced field emission from WO$_3$ nanowires for flat-panel X-ray sources", *ACS Applied Nano Materials.*, vol. 2, pp. 5206-5213, 2019.

[6] Y. Yao, D.D. Sang, L.R. Zou, et al "A review on the properties and applications of WO$_3$ nanostructure-based optical and electronic devices", *Nanomaterials*, vol. 11, p. 2136, 2021.

[7] S. Saygi, A. Koudymov, V. Adivarahan, et al "Real-space electron transfer in III-nitride metal-oxide-semiconductor-heterojunction structures", *Applied Physics Letters*, vol. 87, p. 043505, 2005.

Design of Setup for Laser Induced Plasma Etching

Lukáš Šilhan[1,2*], Jan Novotný[1], Tomáš Plichta[1], Jan Ježek[1], Ondřej Vaculík[1,2], Mojmír Šerý[1]

[1]Institute of Scientific Instruments of the Czech Academy of Sciences, v.v.i., Brno, Czech Republic

[2]Institute of Physical Engineering, Faculty of Mechanical Engineering, Brno University of Technology, Brno, Czech Republic

*Corresponding author: silhan@isibrno.cz

Abstract—Plasma etching introduces a physically activated chemical process highly utilized in the semiconductor industry. However, for the creation of etched structures mask has to be prepared on top of the etched surfaces. This is usually achieved by electron beam lithography, which adds to the complexity and financial demands of the overall process. We present the design of a setup for maskless plasma etching, which utilizes a tightly focused ultrashort laser pulse for the ignition of etching plasma in a custom vacuum chamber with a connection to a gas-containing etching species. In addition, etched structures are written into the surface of sample by scanning with the vacuum chamber with relation to fixed laser focus. This enables maskless 3D etching of samples with a less complicated technological process.

Keywords—*vacuum chamber, plasma etching, femtosecond laser, micromachining*

I. Introduction

Laser Induced Plasma Etching (LIPE) represents a cutting-edge technique in the field of ultraprecision material processing [1], combining the capabilities of laser technology with plasma chemistry to achieve highly controlled etching results. The development of an effective LIPE setup involves intricate design considerations to optimize parameters such as laser wavelength, pulse duration, plasma generation, and material interaction. Advantages of LIPE in comparison to conventional processing techniques introduce true 3D maskless etching, flexibility of design and reduction of the cost of overall system. Etching of various materials has been reported such as germanium [2], silicon [3], silicon carbide [4] and polymers [5]. Mostly the reported setups utilize a focusing lenses with long focal lengths. This paper presents the design and implementation of a novel setup for LIPE, aimed at enhancing plasma confinement by using a focusing lens with high NA and shorter focal length. This research contributes to the advancement of LIPE technology, paving the way for its broader application in fields such as semiconductor manufacturing, photonic structures manufacturing, and surface engineering.

II. Opto-Mechanical Design

Fig. 1 depicts the schematic layout of the experiment. The pulsed laser source is delivered to the vacuum chamber via reflective mirrors. An optical shutter is placed in front of the laser source for control during the etching process. The laser then travels through a beam expander to achieve optimal focus. In this work, a high NA 0.45 aspherical lens is placed near the vacuum chamber window viewport to deliver a diffraction-limited focus of the laser source inside the chamber. The combination of sufficient numerical aperture and focal length of the lens was considered to overcome mechanical constraints. Tightly focusing the laser beam creates an intense light field [6], which ionizes the gas molecules via multiphoton absorption [7]. The vacuum chamber is located on an XY micropositioning stage controlled via a PC. To control the position of the laser focus in relation to the sample, the focusing objective is mounted on a Z linear stage. The optical setup further utilizes simple epi-illumination of the sample via a collimated LED source, which is delivered by a pair of beamsplitters and imaged via a CCD camera.

III. Vacuum Chamber

To provide an atmosphere of gas-carrying reactive species, an innovative vacuum chamber was designed. The chamber consists of two viewports: one for delivering the laser beam inside the chamber and a side viewport for determining the plasma position in relation to the sample and for spectroscopic control of the etching process. Additionally, the chamber includes two KF25 connectors, one for connection to the gas system and the other for connection to a thermistor and heating element, enabling further control of the sample during the etching process. Furthermore, a connection from a stainless steel tube is included to deliver reactive gas inside. The entire assembly is machined from 1.4044 stainless steel, which is known for its durability in highly corrosive environments.

Furthermore, a simple load-lock system is created on the rear side of the chamber for the convenient insertion and extraction of tested samples. The gas system consists of a cascade turbomolecular pump and a rotary pump, which first evacuates the ambient atmosphere and then help to maintain a constant flow of reactive gases while pumping out the etching products. The mixture of gases is delivered via mass flow controllers, and the gas system also includes accessories such as valves to ensure precise control.

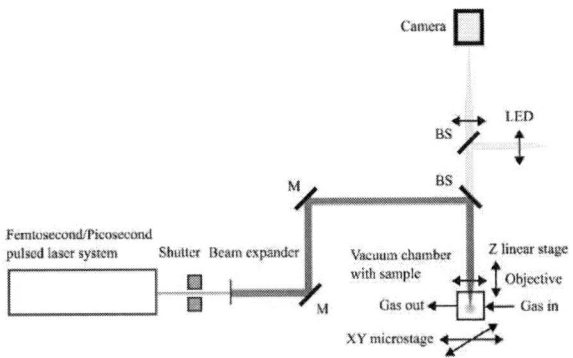

Fig. 1. Diagram of optical layout and vacuum chamber connected to micropositioning stage.

IV. METHODS

The initial tests for plasma ignition utilized a Ti:sapphire femtosecond laser oscillator (Mira HP, Coherent), which produces 130 fs pulses at a repetition rate of 76 MHz and an energy per pulse of 52 nJ, operating at a wavelength of 800 nm. This laser system is capable of writing optical waveguides into bulk of glass via refractive index change. However, it was insufficient to ignite plasma in the gases. Therefore, a more powerful laser system, PERLA 100, was used. This system is based on a Yb:YAG thin disk regenerative amplifier, delivering pulses at a fundamental wavelength of 1030 nm with energies up to 20 mJ operating at a repetition rate of 1-200 kHz and a pulse duration close to 1 ps. Fig. 2 depicts the preliminary test of an ignited plasma plume in an ambient atmosphere.

V. OUTLOOK

Further work will include an ultrafast laser source based on Yb:KGW regenerative amplifier Pharos 10 W, which provides a tunable range for pulse length from 290 fs to 10 ps with energy up to 0.4 mJ at repetition rate 20 - 200 kHz on the fundamental wavelength of 1030 nm. The setup will be used for writing photonic structures into optical materials.

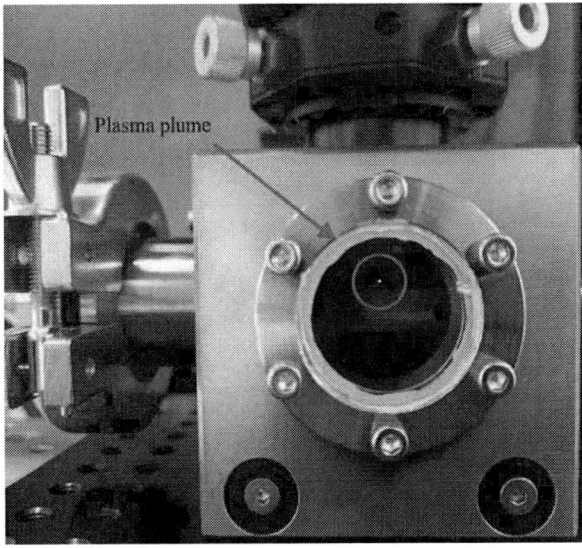

Fig. 2. Preliminary test of ignited plasma in the ambient atmosphere via pulsed picosecond laser with pulse energy of just 80 µJ.

ACKNOWLEDGMENT

The authors would like to thank the institutional support of the Czech Academy of Sciences (RVO: 68081731) and Technology Agency of the Czech Republic TA CR – Center of Advanced Electron and Photonic Optics (TN 02000020/004).

REFERENCES

[1] TANIGUCHI, Norio. Current status in, and future trends of, ultraprecision machining and ultrafine materials processing. *CIRP annals*, 1983, 32.2: 573-582.

[2] EHRHARDT, Martin, et al. Dry etching of germanium with laser induced reactive micro plasma. *Lasers in Manufacturing and Materials Processing*, 2021, 8.3: 237-255.

[3] HEINKE, Robert, et al. Dry etching of monocrystalline silicon using a laser-induced reactive micro plasma. *Applied Surface Science Advances*, 2021, 6: 100169.

[4] ZIMMER, Klaus, et al. Etching of SiC–SiC-composites by a laser-induced plasma in a reactive gas. *Ceramics International*, 2022, 48.1: 90-95.

[5] STREISEL, Leon, et al. Ultrahigh Precision Machining of Polymer Surface using Laser-Induced Reactive Micro-Plasmas. *Journal of Laser Micro/Nanoengineering*, 2022, 17.1.

[6] BUKIN, Vladimir Valentinovich, et al. Formation and development dynamics of femtosecond laser microplasma in gases. *Quantum Electronics*, 2006, 36.7: 638.

[7] HOSSAIN, Afaque M., et al. Time-and position-dependent breakdown volume calculations to explain experimentally observed femtosecond laser-induced plasma properties. *ACS Photonics*, 2023, 10.5: 1232-1239.

Evaluation of Virtual Source Size Measurement System for a Field Emission Electron Gun

Erina Kawamoto*, and Soichiro Matsunga

R&D group, Hitachi Ltd, 1-280 Higashi-koigakubo, Kokubunji, Tokyo, Japan.
*Corresponding author: erina.kawamoto.rm@hitachi.com

Abstract—Abstract— Size of the effective electron source is one of the key parameters affecting the special resolution of electron microscopy. The effective source of a field emission (FE) gun is not equal to the actual electron emission surface but is formed on the backside of the emission surface. Since this virtual source is about tens of nm in size, we proposed a method of enlarging the virtual source by projecting it onto a detector using a double lens system. The double-lens system facilitated obtaining sufficient magnification and highly accurate evaluation of the magnification factor which determines the accuracy of the measurement. In this magnification projection system, the accuracy of the evaluation of the virtual source position affects the evaluation of the magnification factor. Therefore, we evaluated the virtual source position with an accuracy of 51.07±0.05 mm. This high accuracy evaluation allowed us to obtain an error of less than 2% between the simulated magnification factor M_{sim} and experimental magnification factor M_{exp}.

Index Terms—virtual source, electron microscope.

I. INTRODUCTION

Resolution of high-performance scanning electron microscope (SEM) has now reached the sub- nanometre range. Field emission electron sources (FE), which emit electrons from needles sharpened to tens of nanometres in diameter, are used in these high performance SEMs, and the performance of FEs is being improved to achieve even higher resolution. In order to fairly evaluate the performance of these new FEs, standardisation of evaluation methods is desirable. The effective source is one of the key parameters affecting the special resolution of electron microscopy. This effective source of FE is an imaginary virtual source formed on the back surface of the source (not the electron emitting surface), which is several tens of nanometres in size [1]. Therefore, it is necessary to measure the small virtual source by magnification projection; Kawasaki et al [2] used an apparatus with a height of 7.4 m and a large magnification factor by increasing the distance between the main plane of the magnifying lens and the image plane; Veen et al [3] used an apparatus with a shorter distance between the virtual source and the main plane of the magnifying lens to obtain a large magnification factor. In this measurement, the accuracy of the virtual source size depends strongly on the distance from the main plane of the magnifying lens to the virtual source. However, the reported equipment cannot directly measure the position of the virtual source. Therefore, it is difficult to judge the measurement accuracy of the virtual source size.

Based on this background, this study proposed a novel system with a double lens, which can measure the size of the virtual source on a laboratory scale and guarantee its accuracy. Two-step magnification projection using double lenses has two advantages. First, sufficient magnification can be obtained even with laboratory-scale equipment. Second, the measurement accuracy can be ensured by controlling the magnification of the double lens. In this study, we proposed a method for precisely measuring the positions of virtual sources, which can be the main cause of errors in the magnification factor.

II. EQUIPMENT AND OPTICAL DESIGN

Figure 1 shows a schematic of the proposed double lens optical system. The height of this equipment implementing this optical system is approximately 1800 mm, which is large enough to be installed in a laboratory. The virtual source is magnified by the first lens. This magnified virtual source is further magnified by the second lens and projected onto the detector. The magnification factor was calculated by multiplying two lens magnifications determined by the geometric arrangement of the virtual source, the first lens, the first image plane (first cross), the second lens and the sample position. The magnification factor was controllable based on the position of the first image plane, i.e. the intensity of the first and second lenses. The optical conditions were such that the aberrations of the double lens system were negligible.

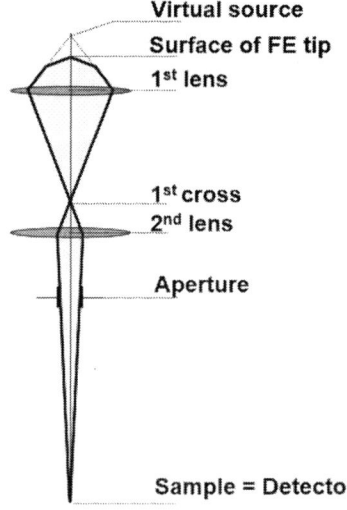

Fig. 1. Schematic of magnification projection optics using double-lens system.

979-8-3503-7977-8/24 $31.00 © 2024 IEEE

The acceleration voltage of the electron beam and the diameter of the aperture were adjusted to obtain suitable optics with negligible aberrations.

The magnified beam size (Rbeam) was measured with the knife-edge method. As a knife-edge, a mask with a cross-shaped aperture in a 1.6 m-thick Si film was used. This mask was prepared by the micro electro mechanical systems (MEMS) process.

III. PRECISE MEASUREMENT OF MAGNIFICATION

The position of the virtual source is the main cause of magnification error. Figure 2(a) shows the optical system used for the virtual source position measurement.

In this measurement, the second lens was not used and only the first lens was used to focus the beam, and the relationship between the intensity of the first lens and the probe current (I_p), which is the amount of current passing through the aperture, was investigated.

Figures 2(b) and 2(c) show the magnified views of the virtual source and the focusing spot, respectively. As shown in Fig. 2(b), I_p is described using the angular current density J_Ω from the virtual source and the beam acquisition angle at the virtual source plane α_0 as follows,

$$I_p = J_\Omega . \pi . \alpha_0^2 \tag{1}$$

and α_0 is described using a diameter of aperture ϕ, a distance from the virtual source and the first lens A_1 as follows,

$$I_p = J_\Omega . \pi . \left(\frac{\frac{\phi}{2} . f_1}{A_1 . f_1 . 285.2 . (A_1 - f_1)} \right)^2 \tag{2}$$

The focal length of the first lens (f1) is shown as

$$\frac{1}{f_1} = \frac{1}{A_1} + \frac{1}{B_1'} \tag{3}$$

The unknown variables J_Ω and A_1 can be measured accurately by fitting Eq.2 with the experimentally obtained relationship between I_p and f_1.

Using the method described above, the position of the virtual source position A_1 was set to 51.07±0.05 mm.

IV. ASSESING MAGNIFICATION SETTING

Figure 3 shows that the relationship between the simulated magnification factor M_{sim} and projected beam diameter R_{beam}. R_{beam} is expressed as the virtual-source diameter r_{source} multiplied by the experimental magnification factor M_{exp}

$$R_{beam} = r_{source} . M_{exp} \tag{4}$$

Since, R_{beam} is proportional to M_{exp}, if M_{sim} coincides with M_{exp}, the graph should be linear through the origin. The black line in Fig.3 shows the result for M_{sim} using the A_1 value of 50.1 mm which is the position of the FE tip surface. The graph is not linear through the origin because the value of A_1 is not set correctly.

On the other hand, the blue line in Fig.3 shows the result for M_{sim} using the A_1 value of 51.07 mm. The data lie on a straight line through the origin. It indicates that the value of A1 has been set correctly.

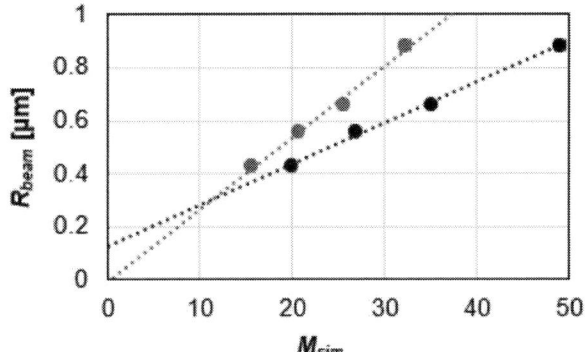

Fig. 3. The relationship between M_{sim} and R_{beam}.

REFERENCES

[1] L. Reimer, Scanning electron microscopy: Physics of image formation and microanalysis, 2nd ed., Springer New York, 1998.
[2] T. Kawasaki, T. Akashi, K. Kasuya, and H. Shinada, Ultramicroscopy 202, 2019, 107.
[3] A. H. V. van Veen, C. W. Hagen, J. E. Barth, and P. Kruit, J. Vac. Sci. Technol. B 19, 2001, 2038.

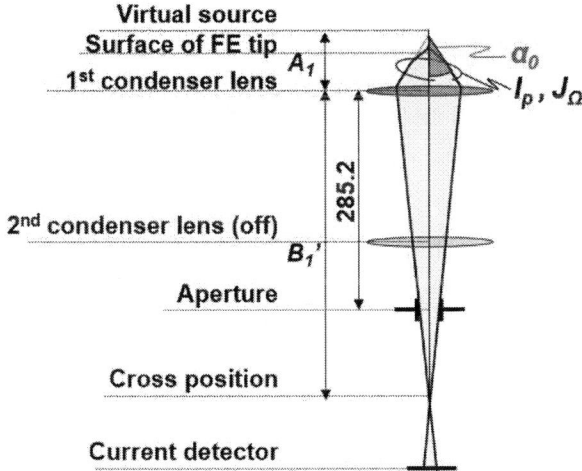

Fig. 2. (a) Schematic of optics for virtual source position measurement, (b) Zoomed-in view of the virtual source and (c) the focused image plane.

Electron Emission Energy Spread Analysis under Plasmon Resonance

Yuyue Ding, Yinyao Chen, Tao Cui, Zheyu Song, Yan Shen*, and Shaozhi Deng

State Key Laboratory of Optoelectronic Materials and Technologies, Guangdong Provincial Key Laboratory of Display Materials and Technologies, School of Electronics and Information Technology, Sun Yat-sen University, Guangzhou 510275, China

*Corresponding author: shenyan7@mail.sysu.edu.cn

Abstract—**The multi-modal electron emission stimulation using the plasmon properties of cathode has been widely concerned. This study elucidates the occurrence of photo-electric co-excited field emission originating from an Au-on-W needle nano-cold-cathode, which was stimulated by a laser of tunable wavelength operating at low intensity. Notably, within the context of plasmon resonance enhancement, the pattern of electron emission energy spread, specifically within the constrained near-infrared wavelength region (CWL~1550 nm), demonstrated a significant increase in tandem with the augmentation of the incident laser power.**

Keywords— plasmon-enhanced electron emission, nano-cold-cathode, electron emission energy spread

I. INTRODUCTION

Cathodes with plasmonic characteristics enhance the electron emission performance by demonstrating their plasmon resonance suitable for an efficient photo-electric co-excitation [1]. Gold (Au), a conventional plasmonic precious metal, has been highlighted for its capacity to significantly enhance the near electromagnetic field, and has exhibited pronounced spatio-temporal features in electron emission from an Au needle nano-cathode [2]. Tungsten (W), one of the most commonly employed cathode materials, characterized by its point electron source design, has been validated to support tunable electron emission dynamics and spatial-temporal distributions, leveraging the inherent properties of incident laser excitation [3]. In this study, we fabricated an Au-on-W composite needle nano-cold-cathode, which benefits from the inclusion of Au plasmons, allowing to achieve emission currents at the level of 100 nA. More importantly, our findings indicate that an increase in the laser excitation intensity results in a well-marked broadening of the electron emission energy spread, particularly for those excited wavelengths at plasmon resonance.

II. RESULTS AND DISCUSSION

Fig. 1 presents measurement illustration of the photo-electric co-excited electron emission of the Au-on-W needle nano-cold-cathode, which boasts a curvature radius (*R*) of 125 nm and cone angle (*α*) of 3.8° (see the inset in Fig. 1). The manufacturing of W needle has gone through the W wire welding and the needle-like electrochemical etching. Before the Au plasmons decoration, the surface of the W needle needs undergo purification *via* a thermal flashing step. Subsequently, a solution containing Au nanoparticles, each with an approximate radius (R_{Au}) of 45 nm, undergoes steps of ultrasonication, centrifugation and dilution before being deposited onto the W needle through a carefully controlled droplet application process.

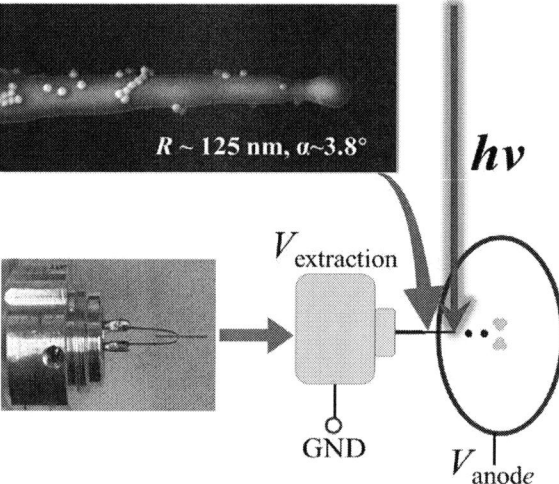

Fig. 1. Schematic diagram of photo-electric co-excited electron emission measurement of the Au-on-W needle nano-cold-cathode. The insets are SEM image of the needle tip and photograph of the cathode.

The Au-on-W needle cathode was mounted on a four-dimensional manipulator, with a flat metal anode precisely adjusting the vacuum gap. During the measurement, a supercontinuum white-light laser emitting wavelengths from 400 to 2400 nm, with a repetition frequency of 4 MHz and pulse width of ~100 ps, was served as the excitation source. The electron emission current under white light excitation was measured using a picoammeter, assisted by an anode bias electrostatic field (E_{bias}). The correlation between the emission current and the electrostatic field under different laser intensities is depicted in Fig. 2a. Notably, at a low average laser intensity (~6 W cm^{-2}), the achievable maximum emission current reaches up to 780 nA, marking a 26-fold increase compared to the pure W needle nano-cold-cathode with similar germetrical features [3]. This demonstrates that the presence of Au nanoparticles, coupled with their plasmonic enhancement effect, substantially elevates the electron yield from the W needle cathode.

In addition, according to the F-N curves in the inset of Fig. 2a, the field emission F-N characteristic curves of the Au-on-W needle cathode are linearly distributed whether under pure electroluminescence (laser off) or photo-electric co-excited emission, indicating that it follows the classic field emission mechanism of metallic materials. However, these F-N curves show different slopes under different laser intensities. Considering that the geometric appearance of the needle cathode has not changed before and after the test, it can be considered that their electrostatic field enhancement factor (*β*) has never changed. Thus it should be caused by changes in the equivalent work function. Specifically, according to the slope changes of the F-N curves, it can be inferred that under such driving field conditions dominated by the local electrostatic field, the continuous increase of the laser excitation intensity is ultimately reflected in the reduction of the equivalent work function of the cathode surface.

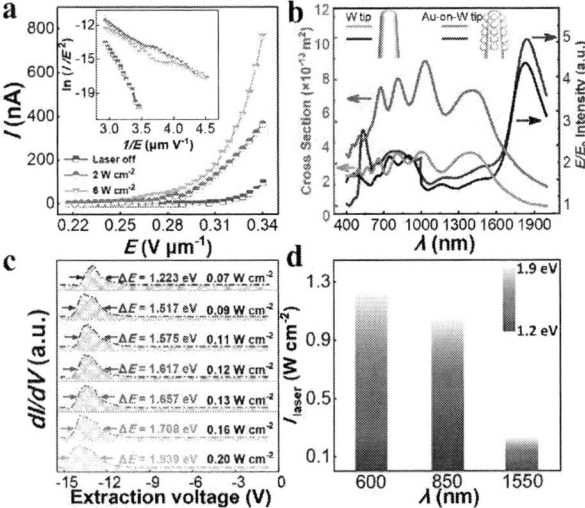

Fig. 2. (a) *I-E* curves and FN plots of the Au-on-W needle under white-light excitation with different intensities; (b) FDTD simulated light absorption cross-sections of the cathode with or without Au plasmon decoration; (c) Electron emission energy spread (Δ*E*) measured under various laser intensities with CWL of 1550 nm; (d) Comparison of Δ*E* and the needing laser intensity for different excitation bands.

To further delineate the optimal band range for laser excitation by the Au-on-W needle, the Finite-Difference Time-Domain (FDTD) was employed to simulate the near-field enhancements and light absorption cross-sections of the cathode with and without Au decoration. The W needle model consists of a spherical cap and a truncated cone, with a fixed tip curvature radius (R) and a cone angle (α). Then, taking 20 Au nanoparticles and orient them along the needle axis. As depicted in Fig. 2b, the participation of Au plasmons enhances the light absorption effect and the near electromagnetic field at the emission edge. The black and orange curves represent the near-field enhancement and light absorption cross section of the pure W needle, respectively, while the blue and red ones reveal those of the Au-on-W needle cathode. It can also be found that compared with the characteristic peaks of pure W needle, the light absorption and near-field enhancement effects of the Au-on-W needle become more significant in the near infrared band (1000-1900 nm). Therefore, we believe that the participation of Au plasmon should be related to both the photon absorption process and the optical-field-induced process of the Au-on-W needle electron emission.

The energy spread (Δ*E*) of the electron beam produced by the Au-on-W cathode was further measured using a retardation potential method. To establish a decelerating electric field for the emitted electrons, a periodic triangle-wave signal ranging from 0 to -15 V was generated by a signal generator in conjunction with a voltage amplifier. With the anode voltage fixed, a Faraday cup attached to the anode collected these emitted electrons through a small hole in the center of a phosphor screen. The electrons were then converted into voltage on an oscilloscope, facilitating the calculation of the electron energy distribution produced by the Au-on-W needle cathode under various excitation conditions. As the intensity of the laser excitation progressively increased, the relationship curves between different electron emission currents and cathode extraction voltages were recorded. Through further differential processing, the corresponding electron beam energy spread was obtained. The distributions of Δ*E* under varying laser intensities at CWL~1550 nm are illustrated in Fig. 2c, where an increase in excitation intensity from 0.07 to 0.20 W cm^{-2} results in a significant increase in Δ*E* from 1.223 to 1.939 eV. This observed trend is attributable to the coupling enhancement of the plasmon resonance effect of Au nanoparticles and W needle in the near-infrared band, leading to a more pronounced light absorption and field enhancement, and consequently, an increase in the electron yield and emission current from the cathode. The resultant elevation in emitted electron numbers with increased excitation intensity further amplifies the electron emission energy spread, under the well-known Boersch effect (a coulomb repulsion).

After a comparative analysis of Δ*E* examined under the other two optical excitation bands at CWL of 600 and 850 nm, as shown in Fig. 2d, one can assess that the average excitation intensity required to achieve similar energy spread across these three different bands. It was discovered that a significantly lower excitation intensity (< 0.2 W cm^{-2}) is needed at CWL~1550 nm to attain energy spread comparable to those at 600 and 850 nm. Therefore, in view of the negative effect of broadened energy spread on the high brightness of electron sources, it is necessary to balance the advantages and disadvantages in the process of introducing plasmons to cathodes.

ACKNOWLEDGMENT

This work was supported by National Key Basic Research Program of China (2019YFA0210201, 2019YFA0210200), National Natural Science Foundation of China (52072416, 51702372), and the Science and Technology Planning Project of Guangdong Province (2023B1212060025).

REFERENCES

[1] P. Dombi, et al. "Ultrafast strong-field photoemission from plasmonic nanoparticles," Nano letters, vol.13, pp. 674-678, January 2013 .

[2] D. J. Park, et al. "Strong field acceleration and steering of ultrafast electron pulses from a sharp metallic nanotip," Physical review letters, vol. 109, pp. 244803, December 2012.

[3] Y. Chen, S. Tang, Y. Shen, H. Chen, amd S. Deng, "A tunable photo-electric co-excited point electron source: low-intensity excitation emission and structure-modulated spectrum-selection," Nanoscale, vol. 15, pp. 8643-8653, April 2023.

Electron Emission Properties of Planar-Type Electron Emission Devices Based On A Graphene/h-BN/Si Heterostructure Fabricated By Inductively Coupled Plasma-Enhanced Chemical Vapor Deposition

Katsuhisa Murakami*, Hiromasa Murata, and Masayoshi Nagao
Device Technology Research Institute
National Institute of Advanced Industrial Science and Technology , Tsukuba, Japan
*Corresponding author: murakami.k@aist.go.jp

Abstract—**A flat-type electron emission devices using a graphene/h-BN/Si heterostructure were fabricated by inductively coupled plasma-enhanced chemical vapor deposition. High emission current density of around 15 mA/cm^2 was achieved from these devices.**

Keywords—Hexagonal Boron Nitride, Graphene, Chemical vapor deposition

I. INTRODUCTION

A metal/oxide/semiconductor (MOS) type electron emission source has the ability of operation even in poor vacuum condition of around 10 Pa, in an atmospheric pressure gas condition and in a solution. The reasons of operable in various circumstance of the MOS type emission device is that Fowler-Nordheim tunnelling through a potential barrier followed by the electron acceleration occur in the device structure. These unique properties will pioneer the various unique applications such as electron microscope in low vacuum conditions, the neutralizer of the ion thruster for the space applications, and direct decompose of the various molecules by electron-beam irradiations. The planar-type electron emission devices, however, have inherent problems of the poor efficiency (typically below 1 %), and wide energy width of the emitted electrons. The primary reasons of these critical problems for the electron beam applications are the energy loss due to the electron scattering during electron acceleration within the device. We have recently overcome the critical problem of inelastic electron scattering by using a gate electrode of around 3-layer graphene and an insulating layer of multilayer hexagonal boron nitride (h-BN) of around 6 nm, resulting in improved the emission efficiency and monochromatic properties of electron beams in MOS-type electron sources. The maximum electron emission efficiency was 48.5 % when multilayer graphene gate electrodes were used [1-5]. Furthermore, to suppress electron energy loss attributed to the electron scattering in the insulating layer by using h-BN insulator, the energy spread of electron beams with 0.18 eV at half maximum width was achieved, which is higher performance compared to that of 0.3 eV of the commercial tungsten field emitter used in a high-resolution electron microscope [6-7]. The h-BN layer of the device was, however, realized by the transferred multilayer h-BN, resulting in leakage current from the cracks generated by the transfer processes. To further improve their electron emission properties, the crack-free multilayered h-BN insulating layer

is demanded. In this study, we fabricated the flat-type electron emission sources utilizing a graphene/h-BN heterostructure directly synthesized by inductively coupled plasma-enhanced chemical vapor deposition (ICP-CVD) and evaluate their electron emission properties.

II. EXPERIMENT

The n-type Si with 300 nm thick SiO$_2$ was used as the substrate. The circle shape with a size of 10 μm diameter was fabricated for the electron emission area patterning using a photolithography followed by a wet etching of the thermal oxide layer. The h-BN insulating layer of approximately 10 nm thickness was deposited directly on the whole substrate by ICP-CVD at 500 °C [8]. The graphene gate electrode with around 3 layers was directly deposited on the h-BN by ICP-CVD at 800 °C. The stacked layers of graphene and h-BN were partly removed by a photolithography and a reactive ion etching using SF$_6$. The contact metal stack of Ni(100nm)/Ti(10nm) was fabricated by an electron-beam evaporation followed by the lift-off process.

The current-voltage curves of the fabricated devices were measured at the base pressure of $10^{-4} \sim 10^{-6}$ Pa. The stainless plate for the anode electrode with a bias of 1 kV was set 5 mm above the device.

III. RESULTS

The optical microscope image of the fabricated device was shown in Figure 1. The h-BN insulating layer was successfully deposited without any cracks and wrinkles, which usually occurs in the transfer process from a Cu foil.

The representative characteristics of electron emission from the fabricated device was shown in Figure 2. High current density of around 15 mA/cm^2 were obtained when the cathode bias was applied to -20 V.

IV. SUMARRY

The flat-type electron source using a graphene and h-BN stack directly deposited on the Si substrate was fabricated by ICP-CVD. We also demonstrate the electron emission from the fabricated device with high current density of around 15 mA/cm^2.

ACKNOWLEDGMENTS

We would like to thank JSPS KAKENHI Grant Numbers JP21H01401, JP22K18855, JP22K18800, and JP24K00954 for financial support of our research.

REFERENCES

[1] K. Murakami, S. Tanaka, A. Miyashita, M. Nagao, Y. Nemoto, M. Takeguchi, and J. Fujita, Appl. Phys. Lett., 108, 083506 (2016).

[2] K. Murakami, S. Tanaka, T. Iijima, M. Nagao, Y. Nemoto, M. Takeguchi, Y. Yamada, and M. Sasaki, J. Vac. Sci. Technol. B , 36, 02C110 (2018).

[3] K. Murakami, J. Miyaji, R. Furuya, M. Adachi, M. Nagao, Y. Neo, Y. Takao, Y. Yamada, M. Sasaki, and H. Mimura, Appl. Phys. Lett., 114, 213501 (2019).

[4] R. Furuya, Y. Takao, M. Nagao, and K. Murakami, Acta Astronaut. 174, 48-54 (2020).

[5] K. Murakami, M. Adachi, J. Miyaji, R. Furuya, M. Nagao, Y. Yamada, Y. Neo, Y. Takao, M. Sasaki, and H. Mimura, ACS Appl. Electron. Mater. 2, 2265-2273 (2020).

[6] K. Murakami, T. Igari, K. Mitsuishi, M. Nagao, M. Sasaki, and Y. Yamada, ACS Appl. Mater. Interfaeces, 12, 4061-4067 (2019).

[7] T. Igari, M. Nagao, K. Mitsuishi, M. Sasaki, Y. Yamada, and K. Murakami, Phys. Rev. Applied, 15, 014044 (2021).

[8] M. Yamamoto, H. Murata, N. Miyata, H. Takashima, M. Nagao, H. Mimura, Y. Neo, and K. Murakmai, ACS Omega 8, 5497-5505 (2023).

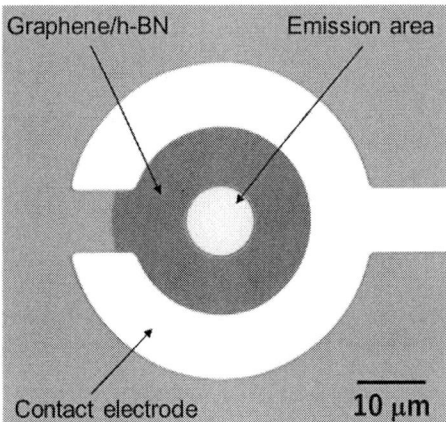

Figure 1: Optical microscope image of the fabricated device using a graphene/h-BN/Si heterostructure.

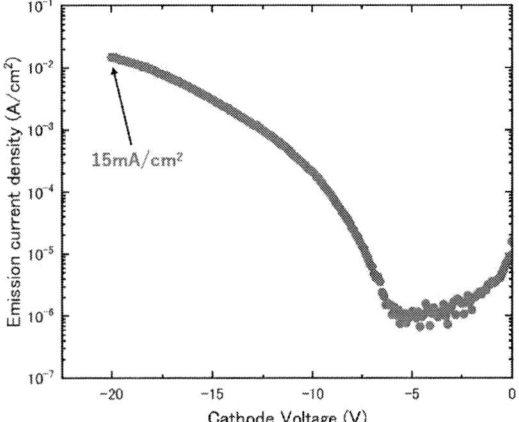

Figure 2: Current density-voltage curve of the fabricated device using a graphene/h-BN/Si heterostructure

Electron-source Investigations with Discharge-resistant CNT Field Emitter Cathodes

Wolfram Knapp
KNAPPTRON GmbH – Vacuum Electronics
D-39291 Moeser, GERMANY
E-mail: dr.who.knapp@t-online.de

Abstract—**Electron sources with thick film CNT field emitter cathodes are resistant to internal discharges of up to several tens of watts. Studies show that in their controlled applications, conditioning for targeted pre-aging, reproducibility and improved long-term stability is achieved. These properties also enable research into the transition from electron field emission to glow discharges and to micro arcs. The presentation will focus on empirical results.**

Keywords—electron source, CNT field emitter cathode, field emission measurement, transition to glow discharge, micro-arc treatment, controlled pre-ageing, long-term stability

I. INTRODUCTION

A discharge breakdown in electron sources with field emitter cathodes is often very problematic for use, as they have a destructive effect on most types of field emitters. As an alternative, we have investigated prototypes of self-developed electron sources with thick-film CNT field emitter cathodes, which are suitable for vacuum electronic applications up to over 10 mA in DC operation.

II. EXPRIMENTAL METHODS

A. Electron Source Design

The focus of our investigations is the development of an electron source with CNT material Carbon Buckypaper, which has a thickness of about 150 μm. Fig. 1 shows the schematic of the electron source (ES) and the circuit set-up for measuring the electron field emission. The active cathode circle area (C) has a diameter of 5 mm and the distance to the micro-grid (G) is ~50 μm, as already described in detail with regard to material [1], in the basic structure [2] and in the field emission measurements [3].

B. Experimental Setup

Fig. 2 is a view of the measurement setup. The oscilloscope of the online control, as shown in Fig. 3, is used for a standard measurement of the FE characteristic [4]. The characteristic curve is recorded at a frequency of 10 mHz and is repeated as often as required for long-term tests.

To determine the electron transfer v_e at the micro-grid, the power supplies PS1 and PS2 are connected in parallel so that all currents of the triode can be measured potential-free, with:

$$v_e = I_A / I_C \qquad (1)$$

and the control for the electron balance in the triode:

$$I_A = I_C - I_G \qquad (2)$$

The anode current is important for the applications, e.g. in X-ray tubes and ionization gauges for vacuum measurements.

In micro-arc treatments, the series resistance R_S has been reduced and the maximum current of PS1 in the grid circuit is controlled (cf. Fig. 1 and measurements in Fig. 3 and Fig. 4).

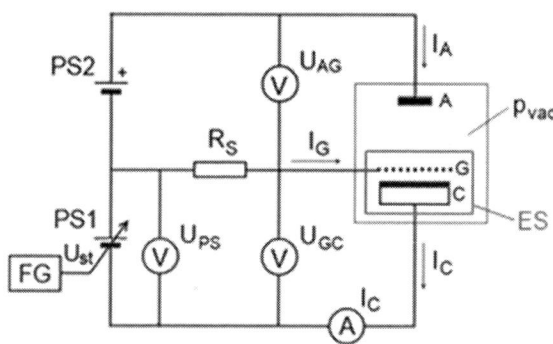

Fig. 1. Circuit diagram for measuring the field emission characteristics.

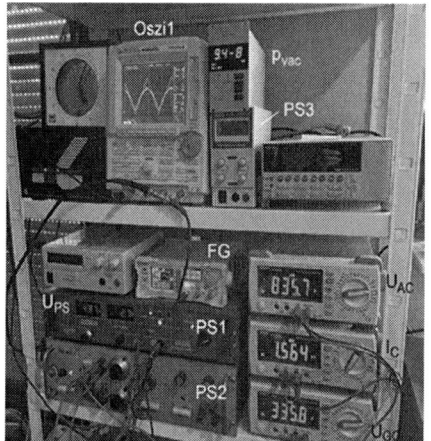

Fig. 2. Measurement setup (cf. Fig. 1) in the KNAPPTRON laboratory.

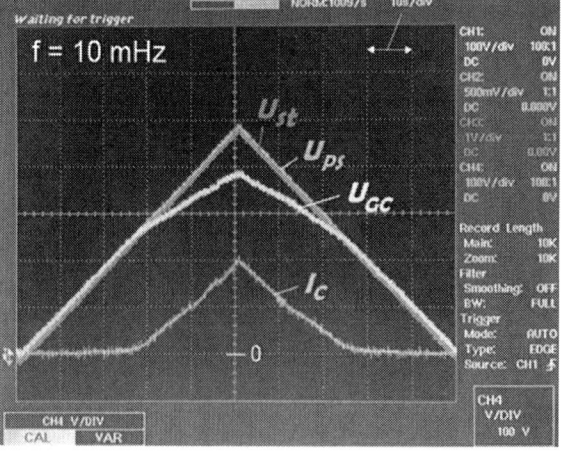

Fig. 3. Oscillogram of typical electron field emission characteristic curves recording over 100 s, I_{C-max} = ~1 mA and without discharges (R_S = 100kΩ).

III. RESULTS AND DISCUSSION

Based on positive experience with discharges, controlled micro-arc treatments are used for rapid pre-aging of CNT field emitter cathodes in order to achieve reproducible long-term stability. See the oscillogram in Fig. 4 as an example.

These robust CNT cathodes are very resistant and therefore particularly suitable for researching the transition from electron field emission to glow discharge [5], which stabilises under suitable conditions and emission currents significantly more than 1 mA [6].

The author calls this transition the "tsunami effect" because the characteristic curve in the three phases:

- normal field emission (see theory of R. Forbes [7]),

- significant decrease due to space charge limitation,

- and sudden increase with a Townsend electron avalanche to a glow discharge (U_{GD} ~ constant)

is reminiscent of this natural phenomenon (cf. Fig. 5, curve 1).

However, important experimental work and theoretical considerations come from plasma physics, as the selection of some relevant references [8], [9] and [10] shows.

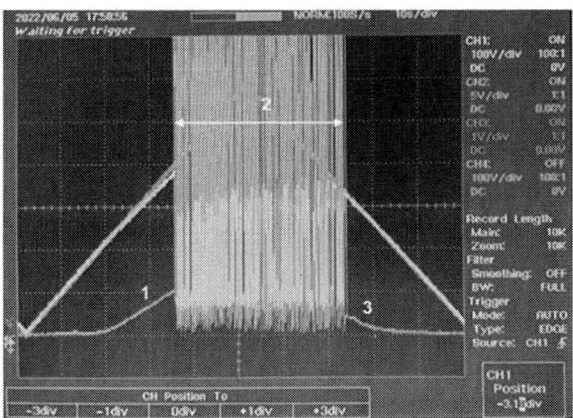

Fig. 4. *U(t)* and *I(t)* curves of our CNT-FE cathode with field emission before (1) and after (3) a controlled micro-arc treatment (2) with many micro-arc discharges about 40 s duration (green curve: $I_C = 5$ mA/div; $R_S = 4.7$ kΩ).

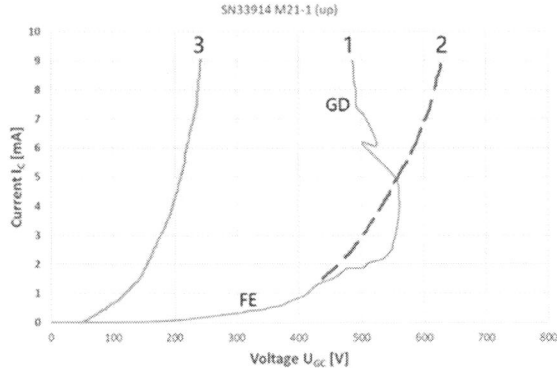

Fig. 5. Measured $I_C = f(U_{GC})$ characteristic curve (1) with "tsunami effect" as transition from field emission (FE) to glow discharge (GD) and the schematic illustrations of FE in the theoretically extrapolation (2) for higher emission current $I_C > 1.5$ mA or at higher field strength (3).

IV. SUMMARY AND OUTLOOK

Since 2000, the author has conducted intensive research on thick-film CNT field emitters of the Carbon Buckypaper type. Regardless of very different electrical loads in a vacuum and years of storage in air, all tested FE cathodes can still be used for electron field emission. The tutorial lecture is a good opportunity to report on the acquired knowledge and empirical experience, with a focus on achieving and stabilizing field emission currents in the range > 1 mA. X-ray tubes and ionization manometers are currently the most important applications.

An important research result on vacuum discharges is that the field emission properties of thick-film CNT field emitter cathodes have deteriorated only slightly, but are latent. We have therefore developed a stress test for the rapid pre-aging of the electron source in order to achieve long-term stable and reproducible FE properties. The time-limited stress test is based on stronger but controlled micro-arc discharges within the electron source (cf. Fig. 4). At higher emission currents, the challenge is that plasma discharge (glow discharge and micro arc) stabilize. Our interesting research outlook: Investigations of micro-plasma cathodes with amplification by CNT field emitter cathodes.

ACKNOWLEDGMENT

Special thanks go to rtw RÖNTGEN-TECHNIK DR. WARRIKHOFF GmbH & Co. KG in Neuenhagen near Berlin, especially to CEO Mr A. Warrikhoff and his colleagues in the FE project team, for the many years of very good cooperation and support for the author's research work.

REFERENCES

[1] W. Knapp, D. Schleussner, "Field-emission characteristics of carbon buckypaper," J. Vac. Sci. Technol. B 21(1), Jan/Feb. 2003, 557-561.

[2] W. Knapp, D. Schleussner, "Special features of emitter sources with CNT field emitter and micro grid," Appl. Surf. Sci. 251, 2005, pp. 164-169.

[3] W. Knapp, "Electron sources with CNT field emitters – results and experience from measurements", 1st Bavarian-Czech-Polish Summer School on Vacuum Nanoelectronics 2023, TAZ Spiegelau, Germany, Sept. 20-21, 2023.

[4] M. M. Allaham, M. S. Mousa, R. G. Forbes, "Comparing the performance of Fowler-Nordheim plots and Murphy-Good plots," 33rd International Vacuum Nanoelectronics Conference (IVNC) 2020, July 6-8, Virtual conf. hosted in Lyon, France.

[5] D. Wenger, W. Knapp, B. Hensel, S. Tedde, „Transition of Electron Field Emission to Normal Glow Discharge", IEEE Transaction on Electron Devices 61 (11), 2014, 3864-3870.

[6] W. Knapp, "Using Fowler-Nordheim plot for the comparison of electron emission efficiency of field emission (FE) and thermal emission (TE) cathodes," 7th ITG International Vacuum Electronics Workshop (IVEW) 2020 and 13th International Vacuum Electron Sources Conference (IVeSC) 2020, May 26-29, Virtual conf. hosted in Paris, France.

[7] G. Gaertner, W. Knapp, R. G. Forbes (Eds.), Modern Developments in Vacuum Electron Sources. Topics in Applied Physics, Vol. 135, Springer Nature Switzerland AG, 2020, Chapter 9, pp. 387-447.

[8] D. B. Go, A. Venkattraman, "Microscale gas breakdown: ion-enhanced field emission and the modified Paschen´s curve (Topical Review)," J. of Phys. D: Appl. Physics 47, 2014, 503001.

[9] A. M. Loveless, L. I. Breen, A. L. Garner, "Analytic theory for field emission driven microscale gas breakdown for a pin-to-plate geometry," J. Appl. Phys. 129, 2021, 103301-1 – 14.

[10] H. Wang, A. M. Loveless, A. M. Darr, A. L. Garner, "Transitions between field emission and vacuum breakdown in nanoscale gaps," J. Vac. Sci. Technol. B 40, 2022, 062805-1 – 9.

Enhanced EUV Lighting with Focusing Electrode Adapted C-beam Irradiation Technique

Iksu Kim, Umesh Balaso Apugade and Kyu Chang Park[*]
Department of Information Display, Kyung Hee Unversity, Seoul, Korea
[*]Corresponding author: kyupark@khu.ac.kr

Abstract— In this study, we investigated the effect of electron spot size on EUV generation by adapting a focusing electrode. We have studied and demonstrated high quality, debris free and low cost of EUV lighting source by C-beam (cold cathode beam) without focusing electrode. By reducing focal spot size (FSS), current density of EUV was improved. By optimizing focusing electrode conditions, we could achieve 32% decrease of Full Width at Half Maximum (FWHM) with minimal loss of anode current.

Keywords— *Focusing electrode, EUV (Extreme Ultraviolet), C-beam, current density, Full width at half maximum (FWHM), Focusing effect*

I. INTRODUCTION

EUV (Extreme Ultraviolet) lithography using LPP (Laser Produced Plasma) or DPP (Discharge Produced Plasma) is important for EUV inspection tool[1]. However, EUV lighting technology faces challenges such as debris, and short lifetimes. To solve these issues, including the short lifetime due to debris and high costs, we developed a novel EUV lighting technique. We successfully demonstrated EUV lighting without encountering debris or cost issues[2][3].

Nevertheless, we observed opportunities for improving EUV intensity and spot size. Therefore, we adjusted the focusing electrode to achieve a smaller spot size without compromising photocurrent. The focuser electrode effectively converges electron trajectories by adjusting the focuser bias[4]. We evaluated the current of a 1x4 array electron source through diode test. Additionally, we measured the anode and cathode currents through triode test to calculate transmittance. By investigate transmittance variations with different focuser biases using this approach, we identified the optimal focuser bias.

Subsequently, we measured changes in sapphire photoluminescence intensity. When the light source of the EUV wavelength reaches, sapphire emits visible light by the photoluminescence[5]. Thus, we could observe EUV irradiation region by sapphire photoluminescence area, indirectly. During this measurement, we analysed grey value and full width at half maximum (FWHM) changes using an EOS camera.

II. Experimental

A. Fabrication of Electron Source

Electron source was grown on the Ni deposited Si wafer with triode PECVD (Plasma Enhanced Chemical Vapor Deposition). We patterned 30 nm Ni deposited on Si wafer by using photolithography and wet etching. CNT's dot size is 3 μm and height is ~40 μm. We used metal mesh gate electrode as a controller of electron's trajectory and current. And insulate all part by ceramic body

B. I-V characteristics of triode module with focusing electrode

We measured the I-V characteristics using a triode module by applying various focusing biases. An anode bias of 15 kV was applied, and the gate voltage was increased up to 2,300 V.

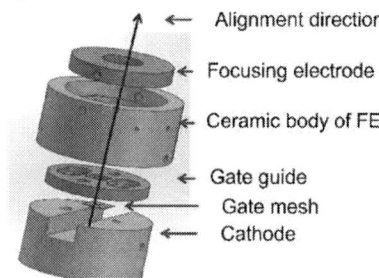

Fig. 1. Schematic of focusing C-beam

Typical operation of C-beam for EUV lighting would be 0.3 mA and 15 kV anode bias. At 15kV of anode bias, we could obtain 0.3 mA anode current. The electron transmittance is very important for low power operation. With the focusing electrode adaption, electron transmittance would be reduced by the loss of focusing interaction. However, the C-beam structure shows very less decreases in electron transmittance due to the directionality of C-beam trajectory. C-beam shows small beam opening angle due to vertically aligned structure.

C- beam's opening angle would be 1.2degree angle, resulting less scattering with gate and focusing electrode. However, the focusing electrode bias increases which more than -700 V, the electron transmittance through focusing dramatically start to reduce by reflection of focusing electrode. At -800 V focusing bias, anode current drop to 23 μA, resulting dramatic reduction of total electron transmittance of 5%.

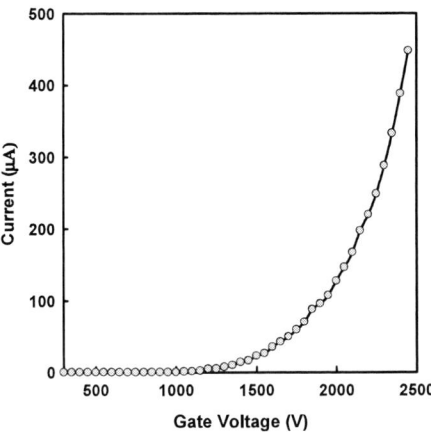

Fig. 2. Anode current of C-beam

Without any focusing bias, the triode system showed an anode current 0.47mA until focusing bias of -500 V, and anode current does not change, significantly. However, at -600 V, the transmittance dropped to about 68% and 50% at -700 V. This drop is due to the repulsive force between the focusing electrode and the electrons, indicating the need to optimize the applied voltage.

C. Light area measurement with photoluminescence (PL) profile on sapphire

We measured the PL intensity profile of sapphire plate for the indirect FWHM measurement of EUV. Sapphire emits a PL of visible wavelength due to the EUV irradiation on the plate. We can indirectly identify EUV FWHM with this sapphire photoluminescence. To optimize of focusing electrode bias, we measured PL intensity profile of sapphire. We used EOS camera to achieve gray value and full width at half maximum of the light on sapphire plate

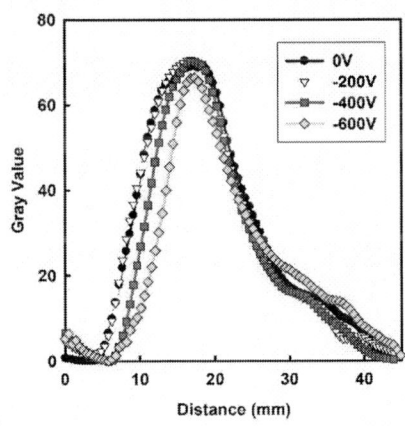

Fig. 3. PL profile on the sapphire by EUV lighting source

III. CONCLUSION

We studied the focusing effect of EUV light area with C-beam irradiation technique. The lighting area measure with photoluminescence area measurement on sapphire plate. Sapphire wafers positioned on the EUV lighting pass and measured the luminescence area compared with various focusing bias. We optimized the focused electrode voltage to increase the photocurrent density without the loss of anode current. An increase in photocurrent density means a more focused generation of EUV photons and a higher concentration of EUV photons in a narrow area. The FWHM value of PL profile changed with anode bias and generation. As a result, we were able to reduce the FWHM about 30%, which also increased the photocurrent density.

ACKNOWLEDGMENT

This work was supported by the Technology Innovation Program (No. 20013595, Extreme ultraviolet light source using nano-electron beam) funded by the Ministry of Trade, Industry & Energy (MOTIE, Korea).

REFERENCES

[1] A. Egbert, B. Tkachenko, S. Becker, and B. N. Chichkov, "Compact electron-based EUV source for at-wavelength metrology," in *High-Power Laser Ablation V*, 2004, pp. 693–703.

[2] B. C. Adhikari, S. T. Yoo, and K. C. Park, "EUV lighting by the cold cathode C-beam irradiation technique," in *2023 7th IEEE Electron Devices Technology & Manufacturing Conference (EDTM)*, 2023, pp. 1–3. doi: 10.1109/EDTM55494.2023.10102964.

[3] S. T. Yoo and K. C. Park, "Extreme Ultraviolet Lighting Using Carbon Nanotube-Based Cold Cathode Electron Beam," *Nanomaterials*, vol. 12, no. 23, 2022, doi: 10.3390/nano12234134.

[4] B. C. Adhikari, B. Ketan, R. Patil, E. H. Choi, and K. C. Park, "Optimization of vertically aligned carbon nanotube beam trajectory with the help of focusing electrode in the microchannel plate," *Sci. Rep.*, vol. 13, no. 1, p. 15630, 2023, doi: 10.1038/s41598-023-42554-8.

[5] J.-I. Lo, Y.-C. Peng, H.-C. Lu, and B.-M. Cheng, "Photoluminescence of optical windows excited with extreme ultraviolet radiation," *Opt. Lett.*, vol. 45, no. 19, pp. 5413–5415, 2020.

Enhanced Field Emission Properties of Titanium Nitride Coated ZnO Nanowires

Xinran Li, Guofu Zhang, Zhuoran Ou, Zhipeng Zhang, Shaozhi Deng, Ningsheng Xu and Jun Chen*

State Key Laboratory of Optoelectronic Materials and Technologies, Guangdong Province Key Laboratory of Display, Material and Technology, School of Electronics and Information Technology, Sun Yat-sen University, Guangzhou 510275, People's Republic of China

*E-mail: stscjun@mail.sysu.edu.cn

Abstract—**ZnO nanowire field emitters have great applications in vacuum microelectronic devices. Surface coating is an effective approach which can enhance field emission properties of field emitters. In this study, titanium nitride layer was coated on ZnO nanowires. The field emission properties were studied in situ using a nanoprobe technique. It is found that the TiN coating lowers the turn-on field and increase the emission current. We attribute the enhanced field emission to the decreased work function after coating.**

Keywords—**ZnO nanowires, TiN coating; field emission; in situ**

I. INTRODUCTION

ZnO nanowires are considered as promising material of large-area field-emitter arrays (FEAs) in applications of flat-panel X-ray source, photodetector and flat panel display [1]. Due to the semiconductor nature of ZnO, uniform emission can be easily achieved for large area ZnO nanowire field emitters. However, the field-emission current needs improvement compared with other more conductive one-dimensional material field emitters, such as carbon nanotubes.

Previous work has shown that titanium nitride (TiN) coating on field emitters can achieve high current and low-voltage operation [2-4]. M. Nakamoto et al. have reported TiN-coated Si FEAs have higher emission current and better thermal stability compared with non-coated Si FEAs [4].

In this work, we employed magnetron sputtering to deposit TiN thin film onto the surface of ZnO nanowires. The field-emission characteristics of single ZnO nanowires were investigated using an in-situ measurement system, and the mechanism for the emission enhancement was discussed.

II. EXPERIMENT

The ZnO nanowires were prepared on silicon substrate using thermal oxidation and the detailed process can be found in our previous work [5]. TiN film was deposited on ZnO nanowires by direct-current (DC) magnetron sputtering of a Ti target using a gas mixture of 87.5% Ar and 12.5% N_2. The sputtering power was set at 3.5 kW.

The characterization of morphology of ZnO nanowires was carried out in a scanning electron microscope (SEM, SUPRA60). Field-emission characteristics of individual nanowires were characterized using in-situ nanoprobe measurement system. In this experiment, we used a Keithley picoammeter for the voltage output and I-V measurement. The nanoprobes were prepared by electrochemically corroding the tungsten filaments. SEM pictures in Fig. 1(a) show the set-up for the field-emission measurement of an individual nanowire.

The sweep voltage range in the field emission characteristic measurement is 0V to 100V with a cathode and anode spacing of 600 nm. The background pressure of the test system is approximately 3×10^{-8} Pa.

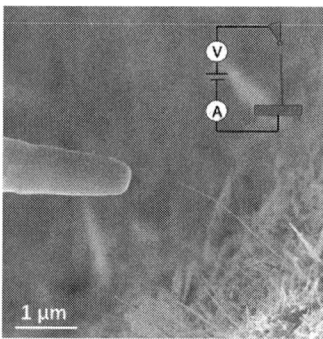

Fig. 1. SEM picture of the experimental set-up during field-emission measurement of an individual nanowire. Inset shows the schematic diagram of the measurement.

III. RESULTS AND DISCUSSION

Fig. 2(a-d) shows SEM images of the prepared ZnO nanowires obtained from the same position of the sample with and without TiN coating, which indicates that the TiN film was successfully deposited on ZnO nanowires.

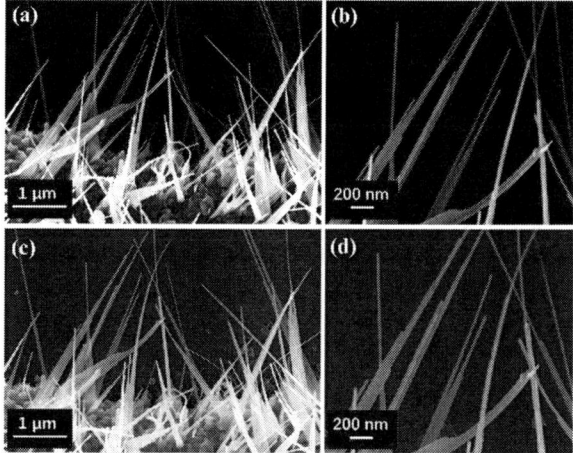

Fig. 2. SEM images of the prepared ZnO nanowires obtained from the same position of the sample before and after coating. (a,b) uncoated, (c,d) TiN-coated.

By the SEM pictures, the diameter of the nanowire tip before coating is estimated to be ~10 nm and it is ~29 nm after coating. The thickness of the TiN coating is estimated to be ~10 nm. The TiN film was characterized using the X-ray photoelectron spectroscopy (XPS). Peaks of Ti, N, and O are observed in the XPS spectrum, which indicated that the prepared TiN film was partially oxidized.

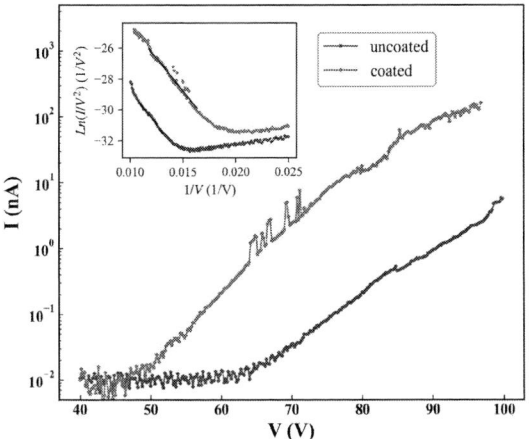

Fig. 3. Typical field-emission I-V curve of single ZnO Nanowires before and after TiN coating. The inset shows the corresponding F-N plot.

TABLE I. COMPARISON OF FIELD-EMISSION PROPERTIES OF SINGLE ZNO NANOWIRE WITH AND WITHOUT COATING

Samples	No.	Maximum emission current (nA)	Turn-on voltage (V)	β (obtained from FN plot)
Uncoated nanowires	1	2.17	84.0	52
	2	0.75	85.8	47
	3	5.70	76.8	34
	4	0.40	93.6	30
	5	5.43	80.4	37
	Avg.	2.89	84.1	40
TiN-coated nanowires	1	3.96	79.2	41
	2	162.00	60.0	51
	3	34.30	69.6	47
	4	11.00	93.0	38
	5	0.24	90.0	58
	Avg.	42.30	78.4	47

TABLE I demonstrates the typical field-emission I-V curves and corresponding F-N plots of single ZnO nanowires with and without TiN coating. Measurement from multiple single nanowires was carried out. TABLE I. summarized the maximum field-emission currents and turn-on voltages obtained from the ZnO nanowires. As shown in TABLE I, the average of maximum field-emission currents of ZnO nanowires with TiN-coating was more than one order of magnitude higher than those of uncoated ZnO nanowires, and the average of turn-on voltages (corresponding to current of 0.1 nA) of TiN-coated ZnO nanowires was lower than those of uncoated ZnO nanowires.

From the F-N plots, upward feature is observed in the high field section for uncoated samples, while downward feature is observed for coated samples. The upward feature is tentatively attributed to the heating-up at high current for uncoated samples, which have high resistance. While the downward feature can often be attributed to the space charge effect. The field enhancement factor (β) was calculated using F-N equation as shown in (1) and the calculated β are presented in TABLE I. :

$$\ln\left(\frac{I}{V^2}\right) = -B\frac{\phi^{\frac{3}{2}}d}{\beta} \bullet \frac{1}{V} + C \, , \qquad (1)$$

where B is a constant equal to 6.53×10^9, β represents the field enhancement factor, d represents the cathode-to-anode distance, and φ represents the work function.

In the calculation, work function of ZnO was adopted for both the coated and uncoated nanowires. Higher average field enhancement factor was obtained for the coated nanowires. However, the SEM inspection shows the coated nanowire tips have larger diameters, which will lower the field enhancement factor. Therefore, the higher β is apparently due to the lowered work function after coating. According to the literature, the work function of ZnO was reported to be 5.3 eV, while the work function of TiN was reported to be 4.2-4.5 eV [6]. Thus, we think the lower work function of TiN may contribute to the obtained lowered turn-on voltage.

IV. CONCLUSION

TiN film was deposited on ZnO nanowires by DC magnetron sputtering. The in situ measurement of field emission characteristics from individual ZnO nanowires revealed that TiN coating lowers the turn-on field and increase the emission current. We attribute the enhanced field emission to the decreased work function. This work indicated that TiN-coated ZnO nanowires have promising applications in field-emission devices.

ACKNOWLEDGMENT

Thanks for the financial support from the National Key R&D Program of China (2022YFA1204200), Key R&D Program of Guangdong (2023B0101200013), National NSFC (82272131 & 62001527), the S&T Department of Guangdong (2023B1212060025).

REFERENCES

[1] D. Chen et al., "Transmission type flat-panel X-ray source using ZnO nanowire field emitters," Appl. Phys. Lett., vol. 107, no. 24, pp. 243105-1–243105-5, December 2015.

[2] S.Y. Kang, J. H. Lee, Y. H. Song, et al., "Emission characteristics of TiN-coated silicon field emitter arrays," J.Vac. Sci. Technol. B, vol.16, pp.871-874, 1998.

[3] M. Nakamoto et al., "Suitability of low-work-function titanium nitride coated transfer mold field-emitter arrays for harsh environment applications," J. Moo. J. Vac. Sci. Technol. B, vol. 29, pp.02B112, 2011.

[4] W. K. Lo, G. Parthasarathy, C. W. Lo, D. M. Tanenbaum, H. G. Craighead, and M. S.Isaacson, "Titanium nitride coated tungsten cold field emission sources," J. Vac. Sci. Technol. B, vol.14, pp.3787-3791 ,August 1996.

[5] C. X. Zhao et al., "Large-Scale Synthesis of Bicrystalline ZnO Nanowire Arrays by Thermal Oxidation of Zinc Film: Growth Mechanism and High-Performance Field Emission," Cryst. Growth Des. 2013, vol. 13, no. 7, pp.2897–2905, June 2013.

[6] Y. Y. Zhuang, Y. Q. Liu, H. Xia, Y. Y. Li, X. Li, and T. Li, "Effective work function of TiN films: Profound surface effect and controllable aging process", AIP Advances, vol.12, Issue 12, pp. 125222, December 2022.

979-8-3503-7977-8/24 $31.00 © 2024 IEEE

Enhancing X-ray Emission of Cold Cathode Flat-Panel X-ray Source by Pulsed Driving

Ruowen Fan, Song Kang, Guofu Zhang, Juncong She, Shaozhi Deng, Jun Chen*

State Key Laboratory of Optoelectronic Materials and Technologies, Guangdong Province Key Laboratory of Display Material and Technology, School of Electronics and Information Technology, Sun Yat-sen University
Guangzhou 510275, People's Republic of China
* Email: stscjun@mail.sysu.edu.cn

Abstract—**The maximum X-ray emission intensity from a cold cathode flat panel X-ray source is limited by the heat produced on the anode. The effect of driving modes on the anode temperature was simulated using finite element simulation software. Pulsed driving is found to effectively lower the anode temperature and increase the X-ray emission of cold cathode flat panel X-ray source. The optimal pulsed driving conditions were predicted by the calculation.**

Keywords—cold cathode; finite element; thermal simulation; pulsed driving

I. INTRODUCTION

Cold cathode flat panel X-ray source (FPXS) using ZnO nanowires and their imaging applications have been reported recently. [1-3] Schemes for the FPXS application in the low dose, high resolution, adaptive stationary computed tomography (CT) was also proposed. [4]

The FPXS composes of large-area field emitters and anode. The X-ray emission is generated by electron bombardment of a large-area anode. Usually, the maximum intensity of X-ray emission is limited by the anode temperature, which is related to the driving conditions of the device. Therefore, it is possible to increase maximum X-ray emission intensity in cold cathode FPXS through pulsed driving while maintaining stable anode temperatures.

In this work the anode temperature of the cold cathode FPXS was simulated under different pulsed driving conditions and anode power density. The increased X-ray emission in cold cathode FPXS through pulsed driving was explored while maintaining a constant anode temperature.

II. SIMULATION METHOD

Fig. 1 presents the schematic of the FPXS, with the anode panel composed of a thin layer of metal Mo deposited on a glass substrate. The cathode panel is nanowire field emitters arrays on electrode.

COMSOL Multiphysics was used to simulate the anode temperature. Fig. 2 depicts the simulation model. The device has cathode and anode glass panels separated by a spacer. In the model the anode is a 1 μm Mo thin film on 3 mm glass. The device is surrounded by air. The anode is loaded with a fixed heat flux and the variation of anode surface temperature with time were obtained. The initial temperature is set to 293.15 K. A representative result of the temperature distribution obtained by the simulation is also shown in Fig. 2.

The influences of direct current (DC) and pulsed driving modes, power loads, and pulse duty cycles on anode surface temperature were studied.

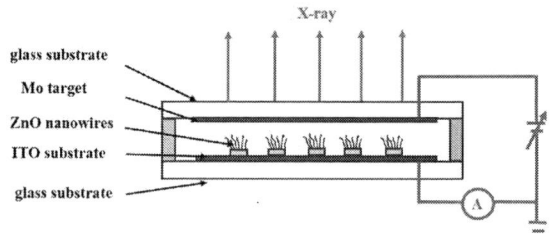

Fig. 1. The schematic diagram of the FPXS.

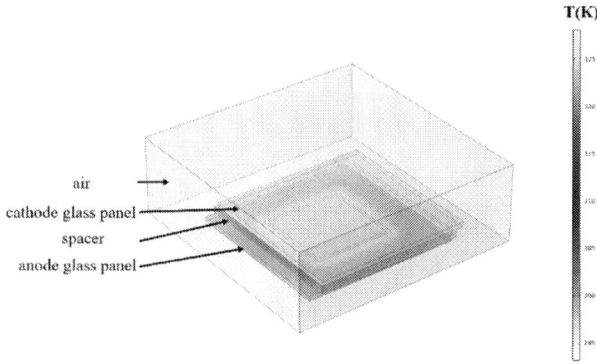

Fig. 2. The model and typical simulation result of the temperature distribution.

III. RESULTS AND DISCUSSION

We first simulated the effect of direct current and pulsed driving on the anode surface temperature under the same heat flux. Fig. 3 (a) illustrates the change with time of anode surface temperature under a consistent power density of 0.3 W/cm² for both DC and pulsed driving. The anode temperature increases rapidly with time and gradually stabilized under prolong operation. The stable temperature is about 358.05 K for DC driving and for pulsed driving it is about 327.90 K. The results indicate that, at the same power density, the anode temperature is lower with pulsed driving. However, the shorter effective operating time of the device under pulsed driving will lower the X-ray emission output.

We further simulated the corresponding power densities for DC and pulsed driving when the device operates under the same stable temperature. Fig. 3 (b) illustrates the simulation results. The calculations show that when the duty cycle of the pulsed driving is 20% and the power density is 1.72 W/cm², the stable anode temperature is the same as that of DC driving with a power density of 0.3 W/cm². Pulsed driving enables higher driving power density, but correspondingly, the effective operating time is reduced. However, considering the effective operating time, it is possible the X-ray emission could be lager by pulsed driving.

979-8-3503-7977-8/24 $31.00 © 2024 IEEE

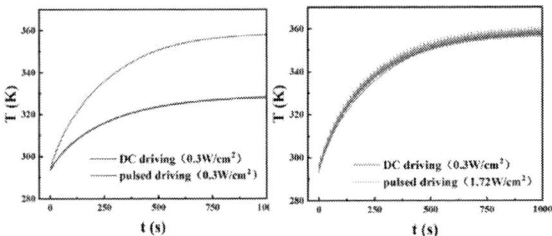

Fig. 3. (a) The temporal variation of anode surface temperature at a consistent power density of 0.3 W/cm² for both DC and pulsed driving. (b)The temporal variation of anode surface temperature when achieving the same stable temperature for both DC and pulsed driving.

When bombarded with electrons, the anode converts roughly 1% of the energy into X-rays. Assuming uniform emission, the device's peak operating power decides the maximum X-ray output. For a certain power density and anode surface temperature, the increment in X-ray emission ($\triangle I$) by pulsed driving can be obtained by using following formula.

$$\triangle I = (P_{pulsed} - P_{DC}) / P_{DC} \times 100\% \tag{1}$$

In which P_{pulsed} and P_{DC} are maximum mean power of the device under pulsed and DC operating, while keeping the same anode temperature.

Therefore, considering the effective operating time, we used (1) to calculate and compare the X-ray output dose for the FPXS under pulsed and DC driving over a fixed period.

Fig. 4 shows the calculated $\triangle I$ under different duty cycle conditions, using a constant pulse width of 1 second at different power densities. The optimal enhancement in X-ray emission is obtained at a duty cycle of about 25%. The increase in X-ray emission reaches 24% for input power density of 0.6 W/cm² while for 0.3 W/cm² it is about 16%.

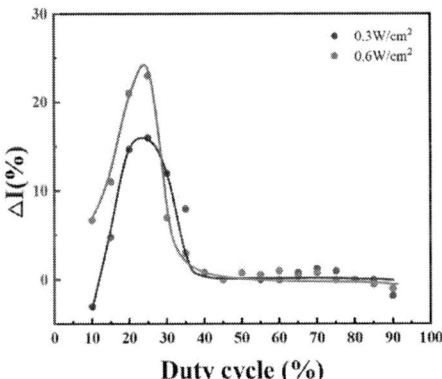

Fig. 4. The increment in X-ray emission calculated under various duty cycle conditions with a fixed pulse width of 1 s at different power density.

In addition to the duty cycle of the pulsed driving, we further calculated the effect of the pulse period (T) on X-ray emission. Fig. 5 illustrates the calculated $\triangle I$ under different duty cycles for two pulse periods (T=5s and 10 s) at a power density of 0.3 W/cm². Shorter pulse periods result in higher X-ray emission. The highest X-ray emission enhancement is also achieved at about 25% duty cycle.

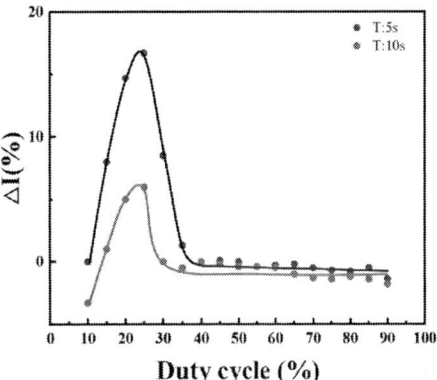

Fig. 5. The increment in X-ray emission calculated under various duty cycle conditions and pulse periods at a power density of 0.3 W/cm².

The above results revealed pulsed driving can significantly boosts the X-ray emission within a fixed operating time under the same stable temperature. As the duty cycle of the pulsed driving varies, the increment in X-ray emission peaks at a ~25% duty cycle. Similarly, a shorter period of the pulsed driving results in a greater increase in X-ray emission. Moreover, the increase in X-ray emission is also related to the operating power of the FPXS.

IV. CONCLUSION

Pulsed driving effectively increases the X-ray emission of cold cathode FPXS. Through calculations of pulsed driving parameters, the increment in X-ray emission under pulsed driving is found to be related to duty circle, pulse period and power density. With a pulsed driving duty cycle of 25%, selecting the smallest possible pulse period can achieve the optimal X-ray output from the FPXS.

ACKNOWLEDGMENT

The authors gratefully acknowledge the financial support from the National Key Research and Development Program of China (Grant No. 2022YFA1204200), Key Research and Development Program of Guangdong Province (Grant No. 2023B0101200013), National Natural Science Foundation of China (Grant Nos. 82272131 & 62001527), the Science and Technology Department of Guangdong Province (Grant No. 2023B1212060025), Fundamental Research Funds for the Central Universities.

REFERENCES

[1] C.Y. Wang, G.F. Zhang, Y. Xu, et al. "Fully Vacuum-Sealed Diode-Structure Addressable ZnO Nanowire Cold Cathode Flat-Panel X-ray Source: Fabrication and Imaging Application", Nanomaterials, 2021, vol. 11, pp. 3115.

[2] K. Wang, Y. Xu, D.K. Chen, et al. "Tungsten Target Optimization for Photon Fluence Maximization of a Transmission-type Flat-panel X-ray Source by Monte Carlo Simulation and Experimental Measurement", IEEE Transactions on Radiation and Plasma Medical Sciences, 2018, vol. 2(5), pp. 452-458.

[3] L.B Wang, Y. Xu, X.Q. Cao, et al. "Diagonal 4-inch ZnO Nanowire Cold Cathode Flat-panel X-ray Source: Preparation and Projection Imaging Properties", IEEE Transactions on Nuclear Science, 2021, vol. 68, pp. 338.

[4] Duan J, Li Y, Mou X, et al, "Coded array beam X-ray imaging based on spatio-sparsely distributed array sources," 17th Virtual international meeting on fully 3d image reconstruction in radiology and nuclear medicine, 2023, pp. 86-89.

Exploring the Future of Field Electron Emission Theory

Richard G. Forbes

Quantum Foundations and Technologies Group, School of Mathematics and Physics,
Faculty of Engineering and Physical Sciences, University of Surrey
Guildford, Surrey GU2 7XH, UK
Permanent e-mail alias: r.forbes@trinity.cantab.net

Abstract—**This presentation relates to the theory of field electron emission (FE), with focus on the theory of current-voltage characteristics and special emphasis on the theory of electron transmission probability. After a survey of where we have got to after about 100 years of investigation, the intention is to present a high-level roadmap for what (in the author's view) needs to be done in order to put the basic principles of FE theory onto a more sophisticated and complete scientific basis. Possible difficulties will be identified.**

Keywords—Field electron emission (FE) theory, FE Systems Engineering, FE current-voltage theory, transmission probability.

I. General introduction

The oral presentation summarised by this extended abstract is a new attempt, in 2024, to offer a high-level roadmap for the future development of "mainstream" field electron emission (FE) theory. By "mainstream" is meant theory relating to FE from metal elements in the "deep tunneling" regime (also known as the "cold FE" regime), including theory relating to measured FE current-voltage and energy-distribution characteristics, and their interpretation.

There are many important/specialized aspects of FE theory outside this mainstream, including temperature-induced and photon-induced effects, FE from materials that are not metal elements, and FE from rough surfaces. Research in these areas is welcome, but we shall not get non-mainstream theory exactly correct if mainstream theory is not exactly correct.

The general structure of the presentation and abstract is as follows. A scientific and technological introduction reviews: (a) the relevance of FE; (b) the current state of "mainstream" FE transmission probability theory; and (c) issues relating to current-voltage measurement. Attention then focuses on the long-term issues, especially transmission probability theory.

To save space, references are given to earlier papers by the author. This presentation is not intended as a review of all relevant work. A copy of the oral presentation slides will be mounted on ResearchGate, and are most easily be found by a websearch on the title. This copy will contain a more comprehensive list of references.

II. Scientific and Technological Background

A. The relevance of field electron emission (FE)

- Many technologies are based on FE. FE theory can assist interpretation of experimental results, and/or guide R&D.
- FE theory is important in electrical breakdown theory.
- FE is one of the paradigm examples of quantum tunnelling. As such, future developments in FE theory might have implications elsewhere in science, for example in quantum biology. (DNA mutation involves thermally activated tunnelling of hydrogen atoms or ions [1].)

B. FE transmission probability theory

FE literature contains very many variants of transmission probability theory, but the commonest variants assume the Sommerfeld model, take the emitting surface as planar and of large lateral extent, and disregard the role of atomic-level wave functions. This has been called *smooth planar metal-emitter (SPME) methodology* [2]. Within this methodology, four often used "levels" of FE theory have been defined [2].

- *Level 1 – FN FE theory.* Theory given by Fowler & Nordheim in 1928 and corrected by Stern et al. in 1929 [3].
- *Level 0 – elementary FE theory*. A modern simplified version of Level 1 theory that omits a pre-exponential factor. Level-0 theory is widely used by experimentalists (albeit undesirably), and is also used in undergraduate teaching.
- *Level 2 – MG FE theory*. Theory given by Murphy & Good in 1956 [4], but best presented using modern SI-type equations. MG FE theory uses an exponent correction factor "v" related to image effects, and makes predictions typically around 100 times larger than Level-1 theory. "v" was originally written in terms of the *Nordheim parameter "y"*.
- *Level 3 – Extended MG (EMG) FE theory*. A variant of MG FE theory, based on modern understanding of the mathematics of "v", developed from 2006 onwards. "v" is recognized as a special solution $v_{FD}(x)$ of the Gauss Hypergeometric Differential Equation, using the *Gauss variable x*. In FE modelling, x is set equal to the *scaled field f* [$=y^2$]. EMG theory is also "experiment facing" in its definition of area. EMG FE theory is described in [5,6].

C. Issues related to current-voltage measurement

In principle, the prediction and interpretation of FE current-voltage measurements is part of a specialized (but under-developed) branch of electronic/electrical engineering recently called [7] *FE Systems Engineering*. Traditional current-prediction and data-analysis methodologies assume that FE systems are *electronically ideal* (i.e., that a characteristic local electrostatic-field magnitude F_C related to the emission process is directly proportional to the measured voltage V_m, with the *local voltage conversion length (LVCL)* $\zeta_C = V_m/F_C$ *constant* (independent of V_m). With modern emitters this is often not the case [7]; thus values of extracted emitter characterization parameters may be spurious.

For SPME-methodology data-analyses, validity tests [7,8] can now check whether systems are electronically ideal, and should now be part of normal research practice.

Other serious difficulties with current technological data-analysis practice, including the widespread use of multiply defective equations, are described in [7–9].

979-8-3503-7977-8/24 $31.00 © 2024 IEEE

III. GENERAL NEEDS FOR FUTURE PROGRESS

• There is inconsistency in the terminology used in FE literature. "Debabelization" and standardization are needed.

• A background difficulty is that some definitions in International Standards are not "fit for purpose for FE", in particular: "Fermi level", "voltage", and "chemical potential".

• FE literature has about 20 different approximations that can be construed as approximations for $v_{FD}(x)$, with some making the false prediction that a FN plot should be linear. "Good" and "high-precision" (error better than 10^{-9}) approximations for $v_{FD}(x)$ are now known [6]. Many older approximations are thus now obsolete, and their use should be discontinued.

• For the present, it would be helpful (particularly in making comparisons) if all current-voltage data-analysis were carried out using a single preferred methodology based on Level-3 theory. This would be "better R&D practice" than commonly used methods, and a good starting place for urgently needed further development of FE theory and data-analysis. Using MG plots [5], rather than FN plots, might be part of it.

• There is a need for more-careful explicit analysis of the experimental conditions and apparatus needed for reliable comparison of FE theory and experiment (see [10,11]). In the last 100 years there have been tens of thousands of experiments on the electrostatics of field emitters, but only one reliable experiment on the basic theory [12]. This generated a result intermediate between Level 1 theory and Level 2 theory, but closer to Level 2 theory. This result needs to be confirmed for metal surfaces other than tungsten (110).

• There are at least 12 reasons [7] why modern FE systems might not be electronically ideal. Some, like space-charge effects, have well developed theory. Other aspects of FE Systems Engineering require significant further development.

IV. SOME IMMEDIATE NEEDS

For emission theory and related data-analysis procedures, two lines of development need to be progressed.

A. Theory for smooth-surface curved emitters

SPME methodology breaks down for emitters that are sufficiently sharply curved, say with apex radius of curvature less than about 20–50 nm. The first obvious advance is to develop analysis methodology (transmission theory, specific statistical mechanics, emission current theory, validity checks, and data-analysis procedures) for electronically ideal FE systems based on "Earthed" spherical emitters. Much progress has been made by Kyritsakis [13]. Theory for more complicated shapes (e.g., [14]) may be needed later.

Data analysis for such systems needs to use numerical multilinear regression techniques. A useful step would be to apply such techniques in SPME methodology and compare extracted parameters with those from FN and MG plots.

B. Getting atoms into planar-surface metal FE theory

An urgently needed step forwards is to "get the atoms into planar-surface metal-element FE theory." Although there have been earlier quantum-mechanical treatments, recent calculations of potential-energy (PE) structures use modern density functional theory (e.g. [15–16]).

It is clear that the future of mainstream FE theory will be "completely numerical". For examples, see [15–17].

An useful first step would be to review comparisons between conventional treatments of MG theory and existing fully numerical treatments, such as that of Mayer [17].

A later need is to explore how to choose "paths of integration" (or a "typical path") through a calculated PE structure. A partial guide to this may be to look again at the theory of field ionization of a hydrogenic ion, as derived from the work of Landau and Lifschitz (LL) [18]. But, here, more fundamental quantum mechanical problems seem to intrude.

V. FUNDAMENTAL QUANTUM-MECHANICAL PROBLEMS

Briefly, it not obvious that existing quantum mechanics is "fully fit for purpose" for all FE theoretical problems. Limited space allows only an indication here. The oral presentation provides longer discussion of the following identified issues.

• The "real" nature of an electron and apparent inadequacy of the Copenhagen interpretation of quantum mechanics. ("What is the length of a tunnelling electron?" and "Can a real electron have negative kinetic energy": [No, in my view.])

• The unduly simplistic nature of "wave-particle duality" and "the uncertainty principle", in the context of FE theory.

• The apparent need to use two *different* interpretations of the wave-function in tunnelling problems involving correlation-and-exchange effects (one to describe "real behavior", the other to discuss the behavior of wave-packets that can split).

• The apparent use by LL of a specialized model of reality in their tunnelling theory. One might call this the "many simultaneous paths" model, in which an electron is considered split into infinitely many infinitesimal elements, each of which follows a different path through the barrier and would have different transmission characteristics. Several issues arise here, relating to coherence and to actual paths.

• Whether final FE emission is incoherent or coherent.

REFERENCES

[1] L. Slocombe, M. Sacchi, and J. Al-Khalili, Comm. Phys. (2002)5:109.

[2] R. G. Forbes, Proc. 2020 29th Internat. Symp. on Discharges and Electrical Insulation in Vacuum. IEEE Xplore, 2021.

[3] T. E. Stern, B. S. Gossling, and R. H. Fowler, Proc. R. Soc. Lond. A 124, 699–723, (1929).

[4] E. L. Murphy and R. H. Good, Phys. Rev. 102, 1464–1473 (1956).

[5] R. G. Forbes, R. Soc. Open Sci. 6, 190912 (2019).

[6] R. G. Forbes, Chapter 9 in: Modern Developments in Vacuum Electron Sources. Switzerland: Springer Nature, 2020.

[7] R. G. Forbes, J. Vac. Sci. Technol. B 41, 028501 (2023).

[8] R. G. Forbes, J. Vac. Sci. Technol. B 41, 042807 (2023).

[9] R. G. Forbes, J. Appl. Phys. 126, 210901 (2019).

[10] A. Ayari, P. Vincent, S. Perisanu, P. Poncharal and S. T. Purcell, J. Vac. Sci. Technol. B 40, 024001 (2022).

[11] S. V. Fillipov, A. G. Kolosko, E. O. Popov, and R. G. Forbes, R. Soc. Open Sci. 9, 220748 (2022).

[12] C. D. Ehrlich and E. W. Plummer, Phys. Rev. B 18, 3767–3771 (1978).

[13] A. Kyritsakis, F. Djurabekova, Comput. Mater. Sci. 128, 15–21 (2017).

[14] V. I. Kleshch, P. A. Zestanakis, and J. P. Xanthakis, Appl. Surf. Sci. 623, 156990 (2023).

[15] B. Lepetit, J. Appl. Phys. 125, 025107 (2019).

[16] Y. Li, J.Mann, and J. Rosenzweig, Instruments 7, 47 (2023).

[17] A. Mayer, J. Vac. Sci. Tchnol. B 29, 021803 (2011).

[18] L. D. Landau and E. M. Lifschitz, Quantum Mechanics, 3rd edition. Oxford: Butterworth-Heinemann, 1977, pp. 295–296.

Extended Definitions of Fermi Level and Fermi Energy and How Lattice Expansion Affects Fermi Energy

Richard G. Forbes

Quantum Foundations and Technologies Group, School of Mathematics and Physics,
Faculty of Engineering and Physical Sciences, University of Surrey
Guildford, Surrey GU2 7XH, UK
Permanent e-mail alias: r.forbes@trinity.cantab.net

Abstract—**Literature and internet discussions about the basic statistical mechanics of electron emitters often display inconsistency in terminology. To counteract confusion, this Poster gives careful "extended"/"adjusted" definitions for the terms "Fermi level" and "Fermi energy". The definitions given are unambiguous, are valid for all materials (metals and non-metals), are valid for both small and large bodies, are valid at all temperatures of practical interest, and are suitable for use in the context of electronic circuits. It is also confirmed that, in the theory of temperature effects in the Sommerfeld model, the total-energy difference $K_F(T)$ between the Fermi level and the Sommerfeld-well base is much more significantly affected by thermally induced lattice-expansion effects than by the constant-volume statistical-mechanical effects discussed in textbooks. To show this, simple formulas for the lattice-expansion effect on $K_F(T)$ are reported. The resulting numerical predictions have not been found in existing literature. In absolute terms the temperature-dependent change in Fermi energy due to lattice expansion is relatively small compared to the Fermi energy itself (typically around 1% or less), so significant practical implications would not be expected.**

Keywords—*Statistical mechanics, Fermi-Dirac statistics, Fermi level, Fermi energy. lattice-expansion effects.*

I. EXTENDED DEFINITIONS OF FERMI LEVEL AND ENERGY

Definitions of the terms "Fermi level" and "Fermi energy" in literature and internet discussions are often inconsistent. This report proposes "extended"/"adjusted" definitions that are useful for the surface electron emission technologies.

For thermodynamic temperature *T*, and electron total-energy E_t, the *Fermi-Dirac distribution function* $f_{FD}(E_t,T)$ is

$$f_{FD}(E_t,T) = 1/[\{E_t - E_F(T)\}/k_B T] + 1]. \qquad (1)$$

where k_B is Boltzmann's constant and $E_F(T)$ is called here the *finite-temperature Fermi level* (FL). $E_F(T)$ is the total-energy level at which $f_{FD}(E_t,T)$ has exactly the value ½. To allocate a value to $E_F(T)$, a *total-energy reference-zero* must be specified. In electronic-circuit contexts, the Fermi level E_{FLE} of the local laboratory "Earth" is often a convenient choice. Fig. 1 illustrates a situation where the electron source of an electron microscope is at high negative voltage.

In the free-electron theory of metals, the base of the Sommerfeld well models the conduction band edge. Denote the total-energy of the well base, measured relative to the same reference-zero as $E_F(T)$, by $E_C(T)$. The *Fermi energy (or energy-difference)* $K_F(T)$ is then defined by

$$K_F(T) \equiv E_F(T) - E_C(T). \qquad (2)$$

These definitions are unambiguous, are valid for all materials (metals and non-metals), are valid for both small and large bodies, are valid at all temperatures of practical interest, and are suitable for use in the context of electronic circuits.

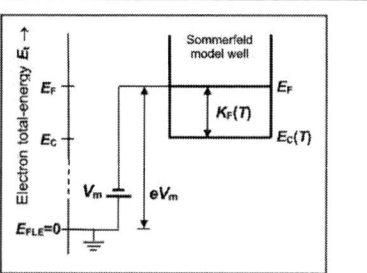

Fig. 1. To illustrate the relationships between relevant energy levels and energy differences when a measured voltage V_m raises the metal-body Fermi level E_F above the Fermi level (E_{FLE}) of the laboratory "Earth".

II. TEMPERATURE DEPENDENCE OF FERMI ENERGY $K_F(T)$

Statistical mechanical considerations as found in textbooks (e.g. [1]) yield the approximate formula

$$(\partial K_F/\partial T)_1 \approx -(\pi^2/6)(k_B/K_{F0})^2 T, \qquad (3)$$

where K_{F0} denotes $K_F(T= 0 \text{ K})$. In laboratory situations lattice expansion also occurs. This provides a second cause of temperature dependence and yields the formula [2]

$$(\partial K_F/\partial T)_2 \approx -2\alpha K_{F0}, \qquad (4)$$

where α is the coefficient of linear expansion. It is also possible to derive expressions for the change ΔK_F in Fermi energy between 0 K and room temperature, taken as 300 K (see [2]).

Table I shows a comparison made at room temperature (taken at 300 K) for selected metal elements, using data available on the internet (see [2]). Clearly the lattice expansion effects are much larger in magnitude than the statistical-mechanical effects discussed in textbooks, and thus should be included in discussions of Sommerfeld model theory. (But note that additional effects may occur in band-structure calculations (e.g., [3]).

TABLE 1. Finite-temperature Fermi energy-difference $K_F(T)$ and its temperature dependence in the Sommerfeld model, for some metal elements, at 300 K. For meaning of column labels, see text.

Metal	K_{F0} (eV)	α (K^{-1})	$(\partial K_F/\partial T)_1$ (meV/K)	$(\partial K_F/\partial T)_2$ (meV/K)	$(\Delta K_F)_1$ (meV)	$(\Delta K_F)_2$ (meV)
Ag	5.49	1.89×10^{-5}	−0.67	−208	−0.100	−45
Au	5.53	1.42×10^{-5}	−0.66	−157	−0.099	−36
Cu	7.00	1.65×10^{-5}	-0.52	−231	−0.079	−46
W	9.75	4.5×10^{-5}	−0.38	−88	−0.056	−17
Pb	9.47	2.89×10^{-5}	−0.39	−547	−0.058	−135
Fe	11.10	1.18×10^{-5}	−0.33	−262	−0.050	−45
Al	11.70	2.31×10^{-5}	−0.31	−541	−0.047	−98
Ag	5.49	1.89×10^{-5}	−0.67	−208	−0.100	−45
1	2	3	4	5	6	7

979-8-3503-7977-8/24 $31.00 © 2024 IEEE

III. WIDER DISCUSSION AND WIDER CONCLUSIONS

The results reported above are the main algebraic and numerical conclusions of a wider investigation into the basic science of the Sommerfeld model and into how it is described and used in the context of the surface emission technologies (field electron emission, thermal electron emission and photoelectron emission). The interim results of this wider investigation will be reported elsewhere in more detail (see [2]), but can be summarized as follows.

• In both literature and internet discussions, significant inconsistency and confusion surrounds the use of the terms "Fermi level" and "Fermi energy". In particular, this inconsistency is noticeable in the contexts of the surface emission technologies and of electronic circuits.

• Existing discussions often contain an inconsistent mixture of terminology drawn from: (a) statistical mechanics; (b) chemical thermodynamics; and (c) electrical/electronic engineering.

• A longer-term aim of the author is to create a properly integrated scientific discussion of the behaviour of the material components in electronic circuits, involving equivalences between the terminology used from all three disciplines.

• The immediate aim, and the objective of [2] and this report, is to create a self-contained and self-consistent description of the basic statistical mechanics of the Sommerfeld model, as applied to the material components used in electronic circuits, in particular a consistent description of the concepts of Fermi level and Fermi energy.

• A distinction needs to be made between (a) the concept of "Fermi level" and (b) the specification of its value.

• Conceptually, "Fermi level" is defined here as the name of the parameter $E_F(T)$ in eq. (1). The Fermi-Dirac distribution function $f_{FD}(E_t,T)$ has exactly the value ½ when the total energy E_t of the fermion under discussion is equal to $E_F(T)$.

• To allocate a quantitative value to a Fermi level, a total-energy reference-zero needs to be specified. In the context of the behaviour of electrons in electronic circuits, the most convenient reference-zero is often the Fermi level of the local laboratory "Earth".

• The name "finite-temperature Fermi energy" is applied here to the temperature-dependent parameter $K_F(T)$ defined by eq. (2). This is the main "adjustment" to terminology proposed in [2] and this report. (Most existing theoretical discussions allocate the name "Fermi energy" to a parameter defined only at 0 K.)

• For additional clarity, the term "Fermi energy-difference" is sometimes used here (instead of "Fermi energy"), but it is not intended that this name should necessarily be used elsewhere.

• Unlike some definitions in the literature, these "adjusted" definitions are unambiguous and apply to all materials, to both large and small bodies, at all temperatures of practical interest, and work in the context of electronic circuits.

• In basic discussions found in textbooks and on the internet, the name "Fermi level" is often allocated to a parameter that is better called the "charge neutrality level (CNL)". For metals at 0 K, the CNL is the "total-energy level of the highest occupied metal electron state at 0 K (HOMES@0K)"; for semiconductors the CNL is the top edge of the valence band; for molecular systems the CNL is the highest occupied molecular orbital (the HOMO). Defining the Fermi level as equal to the charge neutrality level is a convention that works for large metal Sommerfeld bodies (where the density-of-states in total-energy is quasi-continuous), but does not work for intrinsic semiconductors or other systems where there is a band-gap with no electron states close to the Fermi level as defined via eq. (1). Thus, to avoid confusion it is better not to use this form of definition for large metal bodies.

• Thermally induced lattice expansion changes the temperature dependence of the finite-temperature Fermi energy-difference $K_F(T)$. The effect due to lattice expansion is much larger than the effect associated with the detailed statistical mechanics of a constant-volume enclosure. Thus, lattice-expansion effects need to be taken into account in the Sommerfeld model. The need to take lattice expansion—and more sophisticated effects—into account has long been known in the context of detailed metal band-structure calculations.

• These lattice-expansion effects on the Fermi energy(-difference) are predicted to be small in comparison with the Fermi energy(-difference) values themselves (typically around 1% or less) Hence, it seems unlikely that any significant practical implications would be expected.

• There are difficulties with some language used in literature and internet discussions. In particular, the parameter $K_F(T)$ defined here is often called the "chemical potential". The parameter $K_F(T)$ does **not** have either the same scientific properties or the same units as the parameter called "chemical potential" in modern chemistry. The practice of calling $K_F(T)$ a "chemical potential" should cease.

• Calling $K_F(T)$ the "Fermi level" is also best avoided. Since alignment of Fermi levels indicates statistical-mechanical equilibrium between bodies, it is best that quantitative values of Fermi levels also be equal. This requires specification relative to a common total-energy reference zero.

• The author is not aware of any method by which the finite-temperature Fermi energy(-difference) can be reliably measured, but the theory here may be relevant to more complex problems, such as the theory of the static Seebeck effect or the temperature-dependence of local work function.

• The next stage in the wider project should be to examine terminology issues that arise in electrical/electronic engineering descriptions of circuit behaviour when different circuit components are made from different electronic materials.

• There will eventually be a question of how the inter-disciplinary aspects of the above issues are best taught.

[1] U. Mizutani, Introduction to the Electron Theory of Metals. Cambridge University Press (Virtual Publishing) 2003.

[2] R. G. Forbes, "Adjusted definitions of Fermi level and Fermi energy, and the effect of lattice expansion on the temperature dependence of redefined Fermi energy in the Sommerfeld model." To be submitted shortly; find by websearch on title.

[3] P. B. Allen and V. Heine, "Theory of the temperature dependence of electronic band structures," J. Phys. C: Solid State Phys. 9, 2305–2311 (1976).

Fabrication of Double-Gate Zinc Oxide Nanowire Field Emitter Arrays for Achieving High Current

Zhuoran Ou, Chengyun Wang, Guofu Zhang, Xinran Li, Hai Ou, Juncong She, Shaozhi Deng, and Jun Chen*

State Key Laboratory of Optoelectronic Materials and Technologies, Guangdong Province Key Laboratory of Display Material and Technology, School of Electronics and Information Technology, Sun Yat-sen University, Guangzhou, China

*Corresponding author: stscjun@mail.sysu.edu.cn

Abstract—**Large area addressable ZnO nanowire field emitter arrays (FEAs) have important application in vacuum microelectronic devices. However, further work on increasing the current is needed for achieving high brightness flat panel X-ray source. In this work, we designed a double-gate structure ZnO nanowire FEAs. Simulation results show that double-gate structure can realize higher emission current compared with single-gate structure. Then, double-gate ZnO nanowire FEAs were fabricated and their performance were studied. Maximum emission current density (3.32 mA/cm²) and transconductance (2433 nS) achieved in our device are the highest among the gate structures which were previously reported. The results verified that good gate control capability can be achieved from double-gate ZnO nanowire FEAs.**

Keywords—**Double-gate, zinc oxide nanowires, field emitter arrays.**

I. INTRODUCTION

Large area addressable field emitter arrays (FEAs) have attracted significant attention in many areas [1,2]. Especially, they are core components in flat panel X-ray sources, which can achieve selective-area and low-dose imaging for medical and industrial applications [3].

Quasi-one-dimensional (Q1D) nanomaterials [4,5] have been extensively studied for cold cathode applications. ZnO nanowires have been proven to be excellent cold cathode candidate for large-area FEAs, by virtues of easy preparation on large area, stable and uniform emission.

Gate ZnO nanowire FEAs were developed in previous studies. Gate ZnO nanowire FEAs with in-plane focus and non-coplanar focus structure were reported [6,7]. Recently, addressable flat panel X-ray source using gated ZnO nanowire FEAs has been reported and high-resolution X-ray imaging was realized [8]. However, further work on increasing the emission current of ZnO nanowire FEAs is needed for achieving brighter X-ray sources. Therefore, double-gate ZnO nanowire FEAs were designed and fabricated to realize high current in this study.

II. DEVICE STRUCTURE AND FABRICATION

The device structure of double-gate ZnO nanowire FEAs is presented in Fig.1. The nanowires are controlled by inner and outer gate at both sides of the cathode. A SiO₂ layer is located between the upper and bottom electrodes. We etch the SiO₂ layer to obtain via holes for the interconnect of upper and bottom electrodes. Focus electrode is patterned at the upper layer surrounding the outer gate. The intersectional arrangement of the gate and cathode electrodes at the bottom layer is used to realize the column and row addressing function.

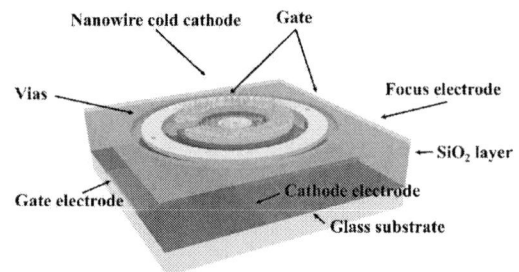

Fig.1. Schematic diagram of double-gate ZnO nanowire FEAs.

We proposed a four-mask process to fabricate double-gate ZnO nanowire FEAs as shown in Fig.2. The procedure begins with magnetron sputtering to form the bottom electrode. Then a SiO₂ layer is deposited by PECVD as the insulating layer. To establish interconnects between the two layers of electrodes, RIE was utilized to etch via holes. Next, we use magnetron sputtering to fill the via holes and form the upper electrode. The process continues with the deposition and patterning of the Zn film through electron beam evaporation. The final step involves the growth of ZnO nanowires in an ambient atmosphere at 470 °C for three hours.

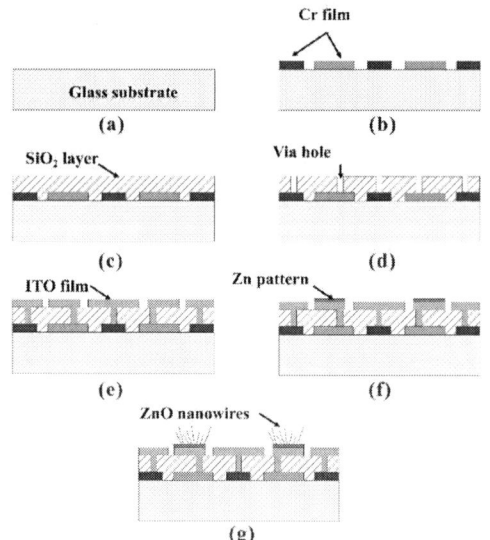

Fig.2. Fabrication process flow for double-gate ZnO nanowire FEAs.

III. RESULTS AND DISCUSSION

The emission current of double-gate ZnO nanowire FEAs was simulated using COMSOL Multiphysics, and were compared with those of the single-gate structure. Simulation of the emission current are illustrated in Fig.3. The gate voltage (V_g) is adjusted from 100 V to 190 V and anode voltage (V_a) is set to 1500 V. The current obtained from double-gate ZnO nanowire is found to be larger than that from single-gate structure. For example, when V_g is 150V, the emission current of double-gate structure is 2.4 times higher than that of single-gate structure. At the same time, the rate of current increasing in the double-gate structure is higher than that in the single-gate. The enhanced emission current is because more nanowire emitters are modulated in double-gate structure. Our simulation results shows that double-gate structure can realize high emission current and transconductance due to the increased modulation capability of double-gate.

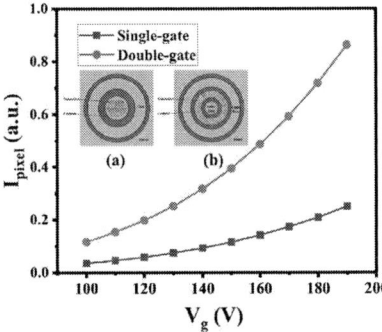

Fig.3. Comparison of the simulated I-V characteristics of double-gate and single-gate ZnO nanowire FEAs. (a) The schematics of single-gate structure; (b) The schematics of double-gate structure.

Fig.4(a) shows the morphology of the device observed by scanning electron microscopy (SEM). The FEAs consist of 160 × 160 pixels with a central spacing of 250 μm. The annular cathode is sandwiched between the inner gate and outer gate. The space between gate and cathode is 10 μm, and the ring width of cathode is 20 μm. Fig.4(b) shows the top-view morphology of the nanowires in our device. The cross-sectional view morphology of nanowires is also shown (Fig.4(c)).

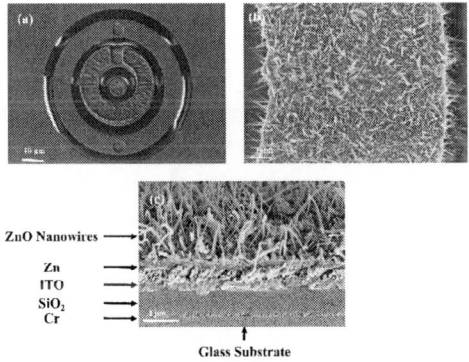

Fig.4. SEM images of the double-gate ZnO nanowire FEAs. (a) Single pixel; (b) Top-view and (c) cross-sectional view of ZnO nanowires on cathode electrode.

The fabricated device was measured using a driving circuit and addressable emission was realized. As shown in Fig.5, I_a-V_a characteristics under different gate voltages were measured. Fig.5(a) shows typical I-V characteristics and the corresponding F-N plots. The maximum emission current of measured from two columns of pixels is 26.8 μA under V_a of

1300 V and V_g of 140 V. The obtained maximum current density is calculated to be 3.32 mA/cm². At the same time, our device has a lower turn-on voltage, which reflects a stronger gate control capability. Furthermore, the transconductance of the device is obtained as 2433 nS. The maximum current and the transconductance are larger than the gated ZnO nanowire FEAs previously reported [5-8].

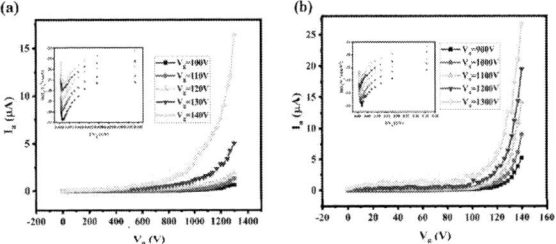

Fig.5: (a)I_a-V_a characteristics under different V_g; (b) I_a-V_g characteristics under different V_a. Inset is the corresponding F-N plots.

IV. CONCLUSION

Double-gate ZnO nanowire FEAs were designed and t fabricated. Maximum emission current density of 3.32 mA/cm² and transconductance of 2433 nS were obtained under anode voltage of 1300 V and gate voltage of 140V. The results verified that high current and good gate control characteristics can be achieved from double-gate ZnO nanowire FEAs.

ACKNOWLEDGMENTS

The authors gratefully acknowledge the financial support from the National Key Research and Development Program of China (Grant No. 2022YFA1204200), Key Research and Development Program of Guangdong Province (Grant No. 2023B0101200013), National Natural Science Foundation of China (Grant Nos. 82272131 & 62001527), the Science and Technology Department of Guangdong Province (Grant No. 2023B1212060025), Fundamental Research Funds for the Central Universities.

REFERENCES

[1] S. Kang *et al.*, "Transparent flat panel X-ray source using ITO transmission anode and ZnO nanowire cold cathode," *IEEE Trans. Electron Devices*, vol. 70, no. 6, pp. 3302–3307, 2023.

[2] Y. Zhao *et al.*, "High current field emission from large-area indium doped ZnO nanowire field emitter arrays for flat-panel X-ray source application," *Nanomaterials*, vol. 11, no. 1, p. 240, 2021.

[3] H. Gao *et al.*, "Field emission of large-area and graphitized carbon nanotube array on anodic aluminum oxide template", *J. Appl. Phys.*, vol. 93, no. 9, pp. 5602–5605, 2003.

[4] L. Zhao, Y.X. Chen and Y. M. Liu, *et al.*, "Integration of ZnO nanowires in gate field emitter arrays for large-area vacuum microelectronics applications", *Cur. Appl. Phys.*, vol. 17, pp. 85-91, 2017.

[5] Y.M. Liu *et al.*, "Fabrication of ZnO nanowire field-emitter arrays with focusing capability", *IEEE Trans. Electron Devices*, vol. 65, no. 5, pp. 1982–1987, 2018.

[6] Y.M. Liu *et al.*, "Fabrication of ZnO nanowire field emitter arrays with non-coplanar focus electrode structure", in *Proc. Int. Vac. Nanoelectron. Conf.*, Regensburg, Germany, pp. 22–23, 2017.

[7] X.Q. Cao, G.F. Zhang and Y.Y. Zhao *et al.*, "Fully vacuum-sealed addressable nanowire cold cathode flat-panel X-ray source", *Appl. Phys. Lett.*, vol. 119, pp. 053501, 2021.

[8] S.Y. Zhang, X.Q. Cao and G.F. Zhang *et al.*, "Optimizing performance of planar-Gate ZnO nanowire field-emitter arrays by tuning pixel density", *Nanomaterials*, vol. 12, pp. 870, 2022.

Field Electron Emission Characteristics of Uncoated Alumel Tips

Marwan S. Mousa[1,*], Enas A. Arrasheed[2], Adel M. Abuamr[1], Ammar Al Soud[3], Dinara Sobola[3]

[1] Department of Renewable Energy Engineering, Jadara University, 21110 Irbid, Jordan.

[2] Department of Communication and Computer Engineering, Jadara University, 21110 Irbid, Jordan

[3] Department of Physics, Brno University of technology, 61600 Brno, the Czech Republic.

* Corresponding email: m.mousa@jadara.edu.jo

Abstract— This research work presents the cold electron emission from uncoated Alumel tips, with apex radii in the micrometer range, produced using an electrochemical etching technique using phosphoric acid (H_3PO_4) solution. The emission characteristics of the Alumel tips were studied in this investigation utilizing current-voltage characteristics (I-V), and Fowler–Nordheim (FN) type plots, electron emission spatial distribution analysis. Energy-dispersive X-ray spectroscopy (EDS) was utilized to examine the chemical composition of the tip, as well as scanning electron microscopy was employed to evaluate the tip surface shape. The results show that the relationship between current and voltage is quasi-Ohmic. The F-N plot operated at low voltages. In addition, emission current exhibiting a bright and stable spatial distribution.

I. INTRODUCTION

Cold-field electron emission (CFE), referred to as cold emission as well as field electron emission, is a phenomenon that happens when electrons escape from the surface of a condensed phase, frequently a solid, into another phase, typically a vacuum[1], [2]. The emitter via cold-field electron emission (CFE) applications frequently develops into a pointed tip[3], [4]. This tip geometry is essential to enhance the electric field at the emitter's surface, allowing electron emission. Apex radius of the tip can be on the nanoscale, varying from a few hundred to a few nanometers[5], [6]. The Alumel tips are mostly composed of nickel (95–96%), with a small percentage of silicon (0.5–1%), aluminum (1-2%), and manganese (1-2%). The specific composition may differ greatly depending on how effectively the product performs [7]. This study examines the field emission characteristics of uncoated Alumel emitters to acquire an understanding of their electron emission properties. Further work includes developing a theoretical framework to justify the results and discussing the real-world uses of these emitters.

II. SAMPLE PREPARATION

A 0.25 mm diameter alumel wire was utilized in this study. Emitters were created by electrochemically etching using a 60% consetation phosphoric acid solution. After etching, the tips were cleaned via distilled water in an ultrasonic device and then examined with an optical electron microscope as shown in figure 1. After that, the emitter was placed in a FEM using an axially tip-to-screen range of 10 mm and a pressure of 10^{-5} mbar. Voltage was applied employing an external power supply, and current measurements were performed with a picoammeter. Spatial electron distribution was recorded using a digital camera.

III. CHARACTERIZATION

SEM-EDS instruments (called MIRA-TESCAN in the Czech Republic) were utilized in this experiment to examine the Alumel sample's composition. The results in figure 1 shows that nickel (91.50%) was the main element with a small percentage of silicon (1.40%), aluminum (1.20%), and manganese (1.80%), along with other low-concentration contaminants from the SEM chamber, such as carbon (1.40%). Furthermore, oxygen (2.20%) is from the apex's oxidized layer, while phosphorus (0.50%) from the etching solution. The spectrum demonstrates that utilizing distilled water for cleaning the samples was successful since the contaminants were in extremely low quantities. SEM micrograph of the prepared tip is shown in figure 2.

IV. RESULTS AND DISCUSSION

The clean Alumel emitter presented in this work have been prepared with sharp apex radii to examine the characteristics of the field electron field emission from this material. Selected outcomes, including the FN type plots of the electron field emission I-V characteristics, were provided. The presented I-V data (Fig. 4) relate to the second cycle of voltage increase through the emitter in order to avoid the voltage switch-on phenomena of coating emitters. The I-V characteristics depends mostly on the size of the emitter's tip radius and to the vacuum level during the experiment. Figure 4 shows the FN plots calculated from the I-V data provided in Figure 3.

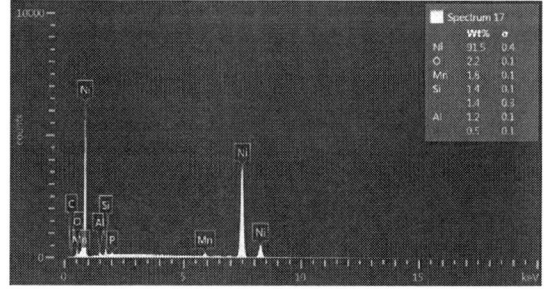

Fig. 1. *EDS spectrum of an alumel field emission apex.*

The electron emission images obtained directly from the phosphorus conductive screen (anode) of the field emission microscope shown in figure 5.

The emission was concentrated into a single very bright spot that indicated no obvious structure within the images .The emission observed at high voltage exhibits a single spot, demonstrating that the electron emission starts from a specific point. The intensity of the emission reduces as the measured current decrease.

Fig. 2. Scanning electron micrograph of the sample.

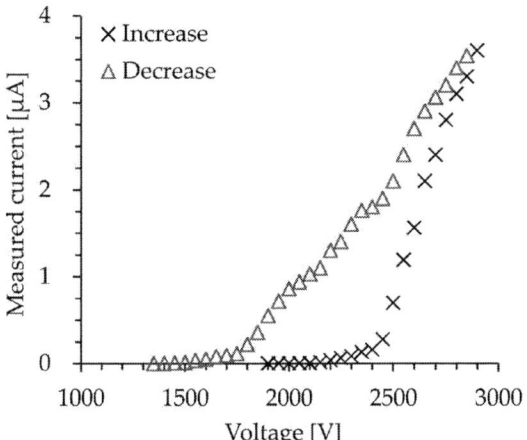

Fig. 3. The current-voltage characteristics for alumel emitter.

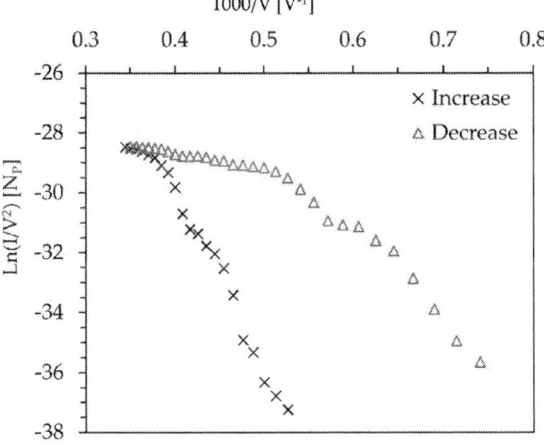

Fig. 4. Shows the F-N plots corresponding to the I-V plots.

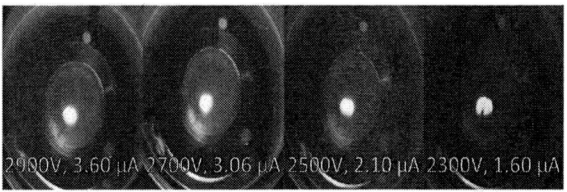

Fig. 5. The field emission pattern images at different voltages as related to the voltage decreasing part of the cycle.

V. CONCLUSIONS

In the present study, an alumel FE cathode tip was examined. The alumel FE cathode tip showed excellent field-emission characteristics, such as high current density. The cathode tip demonstrated reasonable stability and durability during the experimental tests. As a result, alumel appears to be a suitable material for field-emission applications.

REFERENCES

[1]	H. A. Al-Braikat *et al.*, "A Comparative Analysis of Field Electron Emission from Carbon Black Embedded within Insulated Copper Hollowed Wires and Glass Tubes," *Karbala International Journal of Modern Science*, vol. 10, no. 2, Apr. 2024, doi: 10.33640/2405-609X.3346.
[2]	A. Alsoud *et al.*, "Electrical Characterization of Epoxy Nanocomposite under High DC Voltage," *Polymers (Basel)*, vol. 16, no. 7, p. 963, Apr. 2024, doi: 10.3390/polym16070963.
[3]	M. S. Mousa and A. M. Abuamr, "Electron Emission from High-Purity Copper Wires," *Jordan Journal of Physics*, vol. 16, no. 2, pp. 247–252, Jun. 2023, doi: 10.47011/16.2.12.
[4]	H. Yanagisawa, T. Greber, C. Hafner, and J. Osterwalder, "Laser-induced field emission from a tungsten nanotip by circularly polarized femtosecond laser pulses," *Phys Rev B*, vol. 101, no. 4, p. 045406, Jan. 2020, doi: 10.1103/PhysRevB.101.045406.
[5]	M. M. Allaham, M. S. Mousa, D. Burda, M. H. AlSa'eed, S. Y. AlJrawen, and A. Knapek, "Analyses of field electron emission Molybdenum current-voltage data using Fowler-Nordheim and Murphy-Good plots," in *2021 34th International Vacuum Nanoelectronics Conference (IVNC)*, IEEE, Jul. 2021, pp. 1–2. doi: 10.1109/IVNC52431.2021.9600771.
[6]	R. V Latham and M. S. Mousa, "Hot electron emission from composite metal-insulator micropoint cathodes," *J Phys D Appl Phys*, vol. 19, no. 4, pp. 699–713, Apr. 1986, doi: 10.1088/0022-3727/19/4/021.
[7]	L. Wyatt, "Materials Properties and Selection," in *Manufacturing Engineer's Reference Book*, Elsevier, 1993, pp. 1/1-1/119. doi: 10.1016/B978-0-08-052395-8.50005-9.

979-8-3503-7977-8/24 $31.00 © 2024 IEEE

Field Emission Characteristics from Vertical Few-Layer Graphene Growth on Graphite Substrate

Yiming Huang, Shuai Tang*, Mingkai Gou, Haonan Zhao, Yan Shen, Yu Zhang, Juncong She, Jun Chen, Shaozhi Deng

State Key Laboratory of Optoelectronic Materials and Technologies, Guangdong Province Key Laboratory of Display Material and Technology, School of Electronics and Information Technology, Sun Yat-sen University, Guangzhou 510275, People's Republic of China

Corresponding author: tangsh58@mail.sysu.edu.cn

Abstract—The vertical few-layer graphene (FLG) with a 5 mm diameter circular area was synthesized on a graphite substrate using microwave plasma enhanced chemical vapor deposition. The FLG was uniformly synthesized with width of 400-800 nm, height of 1.6-2 μm, thickness of few-layers and with defects. The turn-on field of the FLG is 7.85 V/μm and the maximum current and current density is 700 μA and 89.1 mA/cm² respectively for emission area of 1 mm diameter. It provides a new way to prepare FLG with good conductive interface and would benefit for its large current field emission.

Keywords—field emission, few-layer graphene, graphite substrate, microwave plasma enhanced chemical vapor deposition.

I. INTRODUCTION

The unique two-dimensional structure, high electron mobility of vertical few-layer graphene (FLG) has sparked interest in the field of field emission. Graphene has been reported has a low turn-on field [1]. The emission current from single FLG is 233 μA, which is at the maximum level of the single cold cathode reported [2]. However, the large current characteristics from FLG film [3] was still not satisfied which is 1/4 compared with the carbon nanotube [4] at the same emission area. It has been proved that the amorphous carbon at the interface between substrate and FLG during chemical vapor deposition synthesis would be the key factor to limit its current [2]. Since graphite is a good conductor and with a perfect match of the crystal lattice of graphene, it is believed the graphene can be grown directly on a graphite substrate and amorphous carbon inter-layer could be removed or reduced heavily.

Herewith, a graphite substrate was used for FLG fabrication by the microwave plasma-enhanced chemical vapor deposition (MPECVD). The field emission properties were also measured. Those results offer a fresh avenue to fabricate FLG with a well-conductive interface, promising in achieving high-current field emission performance.

II. EXPERIMENT

The FLG was synthesized on a graphite substrate using MPECVD (CN-CVD 100, ULVAC). Firstly, placing the graphite substrate on a cathode plate within a quartz tube, maintaining an H_2 flow rate of 100 SCCM, and sustaining a 200 Pa air pressure. Then, the graphite substrate was heated for 15 minutes using a 500W microwave plasma generator. Subsequently, maintaining a 5 SCCM CH_4 flow rate to grow FLG with 100 V DC bias applied. The morphology was characterized by SEM (Supra 60, Zeiss). The crystalline properties were analyzed via Raman spectroscopy (InVia Reflex).

Fig. 1 shows the field emission test structure. A metal probe anode with a radius of 0.5 mm was placed at a spacing of 200 μm from the FLG for field emission measurement at a vacuum of 5.8×10^{-6} Pa.

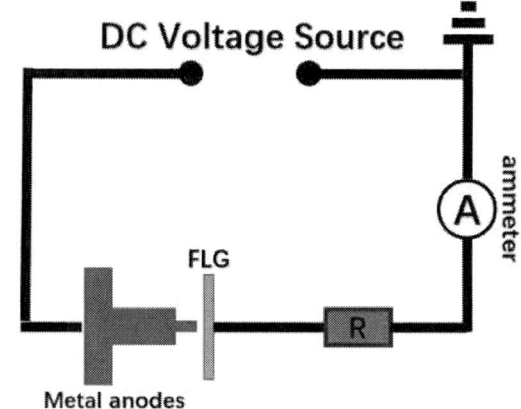

Fig. 1. Schematic diagram of the field emission test structure.

III. RESULT AND DISCUSSION

The top view of SEM image (Fig. 2(a-c)) demonstrates that the FLG grows uniformly on the graphite substrate. FLG has a length of 450-800 nm. The cross-sectional view of SEM image (Fig. 2(d)) demonstrates that the FLG is thin and sharp. FLG is grown vertically on graphite substrate with a height of 1.6-2 μm, which is good for field emission.

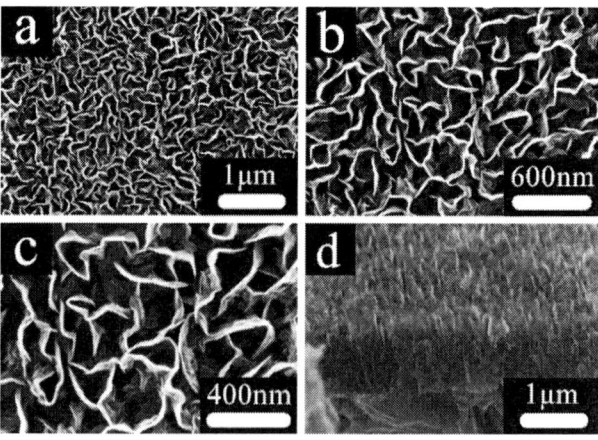

Fig. 2. SEM image of the FLG under (a-c) Top view showing uniform growth and (d) Cross-sectional view showing vertical growth.

The Raman spectrum of the FLG sample and the graphite substrate are plotted in red and black color respectively in Fig. 3. The I_{2D}/I_G is 0.72 for FLG which is much larger than the I_{2D}/I_G of 0.25 for graphite, indicating that FLG has a few-layer structure. Compared to the graphite substrate, the D peak is strong and $I_D/I_{2D}=3$, indicating that the boundary defect is the main defect. In our case, it is probably due to the graphene is in a vertical type and the edge of FLG is regarded as defects.

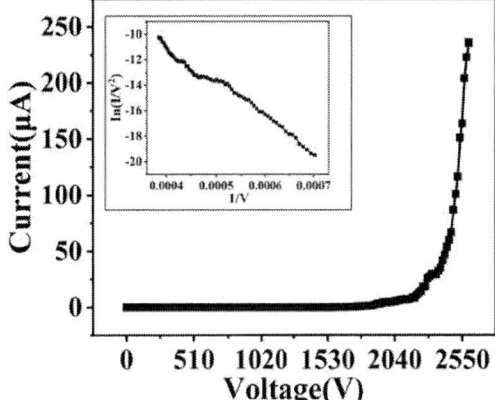

Fig. 4. Field emission I-V and F-N plot of the FLG.

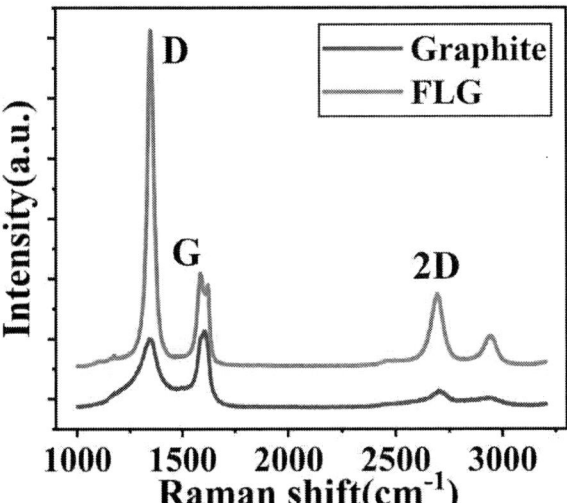

Fig. 3. Raman spectrum of the FLG and graphite substrate.

The Field emission characteristics of FLG grown on graphite substrate was measured using a metal probe anode of 1mm diameter. The field emission I-V plot is shown in Fig. 4. The linear F-N curve (Fig. 4) shows FLG follows the traditional F-N theory. When the applied voltage was set as 1570 V, the current was 77 nA and the corresponding turn-on electric field was 7.85 V/μm. Increasing the voltage to 2600 V, the current increased to 253 μA and the corresponding current density is 32.2 mA/cm². Subsequently, at 2800 V, the vacuum-breakdown occurred with a peak current of 700 μA and a corresponding current density of 89.1 mA/cm².

The field enhancement factor β is expressed from F-N theory:

$$\beta = \frac{-6.83 \times 10^9 d\phi^{3/2}}{k} \qquad (1)$$

where ϕ is the work function of the FLG. d is the gap between cathode and anode. K is the slope of F-N curve. By substitute the d of 200 μm, work function of 5 eV for FLG and slope of -127, β is calculated as 601.

IV. CONCLUSION

Herewith, the MPECVD method was used to grow FLG on the graphite substrate. The SEM and Raman spectra demonstrate it is vertical FLG with defects. The turn-on field of the FLG is 7.85 V/μm and the maximum current and current density is 700 μA and 89.1 mA/cm² respectively for emission area of 1 mm diameter. The optimal parameters for growth of FLG on graphite without amorphous inter-layer and its large current field emission characteristics is still under-investigation.

ACKNOWLEDGMENT

This work was supported by the National Key Basic Research Program of China (Grant no. 2019YFA0210201, 2019YFA0210200, 2023YFF0719004), the National Natural Science Foundation of China (Grant no. 62301619), the Key Field Research Program of Guangdong Province (Grant no. 2022B0303030001), the Special Topic on Basic and Applied Basic Research of Guangzhou (Grant no. SL2023A04J01793), the Fundamental Research Funds for the Central Universities, Sun Yat-sen University (Grant no. 23ptpy04) and Start-up Funds for the Central Universities, Sun Yat-sen University.

REFERENCES

[1] A. Malesevic, R. Kemps, A. Vanhulsel, M. P. Chowdhury, A. Volodin, and C. Van Haesendonck, "Field emission from vertically aligned few-layer graphene," Journal of Applied Physics, vol. 104, no. 8, p. 084301, Oct. 2008.

[2] S. Tang, Y. Zhang, P. Zhao, R. Zhan, J. Chen, and S. Deng, "Realizing the large current field emission characteristics of single vertical few-layer graphene by constructing a lateral graphite heat dissipation interface," Nanoscale, vol. 13, no. 10, pp. 5234–5242, 2021.

[3] Y. Zhang, D. Deng, L. Zhu, S. Deng, J. Chen, and N. Xu, "Pulse Field Emission Characteristics of Vertical Few-Layer Graphene Cold Cathode," IEEE Transactions on Electron Devices, vol. 61, no. 6, pp. 1771–1775, 2014.

[4] W. Zhu et al., "Epitaxial growth of multiwall carbon nanotube from stainless steel substrate and effect on electrical conduction and field emission," Nanotechnology, vol. 28, no. 30, p. 305704, Jul. 2017.

979-8-3503-7977-8/24 $31.00 © 2024 IEEE

Field Emission X-ray System for Online Conveyor Belt Imaging

Rui Zhou[1], Jiaqi Wang[1,2], Zhemiao Xie[1,2], John T.W. Yeow[1,2]

[1]*Systems Design Engineering University of Waterloo* Waterloo, Canada
[2]*Auroray Inc.* Mississauga, CA
jyeow@uwaterloo.ca

Abstract—**In this research, we introduce an innovative field emission (FE) X-ray system for real-time conveyor belt imaging. This innovative system is distinguished by its rapid detection and imaging speed, capable of detecting and capturing X-ray images within 1 ms for objects moving at speeds of up to 1 m/s. This system also incorporates comprehensive functionalities for fully online monitoring and control. The rapid response, lightweight design, portability, and remote accessibility of the system make it particularly suitable for applications where radiation is generated only when required, and human intervention is limited.**

Index Terms—**Field emission, conveyor belt system, X-ray detection, real-time imaging**

I. Introduction

As technological developments expand and application scenarios diversify, the limitations of traditional X-ray technology become increasingly evident, particularly in terms of speed and portability. Field Emission (FE) X-ray technology, known for its rapid response time and low power consumption, addresses these challenges effectively [1]. Central to our innovation is the FE tube, which powers our X-ray system designed for online conveyor belt imaging. The unique cold cathode field-emission mechanism of the tube eliminates the need for heating and cooling processes of our system, enabling extraordinarily fast response times and immediate, high-quality imaging [2], [3]. Moreover, the highly tunable duty cycle of the FE tube enhances energy efficiency and reduces operational costs, making it an ideal solution for portable applications where human intervention is limited. Our online conveyor system adopting the novel FE tube allows fully remote and real-time control and imaging, not only advances technical capabilities but also improves operation convenience and safety, meeting the stringent requirements of advanced medical imaging, manufacturing quality control, and security applications. These advantages demonstrate the essential role of continuous innovation and the need for wider adoption of FE X-ray technology in various sectors.

John T.W. Yeow is the corresponding author and with Department of Engineering, University of Waterloo, Waterloo, ON, CAN, jyeow@uwaterloo.ca

II. System Design

The integration of a FE X-ray tube is instrumental in enhancing the performance of our system. As shown in Fig. 1, the FE X-ray tube comprises a cold cathode tip, a reflective anode, and a vacuum enclosure. Through field emission, electrons are emitted from the cold cathode tip under intense electric fields without the need for heating. These electrons then accelerate toward the anode, where their impact generates X-rays. The vacuum enclosure prevents electron attenuation by air molecules, thereby increasing X-ray production efficiency. This setup allows for precise control over the X-ray intensity and energy, enabling rapid activation while minimizing radiation exposure and energy consumption.

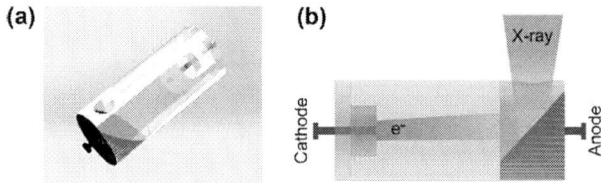

Fig. 1. (a) 3-D view of our FE X-ray tube. (b) Schematic representation of electron emission and X-ray generation process.

At the core of our X-ray system are two primary subsystems: the fast switching system and the fast imaging system. The fast switching system regulates the trigger input to control the X-ray generation, ensuring precise timing for the initiation and duration of the X-ray output. The fast imaging system, activated concurrently with the fast switching system, maintains the same operational window to accurately capture the X-ray output detected by the X-ray detector. This data is then immediately transferred to the PC for real-time imaging. We have meticulously engineered both the fast switching and fast imaging systems to operate synchronously, ensuring identical opening and closing times for complete capture of the X-ray signal. Both subsystems are remotely configurable and automated, enabling the rapid generation and detection of X-rays within 1 ms as objects pass directly beneath the FE X-ray tube. This setup efficiently captures and images objects

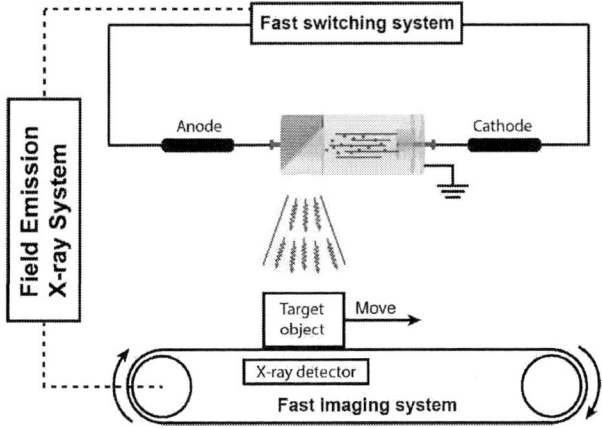

Fig. 2. Schematic illustration of the FE X-ray system for online conveyor belt imaging.

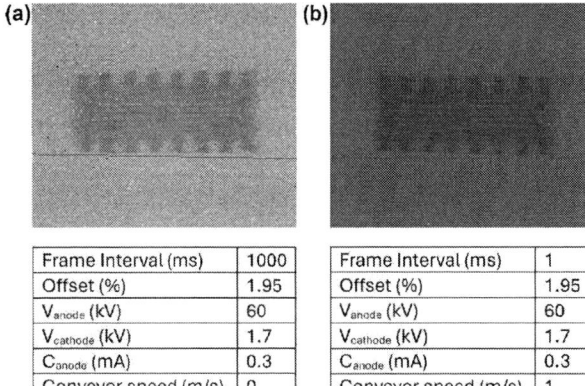

Frame Interval (ms)	1000
Offset (%)	1.95
V_{anode} (kV)	60
$V_{cathode}$ (kV)	1.7
C_{anode} (mA)	0.3
Conveyor speed (m/s)	0

Frame Interval (ms)	1
Offset (%)	1.95
V_{anode} (kV)	60
$V_{cathode}$ (kV)	1.7
C_{anode} (mA)	0.3
Conveyor speed (m/s)	1

Fig. 4. Comparison of the images and parameter settings between (a) regular display mode for the static IC chip and (b) instant imaging of the moving IC chip.

moving at speeds of up to 1 m/s with exceptional precision and clarity.

III. RESULTS

We compared the responsiveness of traditional thermionic versus field emission X-ray systems to trigger inputs, and the results are demonstrated in Fig. 3. Thermionic X-rays exhibit a delayed response due to pre-heating and cooling requirements of the thermionic mechanism, resulting in discrete pulses with noticeable activation and deactivation times. In contrast, FE X-rays respond instantaneously, closely mirroring the trigger frequency with consistent and continuous output. This immediate responsiveness of FE technology, enables more precise emission control, fast response imaging and very low power consumption.

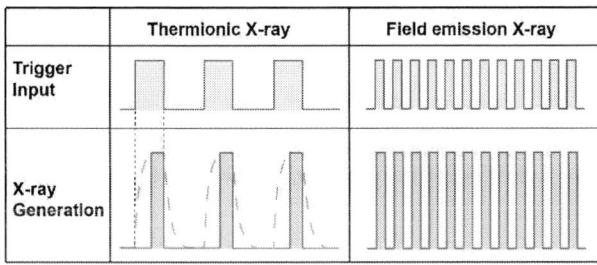

Fig. 3. Thermionic versus field emission X-ray systems: trigger input and X-ray generation profiles.

Our experimental observations have also revealed the online conveyor belt's fast and accurate imaging capability. The scanning results of an integrated circuit (IC) chip, measuring 19 x 8 mm, are illustrated in Fig. 4. The X-ray images clearly depict the chip and its peripheral circuits beneath the packaging. Specifically, Fig. 4a displays the imaging results for the stationary IC chip, while Fig. 4b shows the outcomes for the moving IC chip. Leveraging the rapid response and highly adjustable duty cycle of the FE X-ray system, we successfully captured these X-ray images within 1 ms. This rapid imaging capability enables precise and immediate detection of objects moving at speeds up to 1 m/s. This efficiency highlights our system's advantages over traditional imaging methods, which require longer response times, thereby enhancing both speed and energy efficiency without compromising image quality. Additionally, the experimental procedure is conducted entirely online and remotely, which significantly enhances the convenience and safety of the X-ray system.

IV. CONCLUSION

The development of this advanced real-time FE X-ray system provides significant advantages for applications requiring rapid response times, superior energy efficient and less wear-and-tear. The system presents an indispensable solution for realizing future portable fast X-ray imaging devices, low-power X-ray applications, and mobile X-ray equipment with remote operation capabilities. This study illustrates the transformative potential of FE X-ray technology in delivering efficient and low radiation/power X-ray system for conveyor belt applications.

REFERENCES

[1] Saito, Yahachi, and Sashiro Uemura. "Field emission from carbon nanotubes and its application to electron sources." Carbon 38.2 (2000): 169-182.
[2] Norman, D., et al. "Surface structure of thermionic-emission cathodes." Physical review letters 58.5 (1987): 519.
[3] Gadzuk, John William, and E. W. Plummer. "Field emission energy distribution (FEED)." Reviews of Modern Physics 45.3 (1973): 487.

From Deformation to Performance: WO₃-Coated Tungsten Emitters Created by Anodization

Zuzana Košelová[1,2*], Daniel Burda[1,3], Mohammad M. Allaham[1,4], Zdenka Fohlerová[2,4], Alexandr Knápek[1,2]

[1]*Institute of Scientific Instruments of the Czech Academy of Sciences*, Brno, 612 00, Czech Republic
[2]*Department of Microelectronics, Brno University of Technology*, Brno, 612 00, Czech Republic
[3]*Department of Physics, Brno University of Technology*, Brno, 612 00, Czech Republic
[4]*Central European Institute of Technology Brno, Brno University of Technology*, Brno, 612 00, Czech Republic
*Corresponding author: lili@isibrno.cz

Abstract— **This study investigates the field emission characteristics of tungsten single-tip field emitters coated with a thin tungsten trioxide (WO₃) layer. The WO₃ dielectric barrier at the metal-vacuum interface aims to enhance performance by extending operational lifespan and increasing current levels. High-voltage applications risk deforming the emitter tip due to phase transitions to plasma. However, altered tip shapes, formed from evaporation may provide stability and performance benefits. Our results show that emitters with thicker oxide layers after deformation can exhibit stable emission characteristics and higher voltage thresholds, suggesting that oxide-coated tungsten tips can enhance cold field emission properties.**

Keywords— *Tungsten emitters, cold field emission, FEM analysis, oxide coating, dielectric barrier*

I. INTRODUCTION

This study investigates the field emission characteristics of tungsten single-tip field emitters coated with a thin tungsten trioxide (WO₃) layer, with a thickness of several hundred nanometres. The incorporation of this dielectric barrier at the metal-vacuum interface of the emitter is expected to enhance its performance by extending its operational lifespan and increasing the current levels compared to uncoated tungsten emitters [1][2][3]. However, due to the high voltage and associated heating or explosive electron emission, there is a risk of deforming the original tip shape through a quasi-steady-state phase transition to plasma [4]. In this work, we aim to determine whether the altered shape, resulting from the evaporation of part of the tip, can provide certain advantages and a more stable final cathode shape.

II. METHODOLOGY

The tungsten emitters used in this study were fabricated from polycrystalline wires with 99.9% purity, a diameter of 0.3 mm, and a length of 1 cm using a two-step electrochemical drop-off etching process in a 2 M aqueous NaOH solution. The primary cleaning procedure involved submerging the emitters in 38% hydrofluoric acid for 10 minutes, followed by rinsing in water to remove oxides, hydroxides, and organic contaminants [5]. The creation of the thin oxide barriers was created via anodization conducted at 5 V and 12 V for 7 min. The voltage was gradually increased from 0.02 to 5 V/12 V over 5 minutes, followed by 2 minutes at 5 V/12 V, in a 0.3 M H_3PO_4 solution with a pH of 1.2. The emitter was integrated as the cathode within the triode configuration of a Field Emission Microscope (FEM), under high vacuum conditions ($\sim 10^{-6}$ Pa). An extraction electrode with a 1 mm diameter aperture was positioned approximately 1 mm above the tip, and an Al-coated Ce:YAG scintillator served as the anode. Before measuring characteristics, the tips were subjected to increasing voltage until a sudden increase and subsequent decrease in emission were observed. During measurements, to protect the emitter from further explosion, only the extraction voltage was increased (~1 V per sec).

III. RESULTS

Fig. 1a-d shows detail of the emitter before and after anodization. The oxide layer fully covers the tip, creating a broader "head" on it. Higher anodization voltages, as 12 V, resulted in thicker layers with broader head. Upon applying high voltage during FEM, this peak evaporated/melted (Fig. 1e-g). Fig. 2 illustrates the obtained current-voltage characteristics and subsequent analyses. These characteristics were measured after tips achieved their new shapes (Fig. 1e-g). This observation is not entirely true for the 5 V samples, as the current characteristics suggest continuous evaporation, indicated by observed fluctuations and a relatively small orthodoxy pass region. The characteristic showed a distinct change in slope at approximately 5950 V (of which 1000 V applied at extractor). Beyond this voltage threshold, the current behaviour passes the orthodoxy test. In contrast, the sample with anodization at 12 V passed orthodoxy and had relatively stable behaviour up to approximately 6900 V, attributed to its unique shape. The thick oxide layer protects the tip along its length while ensuring that only the apex emits, and the rounded tungsten tips ensure a stable electron flow. However, these thicker tips have the disadvantage of lower maximum current levels. The emission picture shows more pronounced light areas for the protrusions on the original tip. Orthodoxy was tested using both the Fowler-Nordheim plot and the Murphy-Good plot (Fig. 2b), utilizing a simulation-friendly web tool [6]. A work function of 5.05 eV was used [7]. Both analyses slightly differed in the obtained Voltage Conversion Length value and Formal Emission Area Value, as shown in Fig. 2. Notably, unlike pure tungsten emitters, which tend to melt with increasing voltage and show difficulties in maintaining a stable emission current and measurement characteristic, those coated with oxide, upon acquiring a new stable shape, did not exhibit such high tendencies towards burnout with escalating voltage levels.

IV. CONCLUSIONS

The aim of this study was to demonstrate that the often undesirable process of oxide layer evaporation can, under suitable conditions, yield desirable results. A shape similar to the 12 V tip example would otherwise require complex fabrication using Focused Ion Beam (FIB) techniques. It is shown that creating similar tips necessitates higher voltages (and consequently higher currents) during anodization. The tips formed in this manner enabled the application of

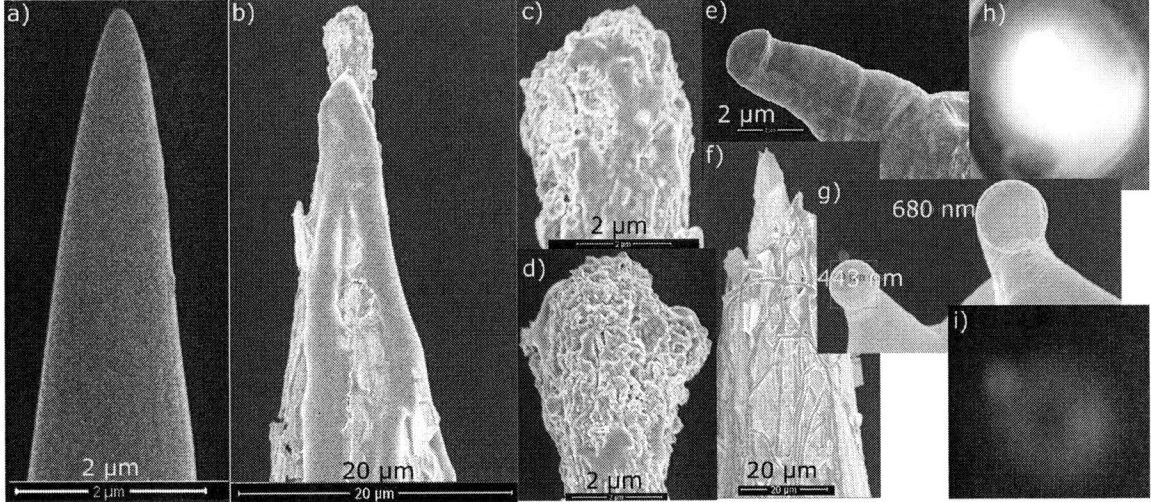

Fig. 1. (a) Cleaned tungsten single tip cold field emitter (b) after anodization at 5 V. (c) detail on created oxide "cap", anodization 5 V and (d) 12 V. (e) Tip after deformation by high voltage, 5 V and (f) for 12V with detail (g). (h) Photo of emission of (e) tip at 6060 V, photo of (f) tip at 6700 V.

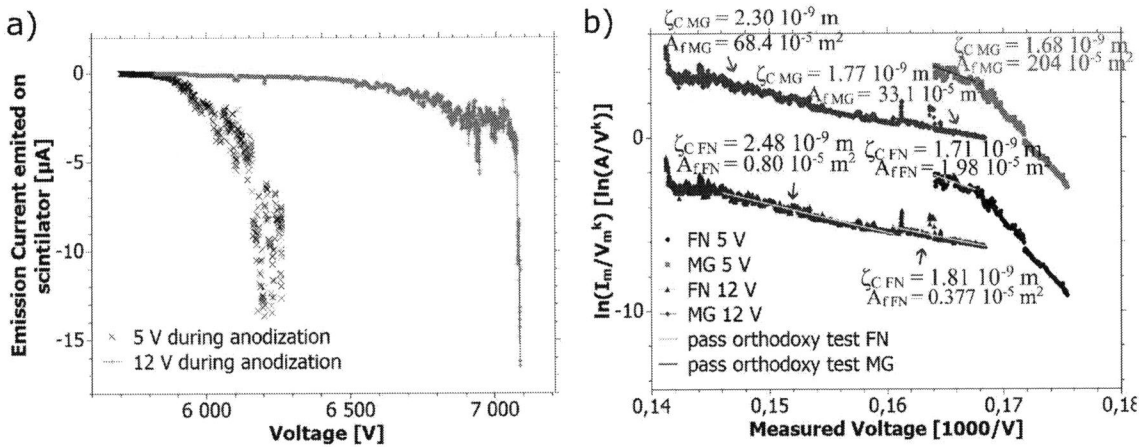

Fig. 2. (a) The example of current-voltage characteristics, (b) the corresponding FN and MG plots

relatively high voltages before further damage occurred, thereby stabilizing current creation. This method of coating tungsten tips with oxide can enhance their effectiveness as cold field emitters without the need for complex facilities.

ACKNOWLEDGMENT

This article was supported by the Czech Academy of Sciences (RVO:68081731) and The Technology Agency of the Czech Republic FW03010504. We acknowledge CzechNanoLab Research Infrastructure supported by The Ministry of Education, Youth and Sports of the Czech Republic (LM2018110), and the project FEKT-S-23-8162 and CEITEC VUT/FEKT-J-24-8567.

REFERENCES

[1] N. Khan, A. Mahajan, A. Arora, K. Sood, S. Kumari, S. Ghosh, M. Jha, "Nanostructured tungsten trioxide developed from environmentally friendly green process as a promising cathode for excellent field emission," *Mater. Chem. Phys.*, vol. 320, p. 129364, Jul. 2024, doi: 10.1016/J.MATCHEMPHYS.2024.129364.

[2] Z. Košelová, L. Horáková, D. Sobola, D. Burda, A. Knápek, And Z. Fohlerová, "Functional Tungsten-Based Thin Films and Their Characterization," pp. 473–479, 2023, doi: 10.37904/METAL.2023.4719.

[3] Z. Knor, S. Biehl, J. Plšek, L. Dvořák, and C. Edelmann, "A contribution to the search for a stable field emission electron source based on W–WOx–Au and W–Al2O3–Au systems," Vacuum, vol. 51, no. 1, pp. 11–19, Sep. 1998, doi: 10.1016/S0042-207X(98)00128-6.

[4] G. N. Fursey, "Field emission in vacuum micro-electronics," *Appl. Surf. Sci.*, vol. 215, no. 1–4, pp. 113–134, Jun. 2003, doi: 10.1016/S0169-4332(03)00315-9.

[5] Z. Košelová, L. Horáková, D. Burda, M. M. Allaham, A. Knápek, and Z. Fohlerová, "Cleaning of tungsten tips for subsequent use as cold field emitters or STM probes," *J. Electr. Eng.*, vol. 75, no. 1, pp. 41–46, Feb. 2024, doi: 10.2478/JEE-2024-0006.

[6] M. M. Allaham *et al.*, "Interpretation of field emission current–voltage data: Background theory and detailed simulation testing of a user-friendly webtool," *Mater. Today Commun.*, vol. 31, p. 103654, Jun. 2022, doi: 10.1016/J.MTCOMM.2022.103654.

[7] G. Halek, I. D. Baikie, H. Teterycz, P. Halek, P. Suchorska-Woźniak, and K. Wiśniewski, "Work function analysis of gas sensitive WO3 layers with Pt doping," *Sensors Actuators B Chem.*, vol. 187, pp. 379–385, Oct. 2013, doi: 10.1016/J.SNB.2012.12.062.

979-8-3503-7977-8/24 $31.00 © 2024 IEEE

Glass-Extraction Electrode for Field Emission Applications

Aleksandra M. Buchta*, Alexander Kassner, Folke Dencker, and Marc C. Wurz
Institute of Micro Production Technology,
Leibniz University Hanover, Garbsen, Germany
*Corresponding author: buchta@impt.uni-hannover.de

Abstract— This work focuses on designing and fabricating an optimized extraction electrode made out of borofloat glass. The electrode is a part of an emitter chip, where the field emitters are manufactured by the wafer dicing technique. The extraction electrode was fabricated by Selective Laser Etch (SLE) process, where the geometry comprises an array of Through-Glass-Vias (TGV) that are positioned concentrically over each emitter tip, a cavity in the area of the TGVs, additional cavities for lowering of the electrode and TGVs for dowel pins for alignment. The current-voltage experiments confirm the enhancement in the measured field emission current dependent on the geometry of the extraction electrode.

Keywords—glass extraction electrode, SLE, silicon field emitters, dicing field emitters, FEA, laser assisted bonding

I. STATE OF THE RESEARCH

Due to its electrical insulation properties and the lack of parasitic eddy currents, glass as a substrate material is a potent alternative for silicon in quantum applications. Our previous work proposed a novel glass-silicon emitter chip comprising diced silicon emitters and an extraction electrode out of 230 µm thick SCHOTT borofloat 33 glass fabricated by LIDE® technology [1]. Glass was used as an alternative to silicon, which is often used as a gate electrode material [2], with the advantage of no need for additional electrical insulation.

The emitter chip geometry has been reduced from 14 mm² to 10 mm² with the same number of emitters (array of 39 x 39 on 8 mm²). The copper metallisation of the electrode (2 µm [1]) has been replaced with a sputtered 50 nm Ti and 500 nm Au coating since it has shown less fluctuation throughout the long-time measurements. Most importantly, due to the relatively high glass thickness (230 µm, [1]), a low transmission was observed in the three-electrode configuration throughout the experiments. Therefore, we have optimised the geometry of the glass electrode.

II. FABRICATION PROCESS

A. Selective Laser Etch

The extraction electrode proposed here was fabricated in-house using the Selective Laser Etch (SLE) process. Within this process, the glass is modified locally by a femtosecond laser and, in the second step, etched with potassium hydroxide (KOH) (Fig. 1A). For the experiments, the laser system LightFab 3D Printer ® was used. The structures were etched for about 24 hours.

B. Geometry of the glass extraction-electrode

The extraction electrode is made out of 500 µm thick borofloat glass. An array of 39 x 39 TGVs corresponds to the number of diced field emitters. Furthermore, the optimized geometry comprises a cavity within the area of the TGVs. The depth of the cavities varies between 400 µm and 460 µm. Therefore, the resulting glass thickness is about 85 µm to 20 µm. This way, a distinctive reduction of the electron path through the TGVs is achieved.

Furthermore, there are cavities in the corner areas of the chip where glass and silicon are bonded by laser-assisted bonding. Through those cavities of about 30 µm depth, the distance between the extraction electrode and the silicon emitters has been reduced. This way, more outbreaks on the flanks of the pyramidical structures are exposed to the voltage, resulting in higher emission currents (Fig. 1 B).

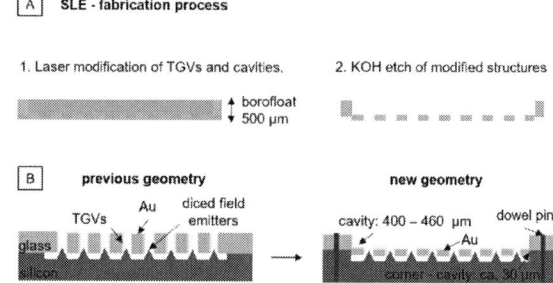

Fig. 1. Schematic image of the SLE fabrication process of the electrode (A) and a cross-section of the previous (B, top, [1]) and new (B, bottom) optimized glass-silicon emitter chip.

C. Alignment of the emitter chip

The emitter chip with diced field emitters was fabricated, as shown in the previous work, by laser assisted bonding [1]. Due to the chip size reduction, the bonding areas have been reduced from 3 mm² to 1 mm² (Fig. 2). Furthermore, we have systemized the alignment of the glass electrode and the emitter tips. The corner-cavities that were initially fabricated to lower the electrode are used for that. Those function as self-allying cavities since quadratic structures in silicon fabricated due to the dicing process fit into the cavities with a misalignment of max. 20 µm. For finer manual, after-alignment dowel pins are used with a diameter of 500 µm. Those enable a finer alignment of the tips within the TGVs, resulting in ca. 10 µm offset (Fig. 2) compared to the time-intensive manual-only alignment shown in the previous work [1].

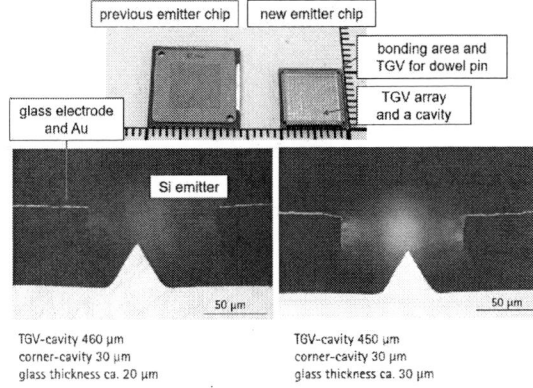

Fig. 2. Upper image shows a top view of the previous emitter chip geometry of 14 mm² (left) and the new with optimised glass electrode (right). Below there are microscopic images of a cross section of emitter chips, with various resulting the TGV thicknesses, after alignment through the chip corner cavities and dowel pins.

III. EXPERIMENTAL RESULTS

A. Test-setup

The glass-silicon emitter chip was characterized in the setup shown in Fig. 3. The holder was fabricated with SLE. The contact areas were sputtered either with Cu or Au and then contacted with screws. A conical spring was used as an electron collector. The CF40 flange was mounted onto an ultra-high vacuum chamber. All of the measurements were conducted at a pressure of about $5*10^{-8}$ mbar. The source measuring unit (SMU) SHR 42 60r iseg was used in all experiments.

B. Current-voltage Characteristics

The efficiency of the emitter chips with high n-doped silicon field emitters was demonstrated through current voltage (I - V) measurements in a three-electrode configuration. The electrodes were: silicon field emitters, glass extraction electrode and the electron collector positioned about 3 mm away (Fig. 3).

The so far tested chips in the diode configuration were: the old geometry (230 µm thickness, no cavity), as well as 400 µm TGV-cavity and 30 µm corner-cavity and 450 µm TGV-cavity and 40 µm corner-cavity. The results with the corresponding residual TGV-height due to the fabricated TGV-cavity are in Table 1. The samples were tested over 30 min.

TABLE I.   I-V-CHARACTERISTICS: DIODE CONFIGURATION

TGV height	230 µm	80 µm	40 µm
corner cavity [µm]	0 µm	30 µm	40 µm
emission I [µA]	135.4 ± 39.4	225.7 ± 70.3	574.52 ± 93

The results show higher emission currents due to the cavities in the chip corners. In the triode configuration two types of samples were tested (Table 2).

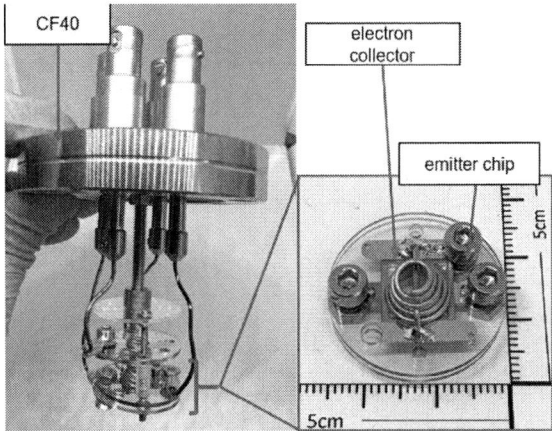

Fig. 3. Test setup for characterisation of the glass-silicon emitterchip in a triode configuration. Left there is the SLE-fabricated fused silica holder with contact pads and a conical spring as an electron collector. On the right there is a zoomed view of the contacting of the electrodes

TABLE II.   TRIODE I-V-CHARACTERISTICS: TRIODE CONFIGURATION

I [µA]	TGV height	
	230 µm	80 µm
emitter chip	135.4 ± 39.4	225.7 ± 70.3
electron collector	10.0 ± 3.2	22.3 ± 12.8

Therefore, we confirm the enhancement of the emission current and electron transmission through the optimisation of the glass chip geometry.

ACKNOWLEDGMENT

This work has been carried out as a part of the joint project "Innovative Vacuum Technology for Quantum Sensors" (InnoVaQ) funded by the German Federal Ministry of Education and Research (BMBF) as part of the funding program "Quantum Technologies – from basic research to market". (Contract number: 13N15919).

REFERENCES

[1] Buchta, A. M. et al., "Novel Glass-Silicon Emitter Chip for Field Emission Applications," 2023 IEEE 36th International Vacuum Nanoelectronics Conference (IVNC), Cambridge, MA, USA, 2023, pp. 207-209

[2] Hausladen, M. et al., "An Integrated Field Emission Electron Source on a Chip Fabricated by Laser-Micromachining and Mems Technology," 2023 IEEE 36th International Vacuum Nanoelectronics Conference (IVNC), Cambridge, MA, USA, 2023, pp. 115-116

HfN and Hf Spindt-type FEA Fabrication Using Triode Reactive High Power Pulsed Magnetron Sputtering

Hiromasa Murata[1,*], Shun Kondo[1,2], Md. Suruz Mian[2], Katsuhisa Murakami[1], Takeo Nakano[2], and Masayoshi Nagao[1]

[1] *National Institute of Advanced Industrial Science and Technology, Tsukuba, Japan*
[2] *Graduate School of Science and Technology, Seikei University, Musashino, Japan*
*Corresponding author: murata.hiromasa@aist.go.jp

Abstract—The triode high power pulsed magnetron sputtering (t-HPPMS) apparatus has been utilized for the fabrication of Spindt-type field emitter array (FEA). t-HPPMS achieves high directionality of sputtered particles through high degree of ionization and by introducing a potential difference between the sample and the plasma. This approach contributes to the downsizing of the apparatus and enables a reactive process for preparing the emitter tips. In this study, we successfully demonstrated the fabrication of HfN and Hf Spindt-type FEA by t-HPPMS.

Keywords—Spindt-type field emitter array, triode reactive high power pulsed magnetron sputtering (t-HPPMS)

I. INTRODUCTION

Spindt-type field emitter arrays (FEA) have been actively studied because of their excellent electron emission performance [1] and have been applied to various applications such as field emitter displays [2] and traveling-wave tubes [3]. In this method, large-size electron beam (EB) evaporation equipment has been used for emitter formation because it requires the directionality of evaporated particles. This resulted in the difficulties large-scale fabrication of FEAs.

Sputtering has been widely employed owing to its advantages in large area coatings. Moreover, sputtering method can easily fabricate compound thin film such as HfN, which has low work function, a task that is difficult with EB evaporation. However, the conventional sputtering methods is unsuitable for emitter formation in Spindt-type FEA because of low directionality of sputtered particles. Improving the directionality of sputtered particles incident on the substrate would help in downsizing the deposition equipment, leading to broader industrial applications of Spindt-type FEA.

High power pulsed magnetron sputtering (HPPMS) efficiently ionizes the sputtered particles by applying short pulse powers to a sputtering target to generate high-density plasma. Furthermore, in the triode HPPMS (t-HPPMS) method, the plasma potential can be controlled by applying a voltage to the cap electrode above the sputtering target, resulting in high directionality of sputtered particles [4]. Thus, t-HPPMS is promising deposition method for Spindt-type FEA. In this study, we fabricated HfN and Hf Spindt-type FEA by using t-HPPMS and evaluated electron emission property.

II. FABRICATION

Fig. 1 shows the fabrication flow of HfN and Hf Spindt-type FEA by using t-HPPMS. SiO_2 (300 nm) was deposited on a Si substrate by plasma-enhanced chemical vapor deposition using tetraethoxysilane gas and then Nb (200 nm) was deposited by direct current magnetron sputtering using Ar gas (Fig. 1(a)). Nb gate holes were formed by using photolithography and reactive ion etching using sulfur hexafluoride gas (Fig. 1(b)). SiO_2 was etched by buffered hydrofluoric acid (Fig. 1(c)). Al sacrificial layer was deposited by EB evaporation in the oblique direction while rotating the substrate (Fig. 1(d)). HfN or Hf were deposited using t-HPPMS as shown in Fig. 2. Here, the time-averaged pulse power, pulse frequency, and duty ratio were set to 150–200 W, 200 Hz, and 5%, respectively. The sputtering target was pure-Hf and the sputtering gases were Ar/N_2 for HfN deposition and Ar for Hf deposition (Fig. 1(e)). Then, Al layer was removed using hydrochloric acid (Fig. 1(f)).

III. RESULTS AND DISCUSSION

Fig. 3 shows the HfN and Hf emitters with high aspect ratio were successfully fabricated, suggesting that t-HPPMS produces highly directional sputtered particles than the conventional sputtering system. The height of the emitters are approximately 600 nm and higher than gate aperture.

We evaluated emission characteristics using the electrical circuit as shown in Fig. 4(a). Fig. 4(b) shows the Fowler-Nordheim (FN) plots calculated from the I-V curves of Hf Spindt-type FEA. The FN plots exhibits a linear behavior, indicating electrons were emitted by FN tunneling.

IV. CONCLUSION

We fabricated HfN and Hf Spindt-type FEA by using t-HPPMS and demonstrated FEA operation. HfN and Hf emitters have relatively high aspect ratio and their height is higher than the gate aperture. Emission property was evaluated by applying a positive voltage to the Nb gate electrode, while the emitter is grounded. FN plots suggests that electrons were emitted from Hf-FEA by FN tunneling.

REFERENCES

[1] C. A. Spindt, "A Thin-film field-emission cathode," J. Appl. Phys. vol. 39, pp. 3504-3505, 1968.

[2] K. Sakurada, M. Kitada, T. Niiyama, M. Namikawa, Y. Takeya and M. Tanaka, "Development of high resolution Spindt-type FED," in Proc. of 13th International Display Workshops, pp.1805-1808, Otsu, Japan, Dec. 2006.

[3] H. Makishima, S. Miyano, H. Imura, J. Matsuoka, H. Takemura, A. Okamoto, "Design and performance of traveling-wave tubes using field emitter array cathodes," Applied Surface Science, vol.146, pp.230-233, 1999.

[4] T. Nakano, H. Taniguchi, N. Dei, Md. S. Mian, K. Oya, K. Murakami, M. Nagao, "Structure optimization of Spindt-type emitter fabricated by triode high power pulsed magnetron sputtering," J. Vac. Sci. Technol. B vol. 40, pp. 063201-1-9, 2022.

[5] Takeo Nakano, Tomoki Narita, Kei Oya, Masayoshi Nagao, and Hisashi Ohsaki, "Fabrication of Mo microcones for volcano-structured double-gate Spindt-type emitter cathodes using triode high power pulsed magnetron sputtering," J. Vac. Sci. Technol. B, vol. 35, pp.022204-1-6, 201

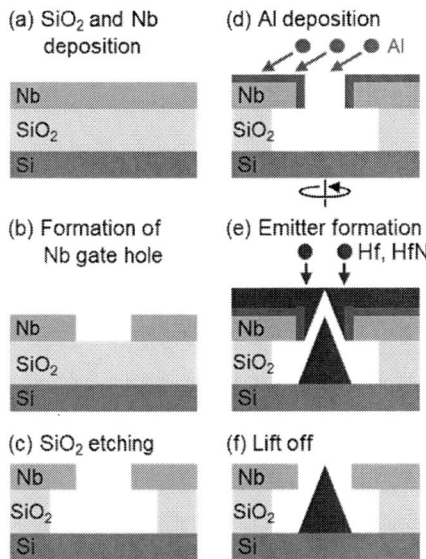

Fig. 1 Fabrication process of HfN and Hf Spindt-type FEA.

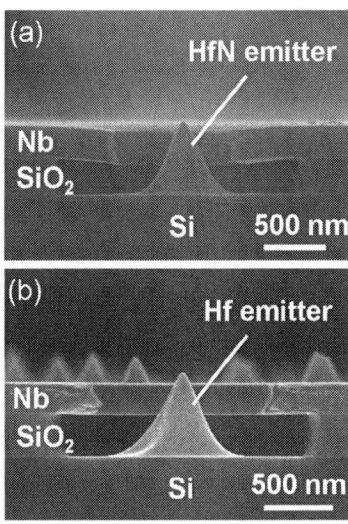

Fig. 3 Cross-sectional scanning electron microscope images of (a) HfN and (b) Hf emitter formed by t-HPPMS.

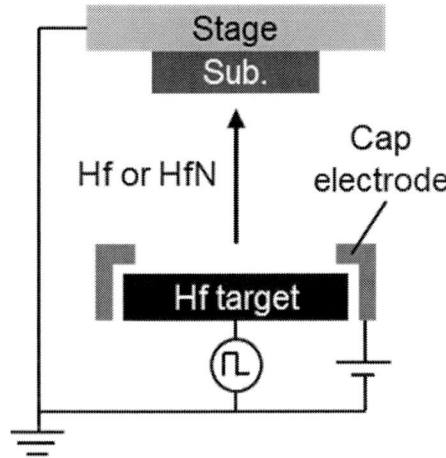

Fig. 2 Schematic of t-HPPMS system.

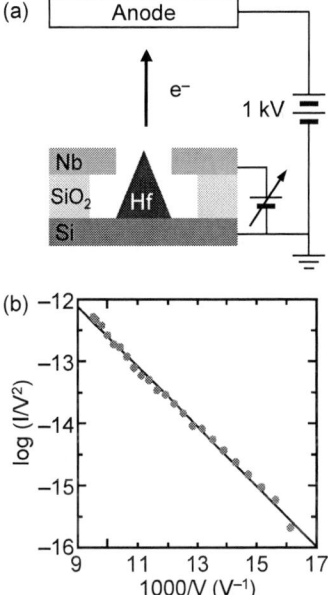

Fig. 4 Evaluation of electron emission property. (a) Schematic of electrical circuit for emission measurement. (b) FN plots calculated from I-V characteristics of Hf emitter.

High Performance Paper-CNT Field Emitters

Michał Krysztof*, Piotr Szyszka, Paweł Urbański, Tomasz Grzebyk

Department of Microsystems, Wroclaw University of Science and Technology, Wroclaw, Poland
*Corresponding author: michal.krysztof@pwr.edu.pl

Abstract—In this paper, technology and preliminary measurements of novel electron emitters are described. The emitters are made of a paper substrate cut in the form of a triangle coated by a layer of carbon nanotubes. Emission characteristics as well as emission current in time were measured for emitters with three different tip angles (30°, 90°, 150°). The emission current is higher for a lower tip angle but presents higher fluctuations. The most stable emission was obtained for the emitter with a tip angle of 150°. The application of such emitters in test structures of miniaturized electron optic devices is discussed.

Keywords—field emitter, carbon nanotubes, electron source, high vacuum MEMS devices

I. INTRODUCTION

Field emitters are widely studied by scientists looking for the best performance of their electron sources in terms of electron beam current, stability of emission, low energy spread, small initial beam spot, suitable for particular applications. At Wroclaw University of Science and Technology we are developing MEMS electron optical devices, such as the MEMS electron microscope [1] or the MEMS point X-ray source [2]. For years we have struggled with the development of suitable electron emitters for those applications, with a high and stable electron current and a long lifetime. During this time, we have developed several emitters that are compatible with MEMS technology, but we are still working on the optimal solution. In this article, we present a reliable, easy to fabricate electron source that can be used in test structures of MEMS devices, when the final emitter has not yet been developed.

II. FABRICATION

The technology of the paper-CNT emitters is very simple (Fig. 1). The substrate for the emitter was cut from laboratory paper in the form of a triangle tip on a rectangular piece of paper.

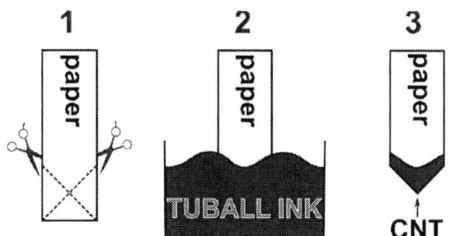

Fig. 1. Paper-CNT emitter technological steps: 1 – substrate forming; 2 – dipping the substrate in CNT solution; 3 – air dry the emitter.

To obtain a high accuracy of the tip angle, a mask with a suitable angle was printed and used as a cutting template. The substrate was then submerged in a CNT solution (TUBALL INK™, OCSiAl, Luxemburg) with weight percentage of components: 0.2% SWCNTs, 2% surfactant, 97.8% deionized

This work was supported by the Polish National Science Centre under Project 2021/41/B/ST7/01615.

water. Since the solution is water-based, the emitter was left to dry in air.

III. MEASUREMENTS AND DISCUSSION

For the experiments, three types of emitters were prepared with different tip angles (30°, 90°, 150°) (Fig. 2). It was done to compare the results with similar experiments presented in [3]. The emitters were mounted to a metal housing, and glass substrates with an ITO layer were used as the anodes. The measurement setup was then inserted into a vacuum chamber ($p = 10^{-5}$ mbar).

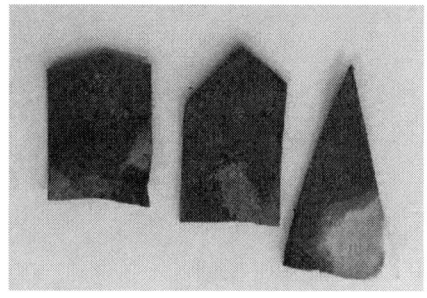

Fig. 2. Paper-CNT emitters before measuremetns.

First, the current-voltage characteristics were measured for a particular emitter (Fig. 3), followed by measurements of electron emission in time (Fig. 4).

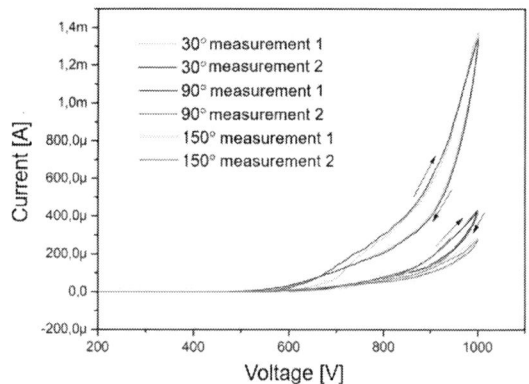

Fig. 3. Current-voltage characteristics for paper-CNT emitters with different tip angle.

The highest current was measured for the 30° emitter tip angle ($I_A = 1.37$ mA) and the smallest for the 150° emitter tip angle ($I_A = 275$ μA). The results are similar to those of emitters made from CNT membranes [3].

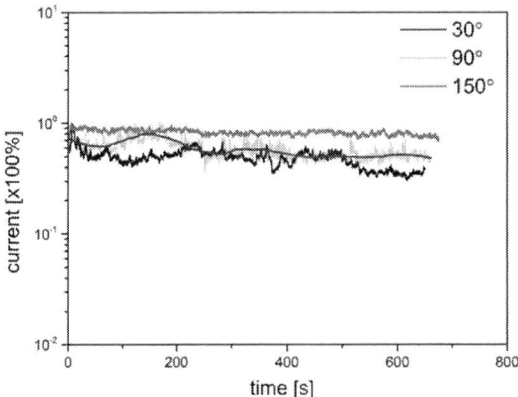

Fig. 4. Normalized current in time for paper-CNT emitters with different tip angle.

The stability of the emitter has the opposite character. The emitter with a 150° tip has the most stable current, whereas the emitter with a 30° tip has the highest current fluctuations. These results are also in agreement with the literature [3].

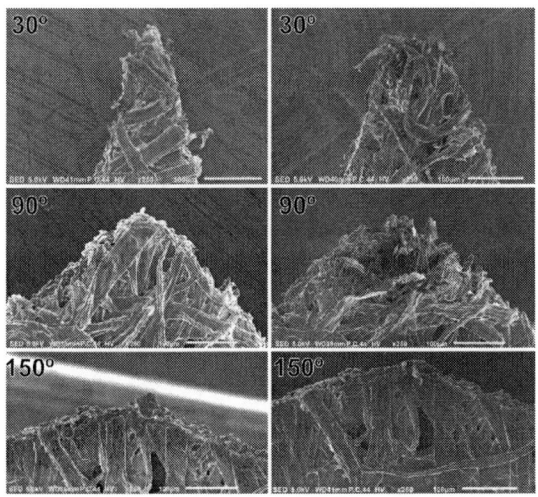

Fig. 5. SEM images of emitters before (left column) and after (right column) measurements.

The fluctuations are a reflection of the damage the emitter suffered by residual ions present in the vacuum chamber. The emitters with lower tip angle emit more electrons, which generate more ions when they collide with residual gas particles, thus causing more damage to the emitter (Fig. 5).

Fabricated emitters work really well, but the question is where they can be used and how? The emitters were designed for use in test structures of MEMS high vacuum devices. To integrate the paper-CNT emitters with silicon electrodes, we have developed a dedicated 3D printed holder. One part of the holder holds and aligns the silicon electrodes, and another one holds the paper-CNT emitter. The second part is design in such a way that emitter with particular tip angle stands out from the holder to a height of ca. 500 μm (Fig. 6a). In this way, we ensure both the constant distance to the electrode and proper alignment with the electron optics axis (Fig. 6b).

Fig. 6. 3D housing for paper-CNT emitters and silicon electrodes: a) 90° tip angle emitter mounted in the housing (inset: SEM image of the 500 μm tip); b) paper-CNT emitter aligned with two silicon electrodes in a test structure.

IV. CONCLUSIONS

The paper-CNT emitters are inexpensive and easy to fabricate alternative to electron emission. They present good parameters with high currents for low emitter angle (30°) or good stability for high emitter angle (150°). This way they can be used interchangeably for applications requiring high currents or high stability. With dedicated 3D printed structure, ensuring very good alignment of the emitter with silicon electrodes, the emitters can be used for testing of electron optical components like electron lenses, octupole scanning systems used in MEMS electron microscope or X-ray targets for MEMS X-ray source.

REFERENCES

[1] M. Białas, T. Grzebyk, M. Krysztof and A. Górecka-Drzazga, "Signal detection and imaging methods for MEMS electron microscope," Ultramicroscopy, 244 (2023) 113653.

[2] P. Urbański, M. Białas, M. Krysztof, T. Grzebyk, "Transmission target for a MEMS X-ray source," Journal of Microelectromechanical Systems, vol. 32, nr 4 (2023) s. 398-404.

[3] D. H. Shin, et al., "High performance field emission of carbon nanotube film emitters with a triangular shape," Carbon, 89 (2015) pp. 404-410.

High Quantum Efficiency Graphene-oxide-semiconductor Electron Source without Negative Electron Affinity Surface

Hidetaka Shimawaki*, Masayoshi Nagao and Katsuhisa Murakami

*Graduate School of Electric, Electronic and Information Engineering, Hachinohe Institute of Technology, Japan

Device Technology Research, Institute National Institute of Advanced Industrial Science and Technology, Japan

*Corresponding author: simawaki@hi-tech.ac.jp

Abstract—**Planar electron emission sources using a graphene-oxide-semiconductor thin film diode structure were fabricated. The photo-assisted electron emission properties were investigated under light irradiation. The emission current proportionally increased to the incident light power. The quantum efficiency of 0.3 % was achieved by the device without negative electron affinity surface.**

Keywords—*photo-assisted electron emission, planar-type, graphene-oxide-semiconductor*

I. INTRODUCTION

Semiconductor photocathodes use III-V semiconductor substrates, such as p-type gallium arsenide. They offer impressive performance compared to metals and other materials, with quantum efficiencies reaching up to ~10% [1]. They can operate with visible laser light with energy higher than the bandgap energy and emit picosecond pulsed electron beams [2]. These outstanding properties can only be achieved with a negative affinity (NEA) surface on the cathode. The NEA surface offers the advantage of low work function, but it is brittle and susceptible to damage from the absorption of residual gases. Therefore, special experienced surface treatment skills and ultra-high vacuum are required to maintain its lifetime.

The GOS electron source is a planar electron source consisting of a graphene-oxide-semiconductor thin film diode structure. It is a type of MOS tunnel cathode [3]. Using graphene as the gate electrode resolves the drawbacks of MOS electron sources, which have an electron emission efficiency of less than 1%, and achieved a high efficiency of more than 10% [4]. This shows that graphene has excellent electron transparency. In addition, graphene has excellent visible light transparency of ~97% [5]. Therefore, GOS electron sources using a p-type semiconductor substrate are expected to emit electrons with high quantum efficiency. GOS electron sources can also operate in low vacuum, allowing them to function without the NEA surface, which is essential for the operation of GaAs photocathodes.

In this study, we investigate the photo-assisted electron emission properties of GOS electron sources and discuss their potential as a semiconductor photocathode operating without a negative electron affinity surface.

II. EXPERIMENTAL

The GOS electron source consists of a thin film diode structure with a thin SiO2 film of ~10 nm thickness

sandwiched between a Si substrate and a graphene film of ~1 nm thickness.

The fabrication process is illustrated in Fig. 1, similar to the standard MOS diode fabrication process. Graphene is directly synthesized on SiO$_2$ by PECVD after the thermal oxidation of a silicon substrate. Further details are available elsewhere [4].

The device was placed in a vacuum chamber with a base pressure of approximately 10^{-6} Pa and was irradiated from the front with a 633 nm CW laser at an incident angle of 45°. The spot size of the laser beam is approximately 0.1 mm in diameter on the device surface of the device.

III. RESULTS

Figure 2 shows the characteristics of the emission and substrate currents and electron emission efficiencies with increasing the gate voltage for our p-type GOS device. The emission area was designed to be 0.1 mm in diameter. In the figure, the solid and dashed lines represent the characteristics with and without 633 nm CW laser irradiation. In the absence of light, the emission current was detected around the voltage of 7 V, and saturation was observed over about 8.5 V due to the limited number of electrons (minority carriers) in the p-type semiconductor. Under irradiation, the threshold voltage

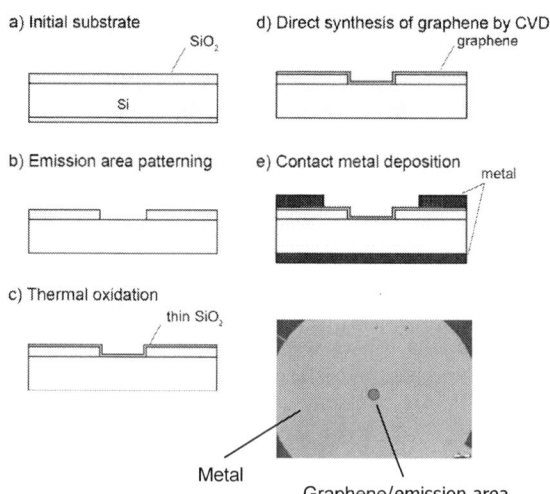

Fig. 1. Fabrication process of graphene-oxide-semiconductor electron source and its optical micrograph.

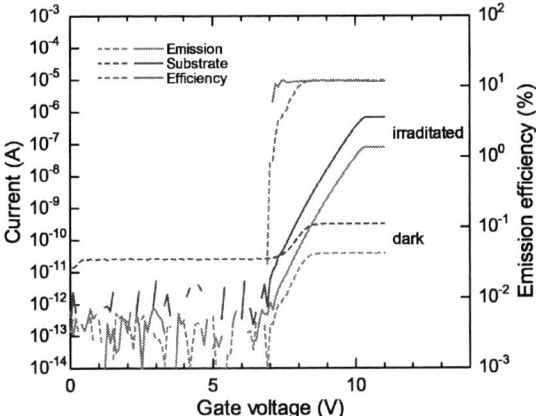

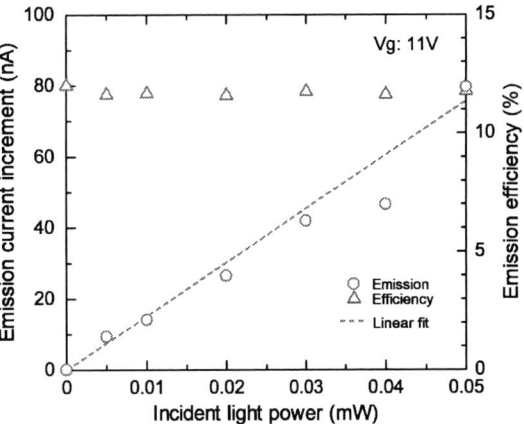

Fig. 2 Typical current-voltage characteristics and electron mission efficiency of a p-type GOS electron source. The dashed line represents the condition without light irradiation, while the solid line represents the condition with irradiation.

Fig. 3 Emission current and electron emission efficiency as a function of the incident light power.

decreased slightly and the emission current increased significantly. This was due to photo-excited electrons from the valence band to the conduction band. The electron emission efficiency was about 12% in both conditions, with and without laser illumination.

Figure 3 shows the correlation between the increase of the emission current, the electron emission efficiency, and the incident laser power. The gate voltage was set to 11 V. The quantum efficiency of the p-type GOS electron source was determined to be 0.3%, estimated from the slope. The electron emission efficiency remains about 12% regardless of the incident light power. This value is similar to that of a typical GOS electron source using an n-type silicon substrate. It indicates that the electron emission efficiency of the GOS electron source is not affected by the substrate type or the incident light power.

IV. CONCLUSION

In this study, the photo-assisted electron emission from a GOS electron source using a p-type silicon substrate has been investigated. Recent n-type GOS devices with a 1 nm thick graphene electrode fabricated using the latest process technology have achieved electron emission efficiencies of

30% or higher [6]. It has been predicted that a device with a single-layer graphene electrode has the potential to achieve a maximum efficiency of about 70%. As a result, the p-type GOS electron sources are expected to increase the quantum efficiency to several percent or higher. The experimental results indicate that GOS electron sources have the potential to serve as semiconductor photocathodes with high quantum efficiency and without a negative electron affinity (NEA) surface.

REFERENCES

[1] W. Liu, S. Zhang, M. Stutzman, and M. Poelker, Phys. Rev. Accel. Beams 19, 103402 (2016).

[2] Aulenbacher K, Schuler J, Harrach D V, Reichert E, Röthgen J, Subashev A, Tioukine V, and Yashin Y., J. Appl. Phys. 92, 7536 (2002).

[3] K. Yokoo, H. Tanaka, S. Sato, J. Murota, and S. Ono, J. Vac. Sci. Technol. B 11, 429 (1993).

[4] K. Murakami, J. Miyaji, R. Furuya, M. Adachi, M. Nagao, Y. Neo, T. Takao, Y. Yamada, M. Sasaki, and H. Mimura, Appl. Phys. Lett. 114, 213501 (2019).

[5] R. R. Nair, P. Blake, A. N. Grigorenko, K. S. Novoselov, T. J. Booth, T. Stauber, N. M. R. Peres, and A. K. Geim, Science 320, 1308 (2008).

[6] K. Murakami, M. Adachi, J. Miyaji, R. Furuya, M. Nagao, Y. Yamada, Y. Neo, Y. Takao, M. Sasaki, and H. Mimura, ACS Appl. Electron. Mater. 2, 2265 (2020).

Characterization and Analysis of Field Electron Emission from Copper Tips

Marwan S. Mousa[1], Adel M. Abuamr[1], Ammar Al Soud[2], Alexandr Knápek[3], Dinara Sobola[2]

[1] Department of Renewable Energy Engineering, Jadara University, 21110 Irbid, Jordan.

[2] Department of Physics, Brno University of technology, 61600 Brno, the Czech Republic.

[3] Institute of Scientfic Instruments of CAS, Královopolská 147, 61264 Brno, the Czech Republic.

* Corresponding email: m.mousa@jadara.edu.jo

Abstract— In the present investigation, field electron emission (FEE) from high-purity copper emitters has been studied under vacuum conditions of approximately 10^{-5} mbar. This research aimed to study the emission properties of copper under low-pressure environments, which is essential for applications in fields such as electron microscopy and vacuum electronics. The current-voltage (I-V) characteristics are analyzed utilizing Murphy–Good (MG) method plots in this study. In addition, a Scanning Electron Microscope with Energy-Dispersive X-ray Spectroscopy (SEM-EDS) was used to visualize the surface of the emitters and analyze sample purity. The spatial distribution of electron emission and current stability were obtained as well to analyze electron emission behavior from the surface of the tips. Electron mapping demonstrated a consistent component distribution throughout the copper tips, showing uniformity in their composition. The current-voltage characteristics showed a notable switching phenomenon that is explained by the existence of a thin oxide layer formed on the emitter's surfaces. This thin oxide layer provides additional quantum barrier for the surface.

I. INTRODUCTION

Field electron emission (FE) is a process where electrons escape from a material's surface into a vacuum caused by a high voltage, producing an extensive surface field, typically a few volts per nanometer[1]. The emitter, commonly in the form of a sharp tip having an apex radius in the range of a nanometer, must be made from superior material in order to achieve its maximum efficiency[2]. Previous studies have extensively investigated FE characteristics across different materials [3], studying both the microstructure as well as theoretical models of field emission [4]. Copper has come to be as an alternative option due to its unique benefits, especially its outstanding electron emission characteristics[5]. Several investigations have been conducted utilizing copper as an electron source for a range of technological and scientific objectives[6], [7]. This research examines copper wires as cold electron sources, targeting to investigate their emission behaviors. Furthermore, this work aims to study the impact of oxidation layers on the current emission characteristics as well as the emission pattern of these emitters.

II. SAMPLE PREPARATION

Copper wires of 99.95% purity and 0.25 mm diameter had been utilized. Emitters were prepared by electrochemical etching using phosphoric acid solution, with a the beginning etching current about ~40 mA and applied voltage of 15V. Following etching process, distilled water was used in an ultrasonic cleaner to clean the copper tips. In addition, scanning electron microscopy (SEM) had been utilized to obtain closer and further detailed images, allowing measurements for calculating the apex radius of the tips. The emitter was installed in a FEM with an axial tip-to-screen separation of 10 mm and a pressure of 10-5 mbar. Voltage was applied employing an external power supply, and current measurements were conducted with a picoammeter. Spatial electron distribution was recorded utilizing a digital camera.

III. RESULTS AND DISCUSSION

The electrochemical etching process produced effectively sharp emitter tips, as shown via the SEM images in Figure 1. The Cu emitter has an apex radius of ~80 nm. The I-V characteristics demonstrated in Figure 2. The emission process current began at 450 V with a current value of 0.02 nA. The applied voltage was increased gradually to 2000 V where the obtained current was 6.48 µA. The voltage was then reduced slowly till the threshold voltage showed up to be achieved at 200 V with emission current of 0.01 nA.

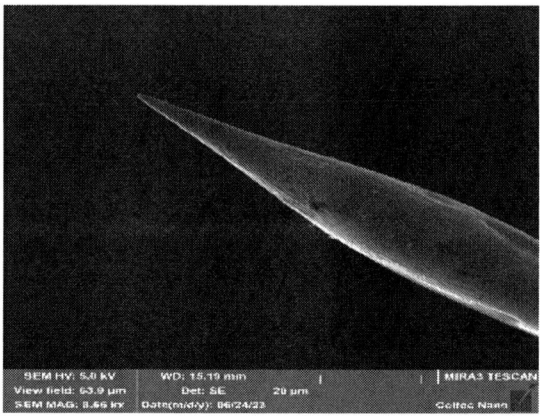

FIG. 1. Shows the scanning electron microscope image of the sample.

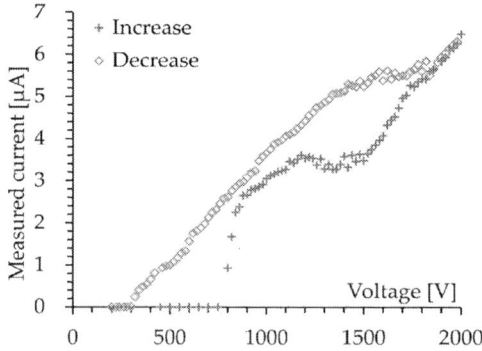

Fig. 2. Shows the current-voltage characteristics of the emitter

The switch-on phenomena can be explained through the existence of a thin oxide layer generated on the emitter surface. This thin oxide layer presents an additional quantum barrier on the surface as it includes traps, which have been separated into two distinct kinds: shallow traps and deep traps[8]. While voltage applies, electrons start traveling from the Fermi level at the tip and accumulate within these traps, and this goes on until the traps have saturated; this phase is known as the charging phase. The electrons are released from the traps through the conductive channels after the applied voltage achieves the switch-on voltage; this phase is known as the discharged phase. It is important to remember that high-energy electrons are more probable than low-energy electrons to escape from shallow traps.

The electron emission pattern as seen in Figure 3, during voltage increase demonstrated a single spot, showing emission from a specific point. The sample's nonlinearity was revealed via Murphy-Goode (MG) plots (Figure 4), which can be attributed to a number of phenomena including space charge absorption, high current density effects, and local electrical field enhancement. Table 1 shows the results of the Field Emission Orthodoxy Tests at every step of the emission process as divided into numerous line segments.

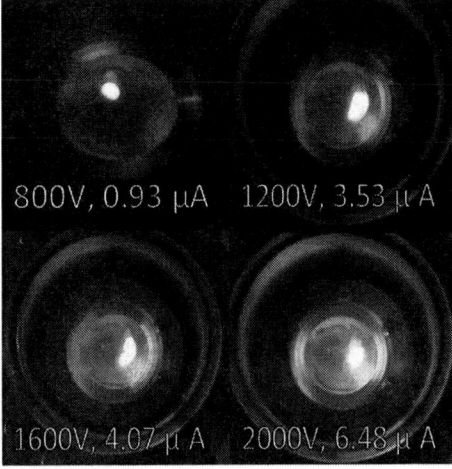

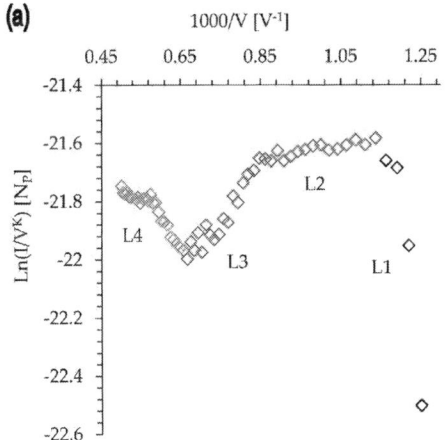

Fig. 3. Shows the electron emission spatial distribution of the sample during the voltage increase. The applied voltage and measured current are provided in each image.

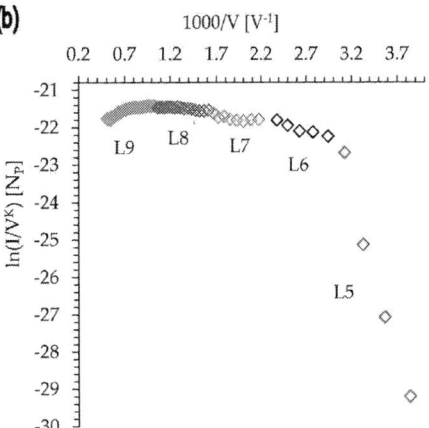

Fig. 4. Shows the Murphy-Goode (MG) plots.

Table 1. Field emission orthodoxy test and field emission analysis results.

Segment index L_i	Test result	$A_f^{SN} \, [m^2]$
L1	P	7.82×10^{-17}
L2	F	N/A
L3	F	N/A
L4	F	N/A
L5	I	6.78×10^{-10}
L6	F	N/A
L7	F	N/A
L8	F	N/A
L9	F	N/A

IV. CONCLUSIONS

The outcomes of this study demonstrate that the emitters show accurate current emission at low extraction voltages, indicating their efficient electron emission characteristics. In particular, the spatial distribution of emitted electrons from the copper emitter remains stable and uniform, demonstrating the reproducibility as well as consistency of the emission process. The oxide layer offered protection to the emitter from ionic shear. In addition, the findings indicate excellent performance in the rectification test at low voltages, thus highlighting promising prospects for further research employing the rectification test and the Murphy-Goode (MG) model for encapsulated emitters.

REFERENCES

[1] R. V Latham and M. S. Mousa, "Hot electron emission from composite metal-insulator micropoint cathodes," J Phys D Appl Phys, vol. 19, no. 4, pp. 699–713, Apr. 1986.

[2] M. S. Mousa and A. M. Abuamr, "Electron Emission from High-Purity Copper Wires," Jordan Journal of Physics, vol. 16, no. 2, pp. 247–252, Jun. 2023.

[3] M. Sreekanth, S. Ghosh, and P. Srivastava, "Highly enhanced field emission current density of copper oxide coated vertically aligned carbon nanotubes: Role of interface and electronic structure," Nov. 2018.

[4] M. M. Allaham et al., "Interpretation of field emission current–voltage data: Background theory and detailed simulation testing of a user-friendly webtool," Mater Today Commun, vol. 31, p. 103654, Jun. 2022, doi: 10.1016/j.mtcomm.2022.103654.

Characterization of Electron Emission from Broad Area Composite Emitter

Ammar Al Soud[1,2*], Marwan S. Mousa[3], Aseel Aljabarat[4], Adel M. Abuamr[3], Ahmad Telfah[5], Alexandr Knápek[6,7], Enas A. Arrasheed[8], Dinara Sobola[1,2]

[1] *Central European Institute of Technology Brno University of Technology Purkynova 656/123,61200 Brno, Czech Republic.*
[2] *Department of Physics, Faculty of Electrical Engineering and Communication, Brno University of Technology, Technická 2848/8, 61600 Brno, Czech Republic.*
[3] *Department of Renewable Energy Engineering, Jadara University, Irbid 21110, Jordan.*
[4] *Department of Physics, Mutah University, Al-Karak 61710, Jordan.*
[5] *Nanotechnology Center, The University of Jordan, 11942, Amman, Jordan.*
[6] *Institute of Scientific Instruments of the Czech Academy of Sciences, Královopolská 147, 61200 Brno, Czech Republic.*
[7] *Department of Microelectronics, Technicka 3058/ 10, 61600 Brno, Czech Republic.*
[8] *Department of communication and computer Engineering, Jadara University, Irbid, Jordan.*

[*] Corresponding author: Ammar.al.soud@vutbr.cz

Abstract— This work studies the emission of electrons from Copper Broad Area Emitter (CBAE) in addition to Copper Broad Area Composite Emitter (CBACE) and provides an explanation of the switch-on phenomenon based on the formation of conductive channels and traps. The samples were examined using field electron microscopy (FEM) at high vacuum conditions (10^{-6} mbar) and a voltage of 15 kV, with the emitter placed 5 mm away from the anode. Scanning electron microscopy (SEM) was used to characterise the epoxy layer after voltage application. The results based on current and voltage characteristics showed that a breakdown of the epoxy layer occurred due to the emission of current at 13 kV.

Keywords—broad-area emitters, Field electron emission

I. INTRODUCTION

Field electron emission (FEM) is a physical phenomenon that has been well explored from the last century to present day. This process involves the emission of electrons from a material through a quantum mechanical tunnelling mechanism, wherein electrons traverse a potential barrier established by the material's surface. The seminal work by Fowler and Nordheim in 1928 laid the foundation for the Fowler-Nordheim theory, up to MG modifications. elucidating the intricacies of quantum tunnelling [1]. Many previous researches have studied the emission of electrons from sharp emitters and broad-area emitters [2], [3]. The aim of this paper is to study the electron emission pattern of CBAE and CBACE after applying 15 kV, to provide an explanation of the electron emission mechanism based on the traps generated in the epoxy layer.

II. SAMPLE PREPARATION

High-purity CBAEs with a diameter of 14 mm and a thickness of 1 mm were used. The CBAE was polished until a surface resembling a mirror was obtained. The polished samples were ultrasonically cleaned with alcohol for 10 minutes. The samples were coated with 300 μm of epoxy resin brand (Epoxy Resin 2301). This involves immersing vertically and gradually of the polished sample into the epoxy/ hardener mixture, ensuring a uniform coating. The CBACE was left for 48 hours in a dry place until the epoxy reached hardness. Next, the sample was placed in a FEM system 5 mm away from a phosphor screen that acts as an anode. The FEM system creates a high vacuum of 10^{-6} mbar. The samples were connected to a high-voltage power supply with an adjustable range from 0 to 30 kV, while the screen was grounded through a micro-pico meter, specifically a Keithley 405 meter, to measure the total emission current. Throughout the experiment, the voltage gradually increased up to 15 kV, resulting in the application of an electric field in the range of 10^6 V/m.

III. RESULTS AND DISCUSSION

A. I-V Characteristic

The I-V values of the CBAE and CBACE were measured for two cycles with each cycle containing two phases: increase and decrease voltage as shown in Figure 1. Figure 1a and b provide a characterization of the emission current from CBAE and CBACE. The current value for CBAE was only 20 nA, due to the difficulty of emitting electrons from flat surfaces. However, this is an important step, as the first cycle helps to clean the surface of any contaminants. The current value increases significantly above 13 kV to around 200 n. The current value increases significantly above 13 kV to around 200 n. The emission mechanism in the voltage increase phase can be characterised into two phases. The charging phase and the discharging phase. Charging occurs in the overvoltage stage up to 13,000 V. In this stage, electrons coming from the Fermi level are continuously injected via nano peaks on the surface of the composite emitter [4]. The electrons begin to accumulate in deep and shallow traps in the epoxy layer formed due to surface

979-8-3503-7977-8/24 $31.00 © 2024 IEEE

roughness and manufacturing defects [6].

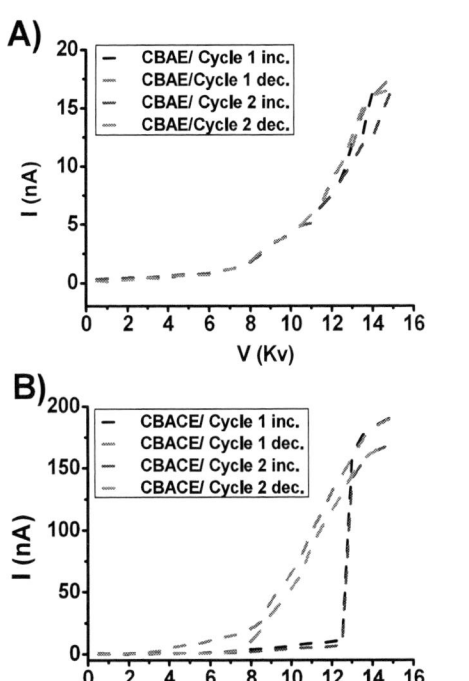

Fig. 1. I-V characteristic for a) CBAE. b) CBACE.

Moreover, electron injection is dominated by Schottky and Fowler-Nordheim injection, and the distribution of electrons in these layers is governed by Fermi-Dirac statistics [7]. After all the traps are filled with electrons, the second stage, branching, begins, where the electrons suddenly rush through the conductive channels that form in parallel with the charging process.

B. Epoxy collapse

Figure 2 shows SEM graph for CBACE after applying 15 kV. The formation of micro-cracks in the epoxy surface (collapse) can be observed starting from the edges and extending to the centre of the sample, due to the large electric field at the edges and the weakening of the epoxy layer at the edges. The formation of these micro cracks can be explained be explained based on the principle of conductive channels. Conductive channels are formed when a high electric field is applied to the epoxy. They contribute to the formation of dipoles that are polarised by the electric field, and when these dipoles come into contact with each other, conductive pathways are formed through which electrons stored in the traps are transferred to the epoxy surface. After the applied electric field is removed, the dipoles cannot return to their original position because they have lost their ability to relax [5]. This means that conductive channels are formed faster in the second cycle, which explains the lower operating current in the second cycle as shown in Figure 1b. As the high voltage continues to be applied, the density of the conductive channels increases and is greater at the edge of the emitter. This causes the epoxy to start corroding. Over time, the depth of corrosion increases, which is referred to

as the partial discharge stage. These discharges appear on the surface of the epoxy. In the next step, the intensity of the partial discharge increases, resulting in the widening of the erosion channel and arborization. Finally, breakdown occurs when the corona discharge imposes electrical stresses higher than the threshold voltage, triggering the onset of discharge and leading to a gradual degradation of the epoxy.

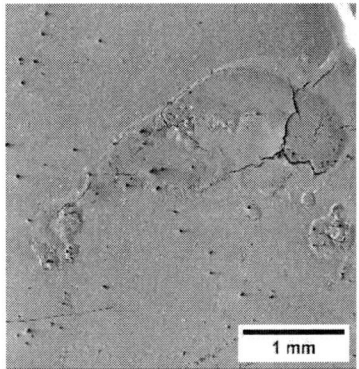

Fig. 2. SEM micrograph for CBACE after applying 15 Kv.

CONCLUSION

The study shows the emission of electrons from large-area copper composite emitters under high voltage and vacuum, using conductive traps and channels. CBAE shows a finite current centred on the edge, while CBACE shows a significant increase due to the release of electrons from saturated traps. This research advances the understanding of conductive traps and channels that can be applied to complex sharp emitters.

ACKNOWLEDGMENT

The research described in this paper was financed by the grant 23-07384S of the Czech Science Foundation (GACR) as well as. Internal Grant Agency of Brno University of Technology, grant No. FEKT-S-23-8228.

REFERENCES

[1] R. G. Forbes, "The Murphy–Good plot: a better method of analysing field emission data," *R Soc Open Sci*, vol. 6, no. 12, p. 190912, Dec. 2019.

[2] A. Al Soud, A. Knápek, and M. S. Mousa, "Analysis of the Various Effects of Coating W Tips with Dielectric Epoxylite 478 Resin or UPR-4 Resin Coatings under Similar Operational Conditions," JJP, vol. 13, no. 3, pp. 191–199, 2020.

[3] S. Bajic and R. V Latham, "A new perspective on the gas conditioning of high-voltage vacuum-insulated electrodes," *J Phys D Appl Phys*, vol. 21, no. 6, pp. 943–950, Jun. 1988.

[4] Y. Gao, X. Liang, W. Bao, C. Wu, and S. Li, "Degradation characteristics of epoxy resin of GFRP rod in the decay-like fracture of composite insulator," *IEEE Transactions on Dielectrics and Electrical Insulation*, vol. 26, no. 1, pp. 107–114, Feb. 2019.

[5] A. Alsoud *et al.*, "Electrical Characterization of Epoxy Nanocomposite under High DC Voltage," *Polymers (Basel)*, vol. 16, no. 7, p. 963, Apr. 2024.

[6] N. Grassie, M. I. Guy, and N. H. Tennent, "Degradation of epoxy polymers: 3—Photo-degradation of bisphenol-A diglycidyl ether," *Polym Degrad Stab*, vol. 13, no. 3, pp. 249–259, Jan. 1985.

[7] S. Maletić, N. Jović Orsini, M. Milić, J. Dojčilović, and A. Montone, "Dielectric properties of epoxy/graphite flakes composites: Influence of loading and surface treatment," *J Appl Polym Sci*, Nov. 2023.

979-8-3503-7977-8/24 $31.00 © 2024 IEEE

Improved Method For Determining the Distribution of FEA Currents by Optical CMOS Sensors

Matthias Hausladen[1, *], Andreas Schels[2], Philipp Buchner[1], Mathias Bartl[1], Ali Asgharzade[1], Simon Edler[2], Dominik Wohlfartsstätter[2], Michael Bachmann[2], Rupert Schreiner[1]

[1]Ostbayerische Technische Hochschule Regensburg, Germany, [2]Ketek GmbH, Munich, Germany

Corresponding author: matthias.hausladen@oth-regensburg.de

Abstract— **CMOS image sensors are utilized to determine the time- and spatially-resolved distribution of the electron emission of silicon field emission arrays. During initial experiments, rather low field emission currents already visibly damaged the sensor surface, altering the system accuracy over the measurement time. Therefore, we coated the sensor surface with copper for protection. In contrast to the original insulating surface, the Cu coating provides a conductive surface for incident electrons and improves heat dissipation in addition. This prevents localized surface charges and surface damages which stabilize the system accuracy.**

Keywords—Field Emission, Field Emission Characterization

I. INTRODUCTION

With field emitting arrays, a uniform emission distribution across all emission sites is desired. Due to slight variations in geometry of the individual emitters, the current distribution is typically unevenly balanced. To observe the current distribution, we developed a method for quasi-simultaneous acquisition of many field emission sites on a cathode array [1, 2]. The pixels of the sensor measure the Bremsstrahlung generated from incident electron beams. The image signals (gray values) are analyzed to calculate a normalized signal share map. Multiplying these separate share factors with an electrically measured total emission current yields optically derived individual emission currents. During initial experiments, rather low field emission currents of a few hundred nanoampere already visibly damaged the sensor

surface, altering the system accuracy over the measurement time (Fig. 1a). Therefore, we coated the sensor surface with a 150 nm thick Cu layer for protection (Fig. 1b). In contrast to the original insulating surface, the Cu coating provides a conductive surface for incident electrons and improves heat dissipation. This prevents localized surface charges and surface damages which stabilize the system accuracy.

II. EXPERIMENTS

For the experiments with the CMOS sensors, we used two different samples: one with a few emitters (2x2 array, Fig. 1c) and one with a large number of emitters (21x21 array, Fig. 1d) [3, 4]. Both were fabricated from n-type Si with a resistivity of 5-10 Ωcm. The sample with few emitters was used to easily detect possible irreversible damage to the sensor surface (Fig. 1a). With the Cu-coated sensor, no damage was observed up to currents of several microampere per individual tip (Fig. 1b). For the further experiments with high integral currents, we then used the array with many tips. Images were taken at different total current values in the range up to 100 µA, where the current was first increased, then kept constant and then reduced again. To evaluate the images, the individual current values of the emitters were calculated and categorized by 80 logarithmic spaced bins (2D-histogram with a range of 1 nA to 150 µA, Fig. 2) The red dashed inset-plot depicts the 1D-histogram of the FE current distribution of datapoint #108 (Fig. 2, red dashed vertical line). The according image of datapoint #108 is

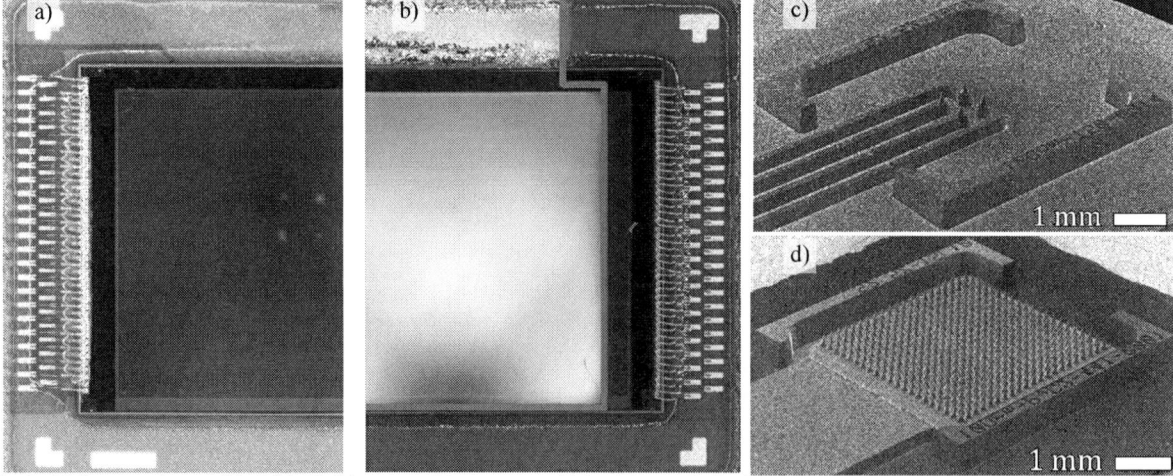

Fig. 1: a) Microscope image of the left half of the uncoated image sensor with local damaged microlens areas. b) Microscope image of the right half of a Cu coated image sensor. The red line indicates the border of the Cu layer. c) Field emission array with 4x4 tips. d) Field emission array with d) 21x21 tips.

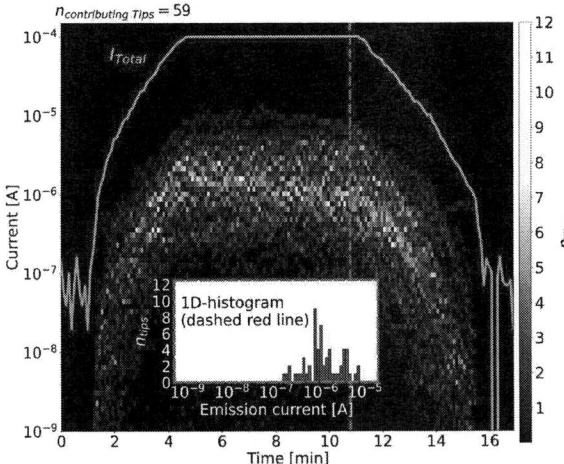

Fig. 2: Distribution of the current contribution of the individual emitters of a field emission array with 21x21 emitters as a function of the total emission current (I_{Total}) using 80 current bins (logarithmic evenly spaced from 1 nA to 150 µA). The colorbar depicts the number of individual tips contributing to the current level bins. The red dashed inset represents the 1D-histogram of datapoint #108 which is indicated by the vertical red dashed line.

depicted in Fig. 3a. All detected emission spots were highlighted with circles. The 4 emitters which exhibited the highest FE current values are colorized. Their individual IV-characteristics (opaque graphs) and current contribution (semi-transparent graphs) to the total emission current (top subplot) were evaluated and plotted in the quadrants of in Fig. 3b. As the total cathode current increases, the individual currents started rising too (Fig. 3b). When the total FEA current reached a value of 100 µA, it was kept constant. During this period, the best performing individual tips

emitted currents of around 10 µA. As we checked the sensor afterwards, no damages or defects were found.

III. CONCLUSION

A Cu layer protects the sensor surface and avoids degradation and modification of the lens material underneath. Moreover, the Cu coating provides a conductive surface to the incident electrons and avoids lens charges, which are locally different on an uncoated image sensor. With a Cu-coated sensor, emitter arrays with higher currents in the microampere range can also be measured without the sensor being damaged.

ACKNOWLEDGEMENT

The research work was funded by the Bavarian Research Foundation under project-number AZ-1583.

REFERENCES

[1] A. Schels u. a., „In-situ quantitative field emission imaging using a low-cost CMOS imaging sensor", Journal of Vacuum Science & Technology B, Bd. 40, Nr. 1, S. 014202, Jan. 2022, doi: 10.1116/6.0001551.

[2] A. Schels et al., "Quantitative Field Emission Imaging for Studying the Doping-Dependent Emission Behavior of Silicon Field Emitter Arrays," Micromachines, vol. 14, no. 11, p. 2008, Oct. 2023, doi: 10.3390/mi14112008.

[3] R. Lawrowski, M. Hausladen, and R. Schreiner, "Individually Addressable Fully Integrated Field Emission Electron Source Fabricated by Laser Micromachining of Silicon," in 2020 33rd International Vacuum Nanoelectronics Conference (IVNC), Lyon, France: IEEE, Jul. 2020, pp. 1–2. doi: 10.1109/IVNC49440.2020.9203470.

[4] M. Hausladen, P. Buchner, M. Bartl, M. Bachmann, and R. Schreiner, "Integrated multichip field emission electron source fabricated by laser-micromachining and MEMS technology," Journal of Vacuum Science & Technology B, vol. 42, no. 1, p. 012201, Jan. 2024, doi: 10.1116/6.0003233.

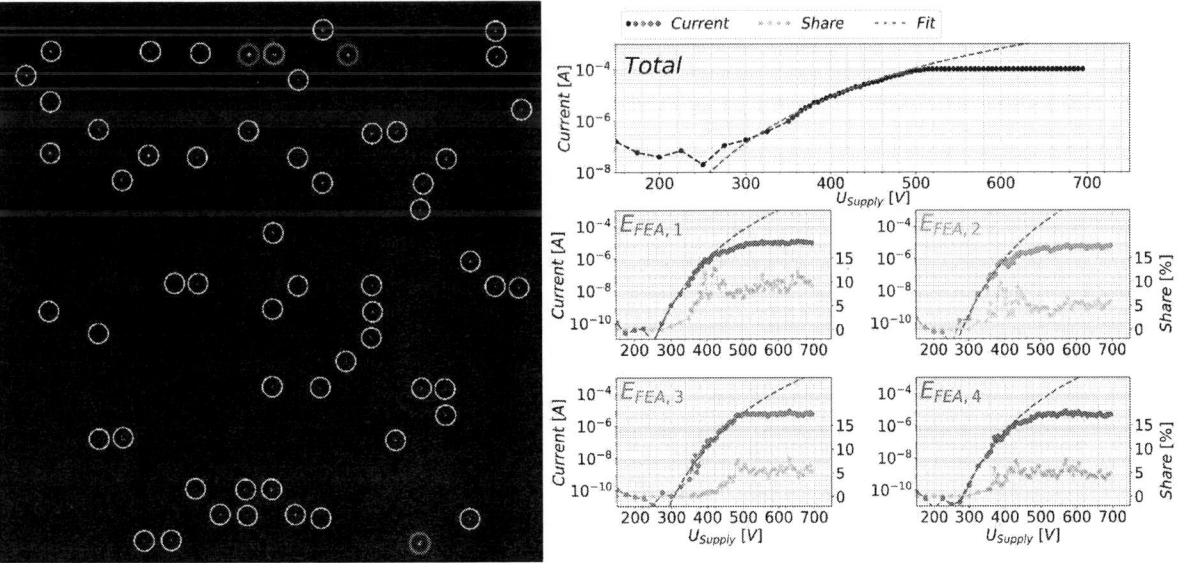

Fig. 3: a) 2800x2800 px (4.34x4.34 mm²) image of datapoint #108, captured by the Cu-coated image sensor during an IV-characterization measurement procedure of the 21x21 cathode with a maximum total current of 100 µA. Each circle indicates a contributing tip ($n_{contributing\ tips}$ = 59). The colorized and labeled circles represent the 4 emitter tips which possessed the highest FE current values. b) Corresponding IV-characteristics of the 4 tip currents (opaque graphs, left axis) as well as their contribution (semi-transparent graphs, right axis) to the total current in percentage.

979-8-3503-7977-8/24 $31.00 © 2024 IEEE

Improved Performance of Planar Vacuum Field Emission Transistors via Angled Gates and Extended Anode

Zelin Yu, Zhenpeng Wang, Zhuoya ZHU, Mei XIAO, Wei LEI, Xiaobing ZHANG[*]

School of Electronic Science & Engineering, Southeast University, Nanjing 210096, China

[*]Corresponding author: bell@seu.edu.cn

Abstract—**Vacuum field emission transistors (VFETs) have recently gathered significant attention. A primary obstacle to the widespread adoption of VFETs is the low electron transmission rate resulting from electrons being intercepted by the gates. In this study, we propose a planar VFET design featuring angled gates and extended anode. Using a particle tracking simulation method, the output and transfer curves of these devices were obtained. Analysis reveals that angled gates effectively reduce the number of electrons intercepted by the gates, thereby increasing the electron transmission rate from 17.96% to 78.03%. Furthermore, the incorporation of an extended anode mitigates the sharp increase observed in the output curves to a certain extent. These findings demonstrate a substantial enhancement in the performance of planar VFETs through structural modification, laying a solid foundation for the future development of VFETs for signal amplifier applications.**

Keywords—*Vacuum field emission transistor, angled gate, electron transmission rate,*

I. INTRODUCTION

Vacuum field emission transistors (VFETs) operate on the principle of field emission, utilizing a vacuum environment for electrons to moving ballistically without lattice scattering.

Reported VFET devices can be categorized into vertical VFET and planar VFET, based on the positional relationship between the cathode and the anode. Vertical VFETs typically require high operating voltages, which often do not match well with low voltage systems in integrated circuit[1]. Conversely, planar VFETs have gathered significant attention due to their processing complexity and voltage requirements[2]. However, a primary obstacle to the widespread adoption of planar VFETs is the low electron transmission rate resulting from electrons being intercepted by the gates.

II. CONFIGURATIONS AND MODELING

A. Configurations

Figure 1 illustrates a planar VFET featuring both flat-bottom gates and angled gates. In this model, the device is constructed on a SiO$_2$ wafer, and the thickness of four gold electrodes is 100 nm. The distance between the cathode and gate measures 50 nm, while the distance between cathode and anode is 400 nm/1 μm.

The Major Research Plan of the National Natural Science Foundation of China (Grant No. 92264103)

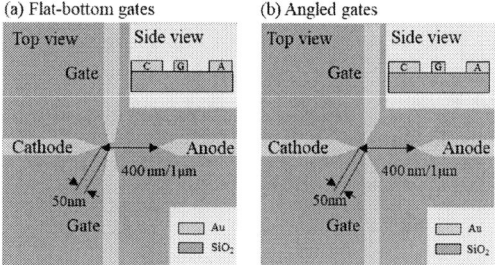

Fig.1. Schematic for planar VFET (a) flat-bottom gates (b) angled gates

B. Field Emission Modeling

VFETs operate on field emission, a process where electrons are extracted from the surface of a condensed phase, such as metal, into vacuum by applying a high electrostatic field[3]. This phenomenon is described by the Fowler-Nordheim (FN) theory. Hence, the emission current density J_{FN} follows[4]:

$$J_{FN} = \frac{e^3 E^2}{8\pi h \phi t^2(y)} \exp\left[-\frac{8\pi\sqrt{2m}\phi^{\frac{3}{2}}}{3heE}\theta(y) \right] \quad (1)$$

where e is the electron charge, E is the electric field intensity of the emitter surface, h is the Plank constant, Φ is the work function of emitter, m is the electron mass, $t(y)$ and $\theta(y)$ are approximate constants. It is worth noting that the emission current is mainly affected by the electric field intensity and the work function of emitter, so the formula can be simplified as:

$$J_{FN} = AE^2 \exp(-\frac{B}{E}) \quad (2)$$

where A = 3.7×10^{-8} A/V^2, B = 5.4×10^8 V/m are the constants of gold cathode. The CST Studio Suite is used to obtain electrical characteristics of devices with four different structures, including devices with flat-bottom gates and a close anode, devices with angled gates and a close anode, devices with flat-bottom gates and an extended anode, and devices with angled gates and an extended anode.

III. RESULTS AND DISCUSSION

The simulation results of output curves for the four structures are depicted in Fig.2. When gate voltage (V_g) is 25 V and anode voltage (V_a) is 30V, the anode current is approximately 1.18×10^{-9} A in Fig.2c, and the corresponding cathode current is approximately 6.57×10^{-9} A, resulting in electron transmission rate of 17.96%. Under the same conditions, the anode current is approximately 5.44×10^{-9} A in Fig.2d, and the corresponding cathode current is approximately 6.92×10^{-9} A, yielding electron transmission rate of 78.03%. These results demonstrate that angled gates can significantly increase the electron transmission rate.

As shown in Fig.2, the anode currents exhibit a sharp increase region as V_a increases to a certain point. Taking the device with angled gates and a close anode as an example, particle trajectory (with V_a =20 V/V_g =25 V and V_a =30 V/V_g =25 V) are shown in Fig.3. The Fig.3a reveals that when V_a is 20 V, some electrons cannot be collected by anode because they possess an upward initial velocity due to the electrical field between the gate and cathode, and V_a is not sufficiently large to attract them to anode. The Fig.3b demonstrates that when V_a is 30 V, the majority of electrons can be collected by the anode. Under fixed V_g, the cathode current remains relatively stable, leading the anode current to saturate when V_a exceeds 30 V. Consequently, when V_a reaches a certain threshold, the anode current undergoes a sharp increase followed by a saturation region. As is shown in Fig.2a, 2c and Fig.2b, 2d, extended anode can partially restrict the sharp increase in output curves. Moreover, due to the greater effect of the flat-bottom gates on the upward initial velocity of the electrons, devices with flat-bottom gates cannot entirely eliminate the sharp increase when the anode is positioned further away from the cathode (from 400 nm to 1 μm).

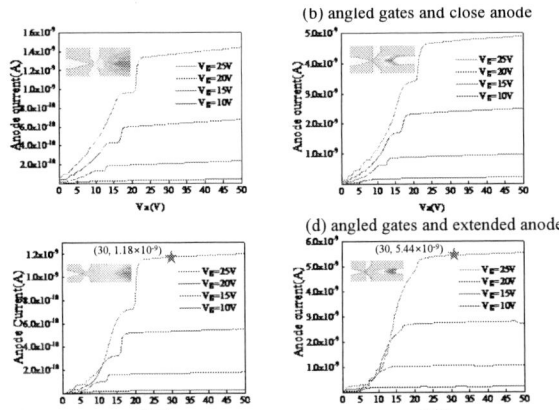

Fig.2. Comparison of output curves with four structures (a) flat bottom gates and close anode (b) angled gates and close anode (c) flat-bottom gates and extended anode (d) angled gates and extended anode

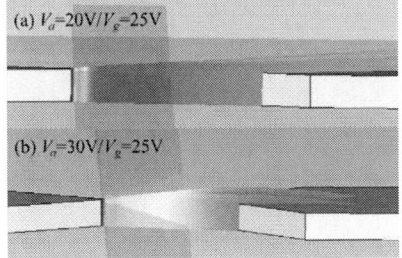

Fig.3. Particle trajectory under two conditions (a) V_a =20 V/V_g =25 V (b) V_a =30 V/V_g =25 V

In Fig.4, the transfer curve of the device with angled gates and extended anode is presented, where V_a is set to 30 V. It is notable that the anode current demonstrates an exponential increase with V_g with a threshold voltage is measure to be 15.4V.

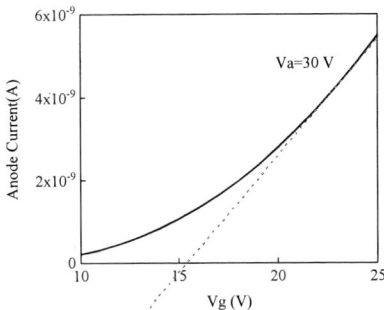

Fig.4. Transfer curve of device with angled gates and extended anode

IV. CONCLUSION

This paper presents a planar VFET featuring angled gates and an extended anode. Simulation results from CST Particle Studio confirm that angled gates can enhance the electron transmission rate, while an extended anode mitigates the sharp increase phenomenon in output curves to a certain degree. Consequently, the modification of the device structure yields a superior output curve, a foundation for future applications in signal amplification.

ACKNOWLEDGMENT

The author thanks the support of the Major Research Plan of the National Natural Science Foundation of China (Grant No. 92264103(Xiaobing Zhang)).

REFERENCES

[1] C. A. Spindt. "A thin-film field-emission cathode." Journal of Applied Physics, vol. 7, 1968, pp. 3504–3505.

[2] J. HAN and J. S. Oh, "Cofabrication of Vacuum Field Emission Transistor (VFET) and MOSFET," IEEE Transactions on Nanotechnology, vol. 13, 2014, pp. 464-468.

[3] A. Kahn, "Fermi level, work function and vacuum level," Materials Horizons, vol. 3, 2013, pp. 7–10.

[4] R. Fowler and L. Nordheim, "Electron emission in intense electric fields," Proceedings of the Royal Society of London. Series A, Containing Papers of a Mathematical and Physical Character, vol. 119, 1928, pp. 173–181.

Improving the Performance of Field Electron Emission from Carbon Fiber Emitters with an Epoxy Resin 478 Coating Layer

Issam Trrad[1,*], Marwan S. Mousa[2], Ahmad M D (Assa'd) Jaber[3], Adel M. Abuamr[2], Ali F. AlQaisi[2], Alexandr Knápek[4]

[1] Department of Communication and Computer Engineering, Jadara University, 21110 Irbid, Jordan

[2] Department of Renewable Energy Engineering, Jadara University, 21110 Irbid, Jordan

[3] Department of Basic Medical Sciences, Aqaba Medical Sciences University, 77110 Aqaba, Jordan

[4] Institute of Scientfic Instruments of CAS, Královopolská 147, 61264 Brno, the Czech Republic.

* Corresponding email: itrrad@jadara.edu.jo

Abstract— The aim of this research is to investigate the differences in behavior and characteristics of electron beams emitted from carbon fiber tips before and after coating their surfaces with a thin layer of epoxy resin 478. Several field electron emission characteristics have been investigated measured under high vacuum (HV) conditions, before and after coating by dielectric materials. The measurements recorded involve current-voltage (I–V) characteristics) as well as spatial current distributions (electron emission images). Furthermore, the surface morphology and chemical composition of the emitters were thoroughly examined utilising a Scanning Electron Microscope with Energy-Dispersive X-ray Spectroscopy (SEM-EDS). The results show significant variations in the emission mechanisms between clean and composite emitters. In particular, the emitters coated with epoxy resin 478 demonstrated an important improvement in the emission characteristics, such as enhanced emission current density and the stability of the emitted current. These results demonstrate how well the insulating layer works to enhance the efficiency of electron emitters fabricated from carbon fiber.

Keywords—Field electron emission; Field emission; Carbon fibers tips.

I. INTRODUCTION

Cold field electron emission (CFE) happens when a strong electric field is applied to metals or semiconductors, leading electrons to tunnel through a potential barrier by quantum mechanical tunneling[1], [2]. This emission can be improved by raising the emitter's temperature or applying powerful electric fields, impacting potential barriers as well as allowing electron emission near the Fermi level contrary to thermionic emission, CFE happens at comparatively low temperatures[3], [4]. The emitted current (I_m) depends mainly on the applied voltage (V_m) as well as the material's local work function (φ)[5], [6], [7]. Carbon-based structures, specifically carbon nanotubes as well carbon fibers, have attracted major attention in both industry and academia for their application as cold cathodes in advanced technologies[8]. These materials demonstrate impressive field emission characteristics, such as low turn on field, high emission currents, current stability, as well as immediate pulse repetition rates. The present investigation studies cold field electron emission from polyacrylonitrile (PAN) carbon fiber type VPR-19, processed at 2800 °C, both before and after Epoxy Resin 478 is coated as an insulation layer. Current-voltage (I-V) characteristics, field electron emission patterns, and SEM images are all included in the study.

II. SAMPLE PREPARATION

Carbon fibre emitters have been produced via an electrochemical etching process utilizing a 0.1 M NaOH solution. The etching procedure was managed by maintaining a starting etching current about 35 µA at 2 volts. Following etching, the carbon fibres were placed in to an ultrasonic cleaning bath in order to remove any NaOH solution residue from the emitter surface. In order coating with Epoxy Resin 478, a two-mole epoxy resin solution has been made by combining the resin with alcohol which allows it to become a homogeneous solution for two hours in total The coating method involved carefully immersing the carbon fibre sample into the Epoxy Resin 478 solution then removing it gradually to ensure that only a thin, uniform film remained on the emitter's surface. Both of the clean and composite emitters were subsequently examined via a field emission microscope (FEM) under high vacuum (HV) conditions to investigate their field emission characteristics.

III. RESULTS AND DISCUSSION

Employing a scanning electron microscope (SEM), the radius of the etched carbon fiber tip (CF-tip) was measured. After that, the coated CF-tip was analyzed with the SEM to determine the thickness of the Epoxy Resin 478 layer. The obtained scanning electron microscope images are shown in figure Fig. 1.

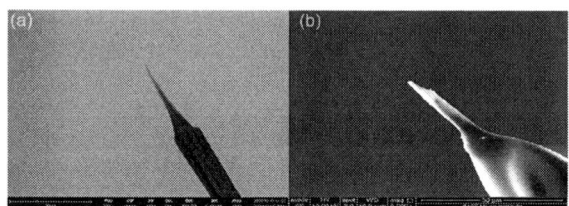

FIG. 1. SEM Micrographs of carbon fiber (CF) emitter (a) before and (b) after coating.

The I-V characteristics for the clean sample are shown in Fig. 2(a) for the full cycle of increasing and decreasing voltages. The threshold voltage was determined at 210 V during the phase of increasing voltage, and it dropped to 140 V during the phase of decreasing voltage. The maximum emission current observed was 2.4 µA at a voltage applied of 600 V.

Following that, the sample was immersed in epoxy resin for twelve times. The coated sample's I-V characteristics, as shown in Fig. 2(b), indicated a switch-on phenomenon at 770

979-8-3503-7977-8/24 $31.00 © 2024 IEEE

V and 2.5 µA of corresponding switch-on current. As the applied voltage raised to 850 V, the emission current attained 3.28 µA. The threshold voltage was observed at 250 V during the phase of decreasing voltage.

(a)

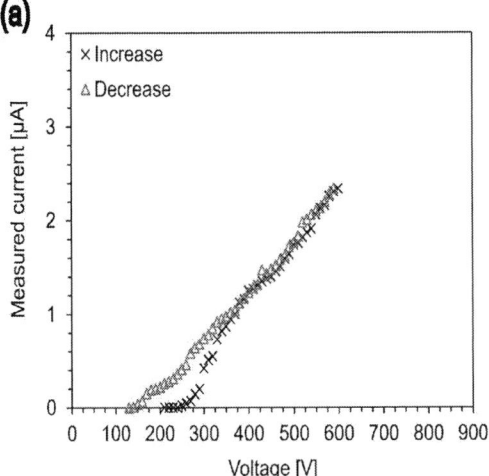

(b)

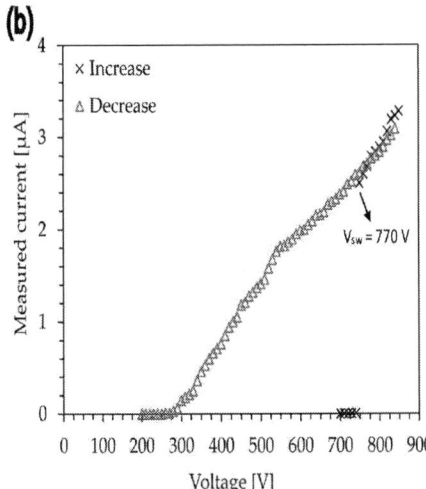

FIG. 2. Shows the current-voltage characteristics of the emitter (a) before and (b) after coating.

The field emission current density distribution patterns are displayed in Figure 3 for the uncoated emitter and Figure 4 for the composite emitter for different applied voltages.

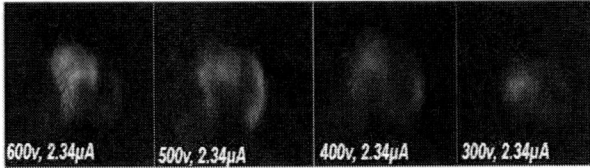

FIG. 3. Field emission patterns at different voltages obtained for the clean emitter.

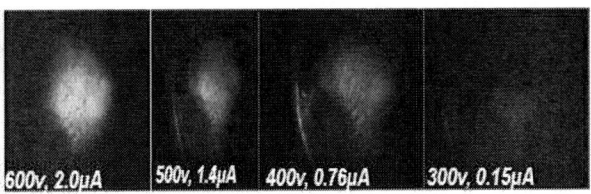

FIG. 4. Field emission patterns at different voltages obtained for the coated emitter.

Because the emission current is widely distributed over a large emission area, the images show low-intensity patterns. This phenomena is explained by the electrons' diffraction at a close distance from the emitter surface.

IV. CONCLUSIONS

In this research, carbon fiber single tip field emitter was prepared utilizing the electrochemical etching method. Comparative study of the I-V characteristics between clean CF-tip and that was coated with Epoxy resin 478 demonstrates significant improvements in performance because of the coating. The composite CF-tip shows higher total emission currents as well as lower field emission responses, demonstrating that the emitter can operate at lower voltages. As a result, the coated emitter is capable of emitting suitable currents within the nanoampere range at significantly lower applied voltages compared to their clean counterpart. This improvement demonstrates how well Epoxy resin 478 works to maximize the carbon fiber tips' field emission characteristics.

REFERENCES

[1] R. V Latham and M. S. Mousa, "Hot electron emission from composite metal-insulator micropoint cathodes," J Phys D Appl Phys, vol. 19, no. 4, pp. 699–713, Apr. 1986, doi: 10.1088/0022-3727/19/4/021.

[2] A. Alsoud et al., "Electrical Characterization of Epoxy Nanocomposite under High DC Voltage," Polymers (Basel), vol. 16, no. 7, p. 963, Apr. 2024, doi: 10.3390/polym16070963.

[3] H. A. Al-Braikat et al., "A Comparative Analysis of Field Electron Emission from Carbon Black Embedded within Insulated Copper Hollowed Wires and Glass Tubes," Karbala International Journal of Modern Science, vol. 10, no. 2, Apr. 2024, doi: 10.33640/2405-609X.3346.

[4] M. S. Mousa and A. M. Abuamr, "Electron Emission from High-Purity Copper Wires," Jordan Journal of Physics, vol. 16, no. 2, pp. 247–252, Jun. 2023, doi: 10.47011/16.2.12.

[5] A. Knápek, M. M. Allaham, D. Burda, D. Sobola, M. Drozd, and M. Horáček, "Explanation of the quasi-harmonic field emission behaviour observed on epoxy-coated polymer graphite cathodes," Mater Today Commun, vol. 34, p. 105270, Mar. 2023, doi: 10.1016/j.mtcomm.2022.105270.

[6] M. M. Allaham, M. S. Mousa, D. Burda, M. H. AlSa'eed, S. Y. AlJrawen, and A. Knapek, "Analyses of field electron emission Molybdenum current-voltage data using Fowler-Nordheim and Murphy-Good plots," in 2021 34th International Vacuum Nanoelectronics Conference (IVNC), IEEE, Jul. 2021, pp. 1–2. doi: 10.1109/IVNC52431.2021.9600771.

Improving the Performance of Nanoscale Vacuum Channel Transistors via Composite Gate

Xin ZHAI, Zhuoya ZHU, Mei XIAO, Wei LEI and Xiaobing ZHANG

School of Electronic Science & Engineering, Southeast University, Nanjing 210096, China

Corresponding author: bell@seu.edu.cn

Abstract— **Nanoscale Vacuum Channel Transistors (NVCT) are positioned to revolutionize the field of electronic devices in the future. However, the elevated operational voltage of planar side gate NVCTs continues to constrain device performance and stability. Overcoming this limitation and enhancing stability through structural design modifications remain formidable challenges. In this study, we propose a composite gate structure that effectively lowers the device's operating voltage and improves its stability. Utilizing composite gate operation markedly reduces the device's operational voltage compared to side gate operation, while concurrently achieving higher anode currents at lower operational voltages (I_a=8.93 nA, V_g=18 V, V_a=35 V). Consequently, these transistors hold promise for a wide array of integrated circuit applications, mitigating the issue of high operating voltages and emerging as a crucial solution in the realm of future electronic devices.**

Keywords— *Nanoscale vacuum channel transistors, composite gate structure, low operating voltage*

I. INTRODUCTION

When the distance between the cathode and the anode is less than the average free path of electrons in air, electrons do not collide with other molecules in the air, allowing for field emission to occur in an air environment (without the need for a vacuum environment)[1]. Nanoscale vacuum channel transistors (NVCTs) operate on this principle. Furthermore, during electron transport from the cathode to the anode, no scattering or collisions occur, resulting in ballistic transport, significantly enhancing the transistor's frequency and efficiency[2,3]. This advancement holds promise for reshaping the design of future electronic devices. However, field emission devices typically require higher operating voltages, and planar NVCTs are no exception. Lowering their operating voltage remains a pressing challenge for the practical implementation of NVCTs.

Here, we propose a composite gate structure, where a combination of biasing applied to both the back gate (heavily doped p-type silicon) and side gate (gold) forms the composite gate. Through a comparative analysis with planar side gate NVCTs, the composite gate structure significantly diminishes the operational voltage of planar NVCTs and attains higher anode currents at equivalent operational voltage levels. Additionally, electromagnetic simulations using CST software were conducted on both structures, revealing that the composite gate structure substantially amplifies the cathode tip electric field, thereby yielding superior performance. The capability of operating reliably at low voltages suggests that NVCTs hold promise for broad application in integrated circuits, thus providing a pivotal technological cornerstone for future electronic devices.

The Major Research Plan of the National Natural Science Foundation of China (Grant No. 92264103).

II. EXPERIMENTAL SECTION

A. Preparation of planar NVCTs

Main fabrication processes of planar NVCTs include electron beam lithography, electron beam evaporation deposition, and ultraviolet lithography. Initially, central electrodes (cathode, anode, and gate electrode) shapes are exposed using electron beam lithography. Subsequently, electrodes are deposited onto the wafer using electron beam evaporation deposition, followed by lift off. Then, ultraviolet lithography is employed to expose pads used for testing. Pads are then deposited using electron beam evaporation deposition, and after detachment, planar NVCTs are obtained.

B. Field Emission Performance Testing

Testing the electrical performance of devices involves utilizing a laboratory-built vacuum testing platform comprising a vacuum chamber, vacuum pump assembly, and source-measurement unit (SMU, Keysight 2400 and Keysight 6487).

C. Electromagnetic simulation

Electromagnetic simulation is conducted using the particle tracing module of the commercial CST software. To simplify the problem, only qualitative considerations of the electric fields between different components of the device are taken into account.

III. RESULTS AND DISCUSSION

After fabrication, the morphology of the device is depicted in Fig. 1, including solder pads, cathode, anode, and gate. Fig. 1(b) magnifies a portion of Fig. 1(a), clearly showing the device structure of planar NVCTs. The distance between the gate and the cathode is approximately 100 nm, while the distance between the cathode and the anode is approximately 180 nm. The composite gate structure is formed through a special wiring arrangement, as illustrated in Fig. 2. Both the side gate and the back gate are controlled using the same digital source meter, applying identical operating voltages. The anode is controlled via another digital source meter.

Electrical performance tests were conducted on both planar NVCTs with side gate and composite gate structures, as depicted in Fig. 3. The composite gate structure demonstrates a clear gate modulation effect. At an anode voltage of 35 V and a gate voltage of 18 V, an anode current of 8.93 nA is achieved. In contrast, the side gate structure exhibits an anode current of only 6.97 nA under an anode voltage of 50 V and a gate voltage of 26 V. The notable difference in the electrical performance of the two types of planar NVCTs underscores the advantages of the composite gate structure.

979-8-3503-7977-8/24 $31.00 © 2024 IEEE

Electromagnetic simulation results from CST software unveil the origin of the low operating voltage in composite gate planar NVCTs. As shown in Fig. 4，by applying identical operating voltages to two gate structure types, the cathode surface electric field intensity of the composite gate planar NVCT surpasses that of the side-gate planar NVCT notably. This enhancement allows the composite-gate planar NVCT to achieve performance comparable to the high operating voltage of the side-gate transistor at a lower operating voltage. Additionally, it demonstrates superior performance at the same operating voltage.

IV. SUMMARY

In summary, we have employed a simple wiring configuration to realize composite gate planar NVCTs, which significantly reduces the operating voltage of planar NVCTs. By comparing the electrical testing performance of side gate planar NVCTs with composite gate planar NVCTs, we have demonstrated the superiority of the latter. Furthermore, electromagnetic simulations conducted via CST software have provided insights into the exceptional performance of composite gate planar NVCTs.

ACKNOWLEDGMENT

The author thanks the support of the Major Research Plan of the National Natural Science Foundation of China (Grant No. 92264103(Xiaobing ZHANG)).

REFERENCES

[1] S. G. Jenning, "The mean free path in air." Journal of Aerosol Science 19.2(1988):159-166.

[2] J. W. Han, J. S. Oh, M. Meyyappan, "Vacuum nanoelectronics: Back to the future? —Gate insulated nanoscale vacuum channel transistor." Applied physics letters, 2012, 100.21.

[3] J. W. Han, M. L. Seol, D. I. Moon, G. Hunter, M. Meyyappan, "Nanoscale vacuum channel transistors fabricated on silicon carbide wafers." Nature Electronics, 2019, 2.9: 405-411.

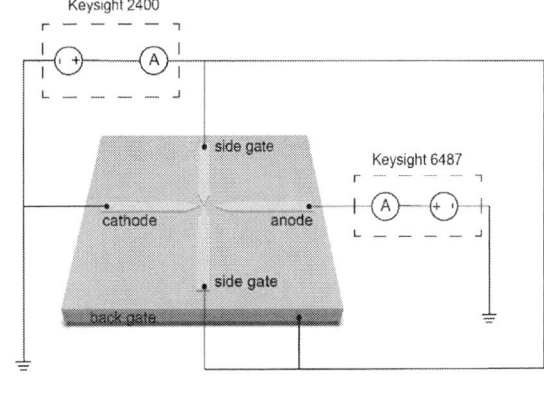

Fig. 2   Diagram of composite gate NVCT.

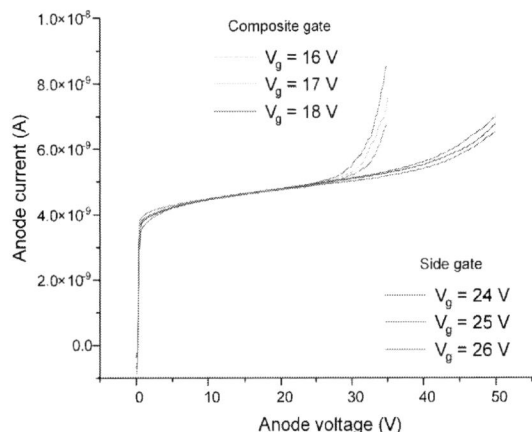

Fig. 3   Side gate and composite gate *I-V* characteristics curves.

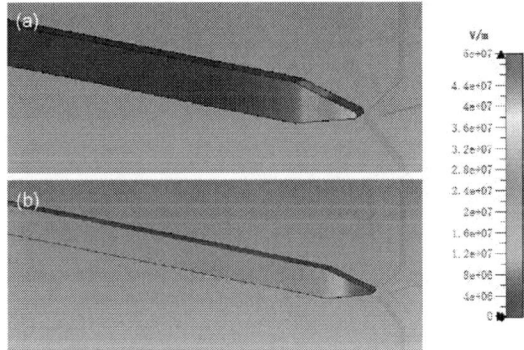

Fig. 4   CST simulation results of electric field distribution on the cathode surface of planar NVCTs. (a) Side gate, (b) composite gate.

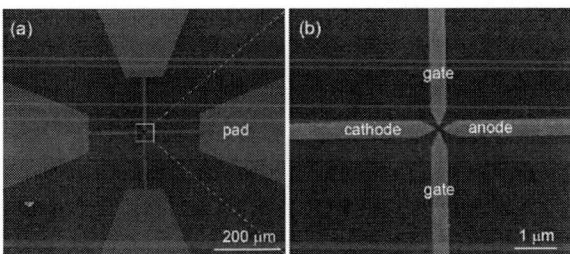

Fig. 1   SEM images of planar NVCTs. (a) Low magnification images of devices including pads, (b) is a locally enlarged image of the box in (a).

In situ Observation of Nanoprotrusion Growth on a Carbon Coated Tungsten Nanotip under Field

Guodong Meng[*], Yimeng Li[*], Roni Aleksi Koitermaa[††], Veronika Zadin[†], Yonghong Cheng[*] and Andreas Kyritsakis[†‡§]

[*]State Key Laboratory of Electrical Insulation and Power Equipment, Xi'an Jiaotong University, Xi'an, China
[†]Institute of Technology, University of Tartu, Nooruse 1, 50411 Tartu, Estonia
[‡]Helsinki institute of Physics and Department of Physics, University of Helsinki, PO Box 43, 00014 Helsinki, Finland
[§]Email: andreas.kyritsakis@ut.ee

Abstract—**Nano-protrusions (NPs) on metal surfaces and their contamination layers can be precursors of vacuum breakdown, detrimental to many devices. We run Field Emission (FE) experiments in-situ a Transmission Electron Microscope (TEM) on an amorphous-carbon (a-C) coated tungsten (W) nanotip. In various cases we observe the FE current-voltage (I-V) curves switching abruptly into an increased current state. Accompanying FE simulations show that the enhanced-current I-Vs are consistent with the growth of an NP at the tip apex. The tip growth hypothesis is confirmed by the direct observation of such an NP and its continued growth during successive FE experiments. We attribute this phenomenon to field-induced biased diffusion of atoms on the surface of the a-C coating.**

Index Terms—**diffusion under field, carbon contamination, protrusion growth, biased diffusion**

I. INTRODUCTION

Applying high electric fields on a metal surface can cause structure instabilities and lead to vacuum breakdown (VBD), severely hindering the stability and performance of electron sources and several other high-field devices. Such issues cannot be solved effectively due to insufficient scientific understanding of the mechanisms that cause them. The migration of adsorbates, such as carbon compounds, etc., under a high electric field may be behind the growth of field magnifying features and subsequent VBD. However, the underlying mechanisms are not fully understood. Processes such as field-induced biased surface diffusion [1] and plastic deformation [2] have been proposed. Yet, none of them has been proved experimentally.

It has been previously shown that the diffusion of surface adsorbates under high electric field can lead to the build-up of NPs during FE [3], which can be exploited to achieve high-performance electron sources [4]. Yet, the physical mechanisms behind this effect remain unclear, prompting in-situ studies. Here, we study the morphological evolution of an a-C coated W nanotip during FE. We measure the evolution of its FE characteristics in situ TEM and record the growth of an NP, which we attribute to field-induced biased surface diffusion.

Funded by the National Natural Science Foundation of China (51977169), the EU's H2020 program ERA Chair "MATTER" (856705), and by the Estonian Research Council's project RVTT3.

II. EXPERIMENTAL SETUP

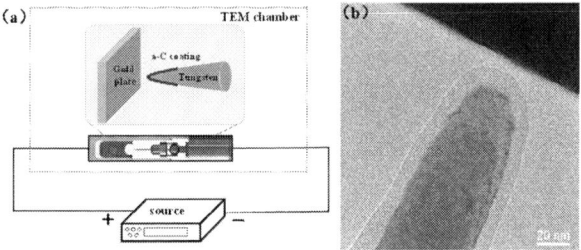

Fig. 1. (a) Schematic of the setup. (b) TEM image of the C-coated W nanotip and the Au plate anode.

The in situ FE setup and electrode morphology imaging are done in a JEOL-2010F TEM (10^{-5} Pa vacuum) with an adjustable electrode gap TEM holder (ZepTools), as shown in Fig. 1(a). Our W tip is coated by electron-beam-assisted deposition of a thin amorphous Carbon (a-C) layer. Fig. 1(b) shows the TEM image of an as-prepared tip. Electron diffraction (SAED) analysis confirms the core-W-shell-a-C structure. Our setup does not allow for an accurate estimation of the gap distance d from TEM images due to the high sensitivity to tilting of the anode plate. Although TEM could not observe the morphological changes of the tip in real time, a careful analysis of the FE behavior can give us significant insights.

III. RESULTS AND DISCUSSION

Fig. 2 gives successive sets of I-V sweeps at different gaps. For $d_{1,2}$, the I-Vs are metastable, evolving through 4 distinct states. State 1 exhibits a stable low-emission current following a Fowler-Nordheim-type curve. State 2 starts with a similar characteristic at low voltages, but when the current reaches $\sim$10 nA, it jumps to high current values. In State 3, the I-V continues in a high current behavior consistent with the appearance of an NP that slightly enhances the field near the tip apex. Finally, in State 4, the current drops back to State 1.

For d_3, the I-Vs have two states. One is similar to State 2, starting at low current and then is enhanced at higher voltages. However, here, the I-V fluctuates instead of jumping to a stable

979-8-3503-7977-8/24 $31.00 © 2024 IEEE

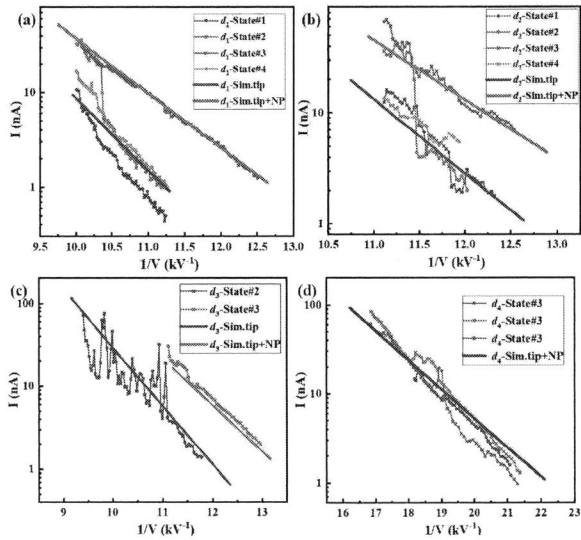

Fig. 2. Measured (dotted) and simulated (solid) I-V curves for various d.

high-current state, indicating an unstable a-C layer. Then, with progressive FE sweeps at d_4, a stable NP was observed at the tip apex, with the current stabilizing at high values (State 3).

To test the NP growth hypothesis, we simulated the FE by the multi-physics software FEMOCS [5], coupled with the general FE calculation software GETELEC [6]. We simulated a tip geometry matching the TEM images. The electrostatic field is calculated around the tip and passed to GETELEC to calculate the current density distribution on the surface. The work function was set to 4.62 eV (graphite). We multiply the current by a fitted pre-exponential factor (0.025) to account for various uncertainties of the FE theory.

For $d_{1,2,3}$, we simulated both a regular tip geometry and one with a small NP at the apex, while for d_4, only the with-NP case was run. The d values were fitted (without NP) to the experimental State 1 curves, yielding d_1=50 nm, d_2=37 nm, d_3=41.5 nm, and d_4=17 nm. The simulated I-Vs are shown with the experimental ones in Fig. 2. The curves without NP agree with the low-current curves (States 1,4), while the one with NP matches with State 3.

These results indicate that the NPs cannot be observed in TEM because they are metastable for $d_{1,2}$. Although the exact mechanism of NP growth is unclear yet, its basic characteristics can be deduced from the observed FE dynamics. The curves in State 2 quickly change when a field that induces high FE current is applied; otherwise, they transit to a relatively stable condition. This implies a field-induced morphological evolution of the surface. Two known mechanisms can be behind this process: 1) plastic deformation due to the Maxwell stress and 2) surface diffusion, biased by the high electric field. Additional experiments ruled out ambient adsorbent deposition, as clean W tips remained clean for a long time under the same conditions. The plastic deformation hypothesis

was ruled out by FEM simulations using our previously developed electrostatic-elastoplastic model [7]. This leaves biased diffusion the most plausible scenario that explains the observed behavior.

IV. CONCLUSIONS

We showed NP growth on the amorphous carbon coating layer of a tungsten nanotip during field emission using experiments inside a TEM accompanied by finite element based simulations. The observed growth is tentatively attributed to field-induced migration of atoms on the carbon surface after we excluded contaminant deposition and field-induced plastic deformation. Our results provide a plausible explanation for the appearance of field-enhancing features that lead to electron emission instabilities and vacuum breakdown.

REFERENCES

[1] G. Meng, Y. Li, R. A. Koitermaa, V. Zadin, Y. Cheng, and A. Kyritsakis, "In Situ Observation of Field-Induced Nanoprotrusion Growth on a Carbon-Coated Tungsten Nanotip," *Physical Review Letters*, vol. 132, no. 17, p. 176201, 2024.

[2] A. Pohjonen, S. Parviainen, T. Muranaka, and F. Djurabekova, "Dislocation nucleation on a near surface void leading to surface protrusion growth under an external electric field," *Journal of Applied Physics*, vol. 114, no. 3, p. 033519, 2013.

[3] V. T. Binh and N. Garcia, "On the electron and metallic ion emission from nanotips fabricated by field-surface-melting technique: experiments on W and Au tips," *Ultramicroscopy*, vol. 42, pp. 80–90, 1992.

[4] M. Duchet, S. Perisanu, S. T. Purcell, E. Constant, V. Loriot, H. Yanagisawa, M. F. Kling, F. Lepine, and A. Ayari, "Femtosecond Laser Induced Resonant Tunneling in an Individual Quantum Dot Attached to a Nanotip," *ACS Photonics*, vol. 8, no. 2, pp. 505–511, 2021.

[5] M. Veske, A. Kyritsakis, F. Djurabekova, K. N. Sjobak, A. Aabloo, and V. Zadin, "Dynamic coupling between particle-in-cell and atomistic simulations," *Phys. Rev. E*, vol. 101, no. 5, p. 053307, 2020.

[6] A. Kyritsakis and F. Djurabekova, "A general computational method for electron emission and thermal effects in field emitting nanotips," *Computational Materials Science*, vol. 128, pp. 15–21, 2017.

[7] V. Zadin, A. Pohjonen, A. Aabloo, K. Nordlund, and F. Djurabekova, "Electrostatic-elastoplastic simulations of copper surface under high electric fields," *Physical Review Special Topics - Accelerators and Beams*, vol. 17, no. 10, p. 103501, 2014.

Lateral Glow Discharge Ion Source for the Integrated MEMS Quadrupole Mass Spectrometer

Piotr Szyszka
Department of Microsystems,
Wroclaw University of Science and Technology
Wroclaw, Poland
piotr.szyszka@pwr.edu.pl

Abstract — **This article presents a glow discharge ion source designed specifically for use in the fully integrated MEMS quadrupole mass spectrometer. The problem that needed to be solved was to design ion source in a form of multi-layer structure from which the ion beam would be extracted and sent towards the analyzer of the lateral MEMS mass spectrometer. In the course of the work, a number of tests structures with varying complexity were proposed and tested. As expected, the trade-off was observed where successive simplifications of structure geometry may result in limiting operating pressure range and limiting the emitted ion currents. The structure, which is optimal in terms of technology consistency with the proposed mass spectrometer, turned out to be efficient enough to operate in the medium vacuum range (from 2×10^{-3} hPa) to high vacuum (below 1×10^{-5} hPa), reaching even tens of µA of extracted ion currents.**

Keywords—ion source, MEMS, mass spectrometry

I. INTRODUCTION

Mass spectrometry is an analytical technique with a wide range of applications and a long history, dating back to the beginning of the 20th century. However, it is only less than two decades when the first attempts to miniaturize these instruments started. The main goal of those efforts is to obtain an instrument that could be delivered to the place where the analysis is to be carried out. This would apply wherever monitoring and detection of hazardous substances is required, or where it is difficult or even impossible to bring the sample to the laboratory (e.g. space exploration). Currently, the most promising technique for reaching the expected degree of miniaturization is the MEMS technology, which, has some limitations related to the specificity of the utilized fabrication techniques. It forces the redesign and often simplification of structures that are miniaturized. This is the case with all MEMS devices and also with MEMS mass spectrometers.

Currently fabricated MEMS mass spectrometers are often made from single silicon wafer in which all electrodes are made. Structures of this kind, referred to as 2.5D, have been reported multiple times as components or mass spectrometers [1–6]. However such configuration pose the limitation what analytical techniques can be applied, reducing available methods to time-of-flight analysis or its variants. To address these issues, it is possible to create 3D structures in the form of multilayer sandwiched structures, where successive layers of electrodes are separated by dielectric layers, enabling on fabrication of more complex analytical tools such as ion traps and filters.

Following this 3D approach we have proposed a concept of MEMS quadrupole mass spectrometer (Fig. 1a), which integrated on one chip modules such as ion source, mass analyzer, detector and can be furtherly revised to include in house developed ion sorption micropump [7] and sampling nanochannels [8]. The utilized mass analyzer was made in the form of silicon rods arranged laterally, connected with glass spacers. The first concept assumed the use of an electron impact ion source based on a carbon nanotube cathode [9]. However, no reliable results in this configuration were obtained, and time degradation of field emission current was observed. On the other hand, we have also developed a glow-discharge ion source, which ionizes gases very efficiently, and is stable over time [10]. The problem with this source was that it emitted an ion beam vertically, thus it could not be integrated with a planar quadrupole analyzer. Therefore, it was necessary to redesign the already existing vertical ion source, in the way it was technologically compatible with the rest of the mass spectrometer structure (Fig. 1b).

However, a few questions arose here: is it possible to recreate a Penning a trap architecture in a planar form? How many layers are needed to do that? Is such a lateral configuration able to sustain stable discharge? and whether is it possible to extract the ion beam towards the analyzer?

II. EXPERIMENT

To answers these questions, several test structures were designed with a different number of layers forming each of

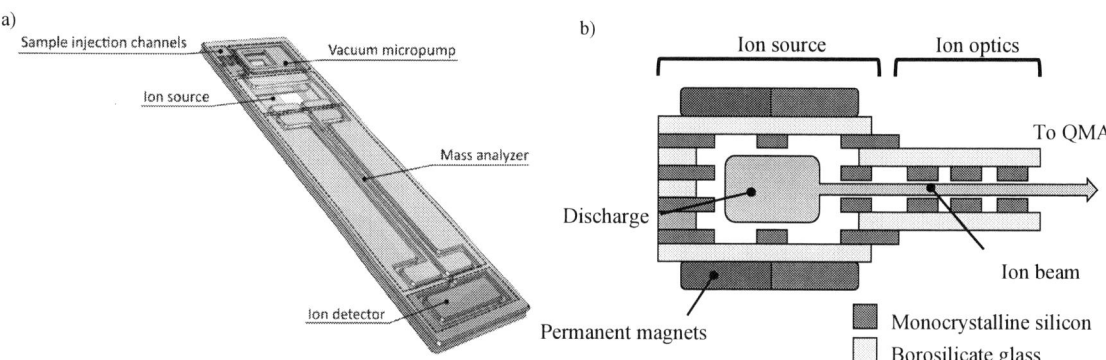

Fig. 1. Lateral integrated MEMS Quadrupole Mass Spectrometer (a) and schematic of the proposed lateral glow discharge ion source (b)

979-8-3503-7977-8/24 $31.00 © 2024 IEEE

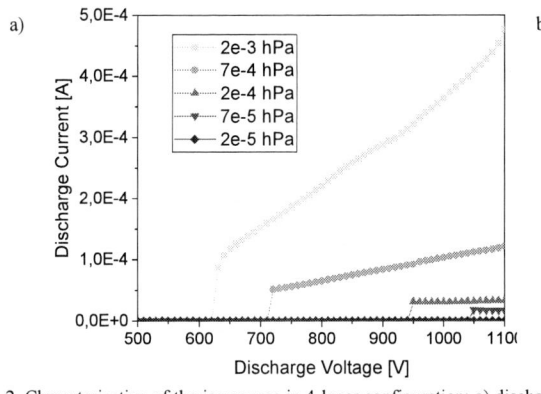 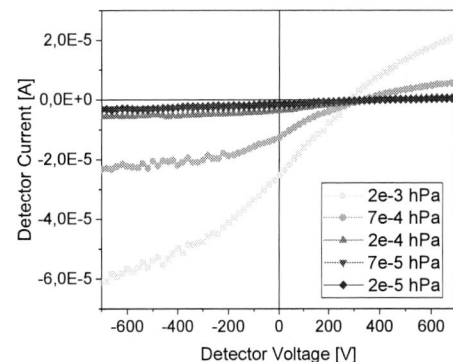

Fig. 2. Characterization of the ion source in 4-layer configuration: a) discharge current in function of voltage and b) detector current in a function of detector voltage; conducted for series of operating pressure

the electrodes. The number of layers was ranging from 7 (for the most extensive solution) down to 3 (for the most simplistic one). Computer simulations showed, that all of them have potential of successful trapping of electrons in crossed electric and magnetic fields, as well as of extracting of the ions from the discharge chamber.

All designed structures took into consideration the possibility of their coo-fabrication with the quadrupole mass analyzer. The fabricated test structures were investigated inside reference high vacuum chamber which allows to perform measurement in pressure down to 10^{-5} hPa. During all performed measurements cathode and anticathode of the source were grounded. Two types of experiments were performed:

   a) Three-electrode experiment (U/I characterization of the discharge).
   b) Four-electrode experiment (extraction of the ions).

A stable discharge was achieved in structures having at least 4 layers (Fig. 2a). The 3-layers structure failed to start up the discharge under any pressure and voltage conditions. When the discharge occurs, in all configurations it was possible to extract the ions from the source (Fig. 2b). It was observed that characteristics of the extracted currents are similar to the ones obtained for well characterized source in vertical construction. The main parameters of the constructed sources were compared with the vertical reference (Fig. 3). It was noticed that source with proposed configuration may operate in a narrower pressure range, while often offering increased discharge currents.

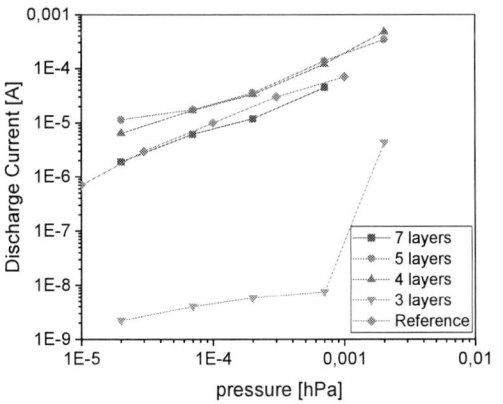

Fig. 3. Comparison of various configurations of the ion source

III. CONCLUSIONS

In this study, we designed and tested a glow discharge ion source for integration with a MEMS quadrupole mass spectrometer. The optimal multi-layer structure demonstrated stable performance in medium to high vacuum ranges (2×10^{-3} hPa to below 1×10^{-5} hPa), achieving ion currents in the tens of μA. A minimum of 4 layers was necessary to sustain stable discharge, with successful ion extraction comparable to vertical ion sources. The results show that the redesigned lateral structures can be efficiently integrated with MEMS mass spectrometers, paving the way for more compact and efficient analytical instruments.

REFERENCES

[1]   C. Freidhoff, MEMS mass spectrometer, Patent US7402799B2, 2005. https://patents.google.com/patent/US7402799B2

[2]   C.M. Tassetti et al., L. Duraffourg, Gas detection and identification using MEMS TOF mass spectrometer, 2013 Transducers and Eurosensors XXVII: (2013) 1162–1165. https://doi.org/10.1109/Transducers.2013.6626979.

[3]   J. Müller et al., A Planar Integrated Micro-mass Spectrometer, in: LC-MS in Drug Bioanalysis, Springer US, Boston, MA, 2012: pp. 423–465. https://doi.org/10.1007/978-1-4614-3828-1_14.

[4]   S. Vigne et al., Gas analysis using a MEMS linear time-of-flight mass spectrometer, Int J Mass Spectrom (2017). https://doi.org/10.1016/j.ijms.2017.03.011.

[5]   J.P. Hauschild, E. Wapelhorst, J. Müller, Mass spectra measured by a fully integrated MEMS mass spectrometer, Int J Mass Spectrom 264 (2007) 53–60. https://doi.org/10.1016/j.ijms.2007.03.014.

[6]   S. Westerdick, B. Walther, P. Hermanns, F. Fricke, T. Musch, Planar Lab-On-A-Chip Micro Mass Spectrometer with Time-Of-Flight Separation, in: 2021 IEEE 34th International Conference on Micro Electro Mechanical Systems (MEMS), IEEE, 2021: pp. 434–437. https://doi.org/10.1109/MEMS51782.2021.9375289.

[7]   T. Grzebyk, A. Górecka-Drzazga, J.A. Dziuban, Improved properties of the MEMS-type ion-sorption micropump, Journal of Vacuum Science & Technology B, Nanotechnology and Microelectronics: Materials, Processing, Measurement, and Phenomena 35 (2017) 062001. https://doi.org/10.1116/1.4994782.

[8]   T. Grzebyk, K. Turczyk, P. Szyszka, A. Górecka-Drzazga, J. Dziuban, Pressure control system for vacuum MEMS, Vacuum 178 (2020). https://doi.org/10.1016/j.vacuum.2020.109452.

[9]   P. Szyszka, T. Grzebyk, A. Gorecka-Drzazga, J.A. Dziuban, Highly effective MEMS gas ionizer-A significant step of development of integrated ion-mass spectrometer, URSI 2018 - Baltic URSI Symposium (2018) 248–249. https://doi.org/10.23919/URSI.2018.8406747.

[10] T. Grzebyk, T. Szmajda, P. Szyszka, A. Górecka-Drzazga, J. Dziuban, Glow-discharge ion source for MEMS mass spectrometer, Vacuum (2020). https://doi.org/10.1016/j.vacuum.2019.109008.

MEMS-based Vacuum Analytical Instruments for Space Exploration

Tomasz Grzebyk, Piotr Szyszka, Michał Krysztof, Paweł Urbański, Marcin Białas, Paweł Knapkiewicz, Jan Dziuban,
Faculty of Electronics, Photonics and Microsystems, Wroclaw University of Science and Technology,
11/17 Janiszewski St., 50-372 Wrocław, Poland
Tel.: +48 71 320 49 77, tomasz.grzebyk@pwr.edu.pl

Abstract—This article summarizes about 10 years of research conducted at Wroclaw University of Science and Technology (in collaboration with partners form other institutions) on elaboration of miniaturized instruments applicable for space exploration. Most attention is paid to gas sample analyzers (mass and plasma spectrometers), but solutions like MEMS electron microscope, and MEMS X-ray source will be mentioned as well.

Keywords-MEMS, field emission elctron source, ion source, mass spectrometry, electron column, space exploration

I. INTRODUCTION

Nowadays, space exploration is driven by few major factors: search for life outside Earth, search for human friendly habitats and search for resources. All mentioned areas require reliable and sensitive analytical methods and instruments. Most interesting is the knowledge about the composition of analyzed samples, but structure and morphology also bring some useful information about processes taking place in space.

There are couple of strategies how to analyze space samples. The first assumes lunching a racket, collecting a sample and bringing it back to a laboratory on Earth. It is the most precise, but at the same time most complex and expensive approach. The second option is conducting measurements from a certain distance, e.g. from an orbit or even from Earth. This method gives averaged and less accurate results. The best option seems to be conducting analysis on site, often using small space vehicles, drones or nanosatellites, equipped with proper analytical instruments. These instruments must by in that case also small, light and preferably cheap to fit the vehicles and if possible – multiplied to be distributed over large area.

II. MATERIALS AND METHODS

Up to now most instruments that were lunched to space were just a smaller versions of classical "Earth" devices, manufactured with precise mechanical methods, weighting several kilograms and occupying many cubic decimeters. To decrease the costs of the missions and to accelerate exploration of different space objects, these instruments had to be further miniaturized.

MEMS (microelectromechanical systems) technology seems to be the best candidate to fulfil all mentioned requirements. In our research we tried to transfer the construction of most popular and most needed analytical equipment to small, silicon-glass devices manufactured by microengineering methods, like photolithography, anisotropic etching, layer deposition and anodic bonding.

Majority of the developed instruments utilize some kind of interaction of the sample with electron or ion beam, thus these devices comprise either miniature field emission sources or small ion sources. In designing process we had to take into account specificity of the technology, limited space for interactions and manipulation of the beam as well as the fact that some physical processes in micro scale are different than in micro scale solutions (e.g. caused by long free electron path compared to inter-electrode distances).

Throughout several years of research, we were able to solve most of existing problems and developed a family of miniaturized instruments, that occupy the space usually smaller than a size of a match box, and weighting from few up to 100 g.

III. DEVICE CONSTRUCTION

First two devices are gas analyzers: miniature mass spectrometer and plasma optical spectrometer (Fig. 1a, b). Both give information about the composition of gas mixtures and the concentration of each element. Both work in reduced pressure, $10^{-7} - 10^{-2}$ hPa in first case, and $10^{-5} - 10$ hPa in the second case. In both gas sample has to be ionized.

MEMS mass spectrometer consists of a glow-discharge ion source, quadrupole filter, faraday cup detector, and ionization vacuum sensor, all placed inside a 3D printed package [1]. It gives the most accurate results, can analyze masses up to 440 AMU, with a unite resolution and sensitivity better than 0,01%. However, it requires quite complex electronic systems and steering.

The second spectrometer, based on a silicon-glass Penning-like ionization cell measures (thanks to the integration with a miniature optical spectrometer) spectrum of the generated plasma [2]. Each gas component gives characteristic lines, present always at the same wavelength. Its intensity can be related to the amount of the component. This method is relatively simple, requires not that complex electronics, and gives very fast results, in one moment whole spectrum and all components are visible. However, it is less sensitive, depending on a gas type it is about 0.1 to 1%. To improve it, some kind of sample preparation system must be adapted (e.g. containing pre-concentrator, molecular sieve, or getter).

Two other devices are similar in construction, but are used for completely different purposes – these are miniature X-ray source and MEMS electron microscope (Fig. 1c, d). They contain carbon nanotube filed emission electron sources, electro-optics column, proper target and micropump responsible for ensuring high vacuum conditions (not always required in space conditions).

The target in the X-ray source is made as 15-μm thick silicon membrane (covered with additional metal layer). It converts the kinetic energy of electrons into radiation [3]. The device works in a constant or impulsive mode, with energies up to 25 keV.

The electron microscope is capped with much thinner silicon-nitride membrane - 100 nm. This membrane separates the vacuum part of the device from a sample which is placed on its surface. If the membrane is thin enough the electrons can pass through it, on the contrary – the gas particles cannot. The image of the sample is obtained by scanning the electron beam with an octopole electrode and detecting secondary or reflected electrons [4].

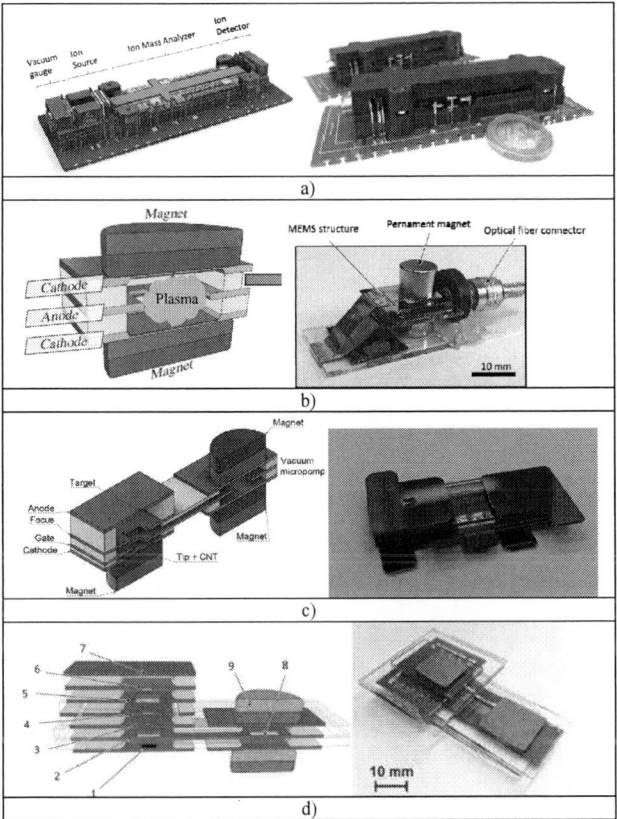

Fig. 1. Family of MEMS and vacuum nanoelelectronic devices for space exploration: a) MEMS mass spectrometer, b) MEMS plasma spectrometer, c) MEMS X-ray source, d) MEMS electron microscope.

In space exploration X-rays can be applied for analyzing solid samples as rocks or regolith present on a surface of different planets. Irradiated samples generate characteristic spectral lines which can be detected by also quite miniature detectors. The second interesting area of utilization is the calibration of cosmic radiation detectors [5]. Microscopes are useful for analyzing the size and morphology of space dust, regolith or maybe even same small biological objects, if ever found outside Earth.

IV. CONCLUSSIONS

Miniaturization is one of the most important research trends in space exploration. Not only satellites and space vehicles are becoming smaller but they also require smaller analytical instruments. In our research we are trying to show that utilization of MEMS and vacuum nanoelectronics devices can be one of the best routs applicable here. We were able to construct very small versions of instruments which can give information on the composition and morphology of space sample on site, without a necessity of bringing them back to Earth. We hope that the application of at least part of these instruments in future space missions will help to answer some of the most bothering scientific questions, e.g. is there any life outside Earth.

ACKNOWLEDGMENT

The work was financed by the projects no. 2016/21/B/ST7/02216 and 2021/41/B/ST7/01615 of Polish National Science Centre and project no. 4000138011/22/NL/KM of the European Space Agency.

REFERENCES

[1] SZYSZKA P., JENDRYKA J., SOBKÓW J., ZYCHLA M., BIAŁAS M., KNAPKIEWICZ P., DZIUBAN J., GRZEBYK T., MEMS quadrupole mass spectrometer. 36th IVNC, 2023: 195-197.

[2] KAPKIEWICZ P., GRZEBYK T., DZIUBAN J., The sensor for H2 content measurements in hydrogenated gaseous fuel, 22nd PowerMEMS Conference, 2023: 27-29.

[3] URBAŃSKI P., GRZEBYK T., Entirely MEMS X-ray source, IEEE Electron device Letters, 2024, in press

[4] BIAŁAS M., GRZEBYK T., Krysztof M., GÓRECKA-DRZAZGA A., Signal detection and imaging methods for MEMS electron microscope. Ultramicroscopy. 2023, 244: 1-10.

[5] GUO Si-Ming, WU Jin-Jie, HOU Dong-Jie, The development, performances and applications of the monochromatic X-rays facilities in (0.218–301) keV at NIM, China, NUCL SCI TECH (2021) 32:65

Miniaturized Rubidium Source for Generating Vapor Phase Atoms for Magneto Optical Traps

Jannik Koch*, Leonard Frank Diekmann, Alexander Kassner, Folke Dencker and Marc Christopher Wurz

Institute of Micro Production Technology, Leibniz University Hanover, Garbsen, Germany

*Corresponding author: koch@impt.uni-hannover.de

Abstract—**Ultracold atoms offer the highest sensitivity for quantum sensors. The industrial use of these systems requires miniaturization of the experimental setups. For this purpose, a concept for a miniaturized atom source using rubidium as atom species for the generation of vapor phase atoms was developed in this work. As an alkali metal, rubidium is highly reactive and reacts directly with small amounts of water or oxygen. In this source, pure rubidium is encapsulated by bonding two micromachined silicon components, a reservoir chip and an active release chip, together in a glovebox under an argon atmosphere. This prevents the rubidium from reacting when the source comes into contact with air. The active release chip had a thin silicon membrane and an additional gold structure that enables the membrane to be heated by Joule heating. After pumping down and baking the vacuum test chamber to an ultra-high vacuum, the release mechanism is triggered by a sharp increase in temperature within milliseconds. After opening the source, a rubidium signal was detected by analyzing the residual gas atmosphere of the vacuum with a quadrupole mass spectrometer.**

Keywords—quantum technology, atom source, rubidium

I. INTRODUCTION

Ultracold neutral atoms can be generated from vapor-phase atoms, which are cooled by laser cooling and trapped by Zeeman splitting in a magneto-optical trap (MOT). Ultracold atoms provide quantum sensors and metrology, such as atom interferometers [1] or atom clocks [2], with sensitivity beyond the classical limit for new measurement standards for seismic monitoring, precise navigation, and fast communication technics [3-5]. As a result, MOTs have stimulated extensive research in atom and quantum physics. The next step is to miniaturize the core components of the MOT for industrial applications. Here, a miniaturized atom source using rubidium as the atom species is shown. Pure rubidium is often used as an atom species in ultracold atom experiments. The main advantage is that the transition line of rubidium is easily accessible for laser cooling. The D2 transition ($5^2S_{1/2} \rightarrow 5^2P_{3/2}$) at 780 nm is the most important [6]. Furthermore, rubidium melts at 39°C and has a high vapor pressure of $5.23*10^{-7}$ mbar at room temperature [6]. A vapor pressure sufficient to load a rubidium MOT is approximately 10^{-8} mbar [7]. Therefore, no additional heating is required to produce vapor-phase atoms. However, pure rubidium is very reactive. On contact with air, it reacts immediately with water to form rubidium hydroxide and hydrogen. Therefore, the oxidation of rubidium by contact with water or oxygen molecules must be prevented throughout the preparation, encapsulation, and storage of rubidium,. The challenge in designing the atom source is to secure the rubidium in an inert atmosphere and to develop an opening mechanism that opens the source in the magnetic optical trap chamber after pumping and baking it to an ultrahigh vacuum. In this study, an approach using silicon-based microtechnology was chosen to realize an atom source.

II. METHODS

The rubidium atom source consists of two components: the reservoir chip and the active release chip, which encapsulate pure rubidium in an argon atmosphere. Thus, the source can be stored under a normal atmosphere and opened inside the vacuum system after pumping and baking the vacuum chamber. For a controlled opening of the active release chip, an external load must generate additional stresses in a membrane, which has a diameter of 8 mm and a membrane thickness about 20 µm. These stresses destroy the membrane and open the source. The stresses required for the opening was achieved using two different types of loading. First, a passive stress due to the pressure difference between the atmosphere inside the source and the vacuum of the MOT chamber, and second, an active stress due to the heating of the membrane. The heating is initialized with Joule heating by applying a voltage to a gold heater structure, which results in additional thermally induced stresses as well as a reduction in the membrane strength. As the membrane temperature increased, the conductance of silicon increased exponentially. If the conductance of the membrane exceeds that of the heater structure, it leads to a sharp increase in the current and membrane temperature, which causes the opening of the source and the release of the rubidium. Both components were manufactured using silicon microtechnology. The processed components are illustrated in Fig. 1. The next process steps take place in a glovebox from MBraun, which enables work under an argon atmosphere and prevents the reaction of rubidium. The rubidium was placed inside the reservoir chip using a heated syringe. Subsequently, rubidium is encapsulated by bonding the components with a transient liquid phase process at 200°C using gold as the base metal and indium as the intermediate layer.

III. RESULTS AND DISCUSSION

The opening of the active release chip is the most important part of this atom source as it enables the release of the rubidium. The experiments to test the opening were carried out with a test setup, in which a controlled pressure load of 600 mbar was applied to the membranes, which had a diameter of 8 mm and a membrane thickness of 19.7 +/- 3.6 µm. The applied voltage was then slowly increased until the membrane ruptured.

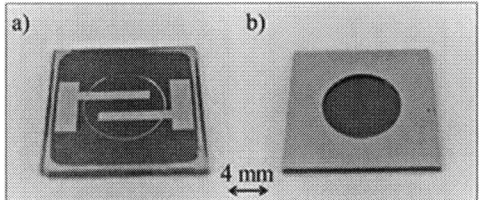

Fig.1 Image of the fabricated atom source components, a) active release chip and b) reservoir chip.

The temperature generated by the heater structure was measured using an infrared camera, which enabled real-time monitoring with a frame rate of up to 125 Hz and measured temperatures of up to 275°C. An example of the opening mechanism is shown in Fig. 2. It should be noted that the temperature measurement does not work directly on the membrane because of the transparency of the thin silicon and directly on the metal due to the difference in emissivity. The current and temperature increase with increased voltage. At temperature of around 200-220°C a sharp increase in current and a high temperature gradient occurred within milliseconds. Here, the current flows directly through the silicon membrane, as the electrical conductance of the silicon membrane increases with temperature. This effect is illustrated in Fig. 2c and Fig. 2d. After opening, the membrane chip was rapidly cooled to below 100°C within ten seconds. The opening test was successful for all membrane chips tested. A total of 18 different chips were tested, requiring a heating power of 26.3 +/- 4.2 W to initiate the opening of the membrane. Thus, this reproducible opening mechanism can be used as release mechanism for the rubidium source. Outgassing of the rubidium source was tested in a baked ultra-high vacuum chamber with a mass spectrometer and a cold cathode gauge to monitor the total pressure and partial pressures of hydrogen, water, argon, and rubidium. At the beginning of the experiment, a total pressure of $3*10^{-8}$ mbar was achieved. The results of the mass spectrometer analysis and the opened source after the experiment are shown in Fig. 3. At t=0 s, the membrane of the atom source was opened with an applied heating power of 20 W. By opening the source, rubidium could be detected by the mass spectrometer with a partial pressure of $5*10^{-8}$ mbar.

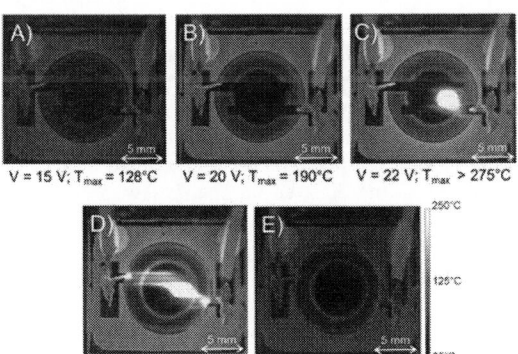

Fig.2 Process of the opening mechanism, IR-images at different times and applied voltages.

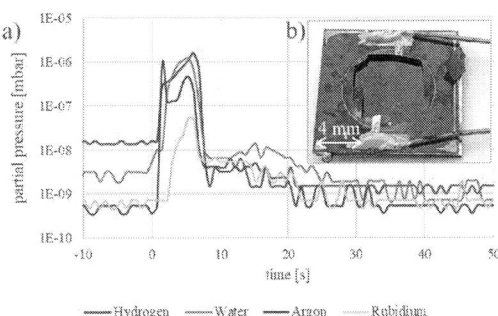

Fig.3 a) Mass spectrometer analysis of the rubidium source during opening and b) opened rubidium source after the experiment.

Unfortunately, the detection signal drops below the noise level after 15 s and, in addition to argon, the hydrogen and water signals increase, indicating insufficient hermetic bonding of the components. Compared to the mass spectrometer, the measured total pressure of the cold cathode gauge was much higher. When the source was opened, the total pressure increased to $1*10^{-4}$ mbar. This was caused by the heating of rubidium during the opening of the membrane. At temperatures of about 100°C, rubidium has a vapor pressure of more than $1*10^{-4}$ mbar [6]. If the source is cooled after the opening of the source, a pressure of about $1*10^{-7}$ mbar turns on after 50 s. This value corresponds to the vapor pressure of rubidium at room temperature. The reason why no more rubidium was detected by the mass spectrometer is due to condensation and adsorption on the walls of the vacuum chamber. A coloured layer was deposited on the vacuum walls and the sample holder due to the rubidium release from the atom source. In addition, rubidium-containing deposits outside the reservoir chamber of the source were detected using EDX.

IV. CONCLUSION

We report the development of a miniaturized rubidium source based on silicon microtechnology. Here, a membrane could be opened by applying heating power after pumping and baking the vacuum chamber. The opening could be initialized at a power of 26.3 +/- 4.2 W. When testing the sources, a rubidium signal was detected with a mass spectrometer during opening, and a constant pressure of about $1*10^{-7}$ mbar was measured after opening, which corresponds to the vapor pressure of rubidium at room temperature.

REFERENCES

[1] Lee J, Ding R, Christensen J, Rosenthal R et al. 2022 Nature communications 13 (1), S. 5131.

[2] Hinkley N, Sherman J, Phillips N et al. 2013 Science (New York, N.Y.) 341 (6151), S. 1215–1218.

[3] Matsuzaki Y, Benjamin S, Nakayama S et al. 2018 Physical review letters 120 (14), S. 140501.

[4] Wang J 2015 Chinese Phys. B 24 (5), S. 53702.

[5] Kitching J 2018 Applied Physics Reviews 5 (3), S. 31302.

[6] Steck, D 2003 Oregon Center for Optics and Department of Physics.

[7] Rushton, J. A.; Aldous, M.; Himsworth, M. D. 2014 The Review of scientific instruments 85 (12), S. 121501

Modeling of Horizontal Integrated Silicon Field Emitter for On-Chip Free Electron Interactions

Goulven Rouillé[1]*, Catherine Weng[1], Anthony Ayari[1], Sorin-Mihai Perisanu[1], Sylvain Combrié[2], Laurent Gangloff[2], Xavier Checoury[3]

[1]institut Lumière Matière, Université Claude Bernard Lyon 1, Villeurbanne, France, [2]Thales research & technology, Palaiseau, France,
[3]Centre de Nanosciences et de Nanotechnologies, Palaiseau, France
Corresponding author: goulven.rouille@univ-lyon1.fr

Abstract— **We present the modeling of an on-chip planar electron gun using COMSOL Multiphysics. The beam coming from this electron gun goes through a trench crossing a photonic crystal. The interaction between the electrons and the photonic crystal generates photon emission by Cherenkov effect [1][2]. An objective is to realize an electron gun working from 30V to 1000V voltage. To obtain enough photons conversion and reach the laser threshold, we aim to reach 100 nA of current in 1 mrad. The electron gun will be fabricated using several steps of e-beam lithography. Modeling of the electron gun and electron trajectories will allow to reach optimal performances while preventing from device failures.**

Keywords—Field emission, COMSOL Multiphysics modeling, Electron beam, Electron gun, Silicon tip emitter

I. INTRODUCTION

The use of relativistic electrons as a gain medium is a long-standing topic in research with the first free-electron laser (FEL) demonstrated in 1971 [3]. FELs offer unique opportunities such addressing spectral windows that are difficult to reach with other kind of lasers. Applications of these lasers are ranging from fundamental physics to medical and industrial applications. However, FELs are still very large instruments, thus the diffusion of such system is limited due to their cost, size, and power consumption. Miniaturization and integration of FELs could lead to a revolution with major implications for free space telecommunications, spectroscopy or for device operating in extreme environmental conditions (e.g., high temperatures). First demonstrations of on-chip interaction between electrons and photons have emerged. Up to now these photonic structures are mainly excited by electrons generated from transmission or scanning electron microscope (TEM or SEM respectively) columns. Integrating an electron source and a photonic structure on a same chip is a real challenge.

II. FIELD EMISSION MODELING

A. On-chip cold emission

On-chip cold emission has already been demonstrated with carbon nanotube grown within an aligned gate [4] but those vertical devices don't allow to align the beam with the trench of the photonic crystal (Fig1). In-plane free electron guns have also been made [5][6] but are not compatible with our project because of the difficulties to align those devices with the crystal photonic. Here, the structure of the electron gun is divided in three parts. The electron gun is composed by a planar n-doped silicon tip and, in order to compensate for asymmetries inherent to field emission, 2 quadrupoles (8 independently polarized electrodes). The first quadrupole is used as an extraction gate while the second quadrupole will focalise the electron beam to increase the percentage of electrons reaching the end of the photonic crystal.

B. Electron gun modeling

The main goal of this work is to model a horizontal electron gun for on-chip light generation. By coupling an electron gun and a photonic crystal, we aim to produce a light emitting device in the near infrared through Cherenkov effect. This electron gun could also be used to excite quantum dots and generate UV light.

In the case of Cherenkov emission, the electron beam will travel along a slot made in a two-dimensional photonic crystal and transfer its energy to the optical modes of the photonic crystal. For efficient photons conversion, and to reach a laser threshold, we target a current of 100 nA at the end of the crystal.

In this poster, we will focus on simulation results of the electron source design using COMSOL Multiphysics but also on the emitted current governed by the Fowler Nordheim law. Impact of focusing electrodes will also be presented according to simulations of the electron trajectories.

Many parameters might influence the electron beam deviation, like tip, electrodes positions and potentials or even the shape of the tip [7]. In COMSOL Multiphysics, several modules can be used and coupled all together. The modules we used for our study are "Electrostatics" and "Charged particles tracing", which are coupled in the "Particle Field Interaction, Non-relativistic". The "Electrostatic" module will use the potentials we define to calculate the electric field and the "Charged particles tracing" will introduce the electron in the modeling and determine their trajectories from the electric environment.

The first part of the work was to define the geometry of the gun tip. Then we evaluated the emitted current density according to the Fowler-Nordheim law and finally we optimised the electron trajectories by adjusting quadrupoles parameters (position, design, polarisation). To do so, COMSOL calculates the electrostatic field from the potential we apply between the tip and the electrodes. Then, we defined the Fowler-Nordheim equation in the global definition to calculate the current density over the tip surface. This current density is used to determine the initial position of the electrons which is closely related to the beam divergence. We

979-8-3503-7977-8/24 $31.00 © 2024 IEEE

finally used the focalisation electrodes to limit the divergence of the beam and maximize the active electrons (Fig 2).

Several tip geometries were under consideration because of the field enhancement factor but also because of physical limitations (tip melting, lithography limits). We initially planned to use a cylinder-shaped tip but to prevent thermal issues, we finally used a conical suspended tip which width will linearly decreases up to 40 nm. Another limitation comes from the fabrication process. As a matter of fact, electrodes of a same quadrupole can't be superposed thus resulting in a dissymmetry in their effect on the electron beam.

Finally, after several optimization steps, we obtained electric fields up to 4.5 V/nm (for 400V applied between gate and tip) and about 1/7000 of the emitted electrons interacting along the 100 μm length of the photonic crystal (efficient electrons).

ACKNOWLEDGMENT

This project is funded by French National Research Agency (OFELIA project, ANR-21-CE24-0007).

REFERENCES

[1] Bézard M., Si Hadj Mohand I., Ruggierio L., Le Roux A., Auad Y., Baroux P., ... & Kociak M. (2024). High-Efficiency Coupling of Free Electrons to Sub-λ3 Modal Volume, High-Q Photonic Cavities. *ACS nano, 18*(15), 10417-10426.

[2] Han Z., Checoury X., Néel D., David S., El Kurdi M., & Boucaud P. (2010). Optimized design for 2× 106 ultra-high Q silicon photonic crystal cavities. *Optics Communications, 283*(21), 4387-4391.

[3] Pellegrini C. (2017). X-ray free-electron lasers: from dreams to reality. *Physica Scripta, 2016*(T169), 014004.

[4] Gangloff L., Minoux E., Teo K. B. K., Vincent P., Semet V. T., Binh V. T., ... & Legagneux P. (2004). Self-aligned, gated arrays of individual nanotube and nanowire emitters. *Nano letters, 4*(9), 1575-1579.

[5] Liu F., Xiao L., Ye Y., Wang M., Cui K., Feng X., ... & Huang Y. (2017). Integrated Cherenkov radiation emitter eliminating the electron velocity threshold. *Nature Photonics, 11*(5), 289-292.

[6] Storeck G., Vogelgesang S., Sivis M., Schäfer S., & Ropers C. (2017). Nanotip-based photoelectron microgun for ultrafast LEED. *Structural Dynamics, 4*(4).

[7] Purcell S. T., Binh V. T., & Baptist R. (1997). Nanoprotrusion model for field emission from integrated microtips. *Journal of Vacuum Science & Technology B: Microelectronics and Nanometer Structures Processing, Measurement, and Phenomena, 15*(5), 1666-1677.

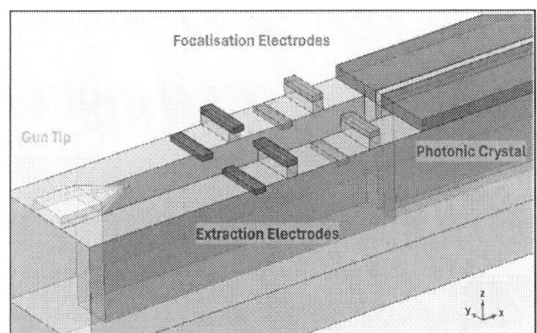

Fig. 1. Electron gun side view showing gun tip (yellow), both extraction (red) and focalisation (green) quadrupoles and the photonic crystal (blue).

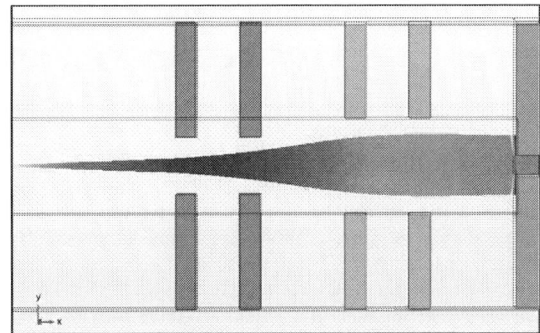

Fig. 2. Top view of the electrodes and their collimation effect on a emitted electron beam. Extraction (red) and focalisation (green) quadrupoles and the photonic crystal (blue).

Modelling of X-Ray Diffraction on Multilayer Objects

Artur Ovcharenko, Serhii Lebedynskyi and Oleksandr Lebed*

Institute of Applied Physics, National Academy of Sciences of Ukraine Sumy, Ukraine
*Corresponding author: lebedO@ukr.net

Abstract—**The work focuses on studying the internal multi-layered structure of optically inhomogeneous objects using the X-ray phase contrast imaging method, which are based on X-ray diffraction phenomenon. Phase contrast method is preferred over conventional X-ray imaging for certain applications, particularly when dealing with materials or biological tissues that do not absorb X-rays strongly. The research employs computer modeling of X-ray diffraction based on the Fresnel-Kirchhoff scalar diffraction theory. It was shown that the modeling result contains quantitative information about the internal structure of the object and its multilayer nature, such as the thickness and refractive indices of individual layers.**

Keywords—multilayer object, X-ray phase contrast imaging, X-ray diffraction, Fresnel-Kirchhoff diffraction theory

I. Introduction

X-ray diffraction is an indispensable tool in materials science, chemistry, physics, and biology for analyzing the atomic structure and properties of materials. By understanding and applying the principles of X-ray diffraction, researchers can gain insights into material compositions, structural phases, and other critical characteristics that inform various scientific and industrial applications.

The study of multi-layer objects is currently a relevant issue for various fields, including medicine (cornea, skin, cells), nuclear energy (TRISO particle nuclear fuel), and accelerator and vacuum technology (thin films, nano coatings, surface treatments). These applications require precise analysis of layered structures to address specific challenges, such as enhancing material properties or improving device performance.

Computer modeling provides an additional research tool that does not require expensive and complex experiments, making it a cost-effective and efficient alternative. It plays an important role in many fields of science and technology by allowing researchers to simulate and study complex systems. Diffraction patterns from micro- and nano-objects with a multilayer structure can be effectively analyzed through simulation. This approach helps in understanding the properties, morphology, and structure of various materials, including metals, polymers, and composites. Such insights are critical for the development of new materials and devices, leading to advancements in technology and industry.

II. Model Development

A. Modeling of a multi-layered object

An algorithm was developed for creating a model of a multilayer object with different values of the refraction decrement, which allows to specify the shape, size, number and thickness of layers, and the refractive index of each layer of the object. A method of creating a 6-layer object (Fig. 1) was proposed for phase contrast image calculation.

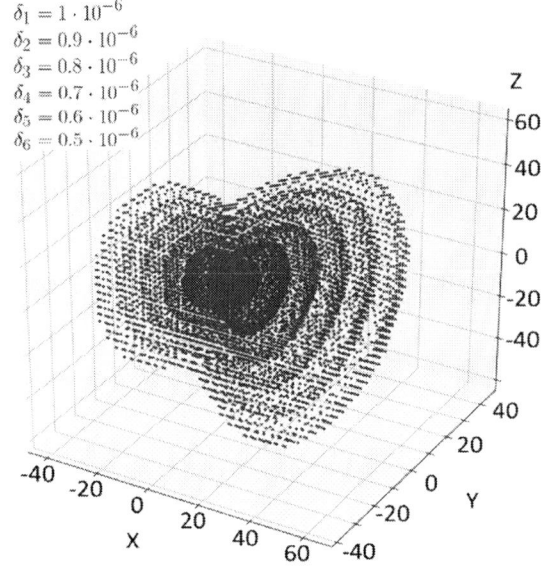

Fig. 1. Multilayer object.

B. Analytical model

Fresnel-Kirchhoff diffraction theory [1,2] was used to simulate formation X-ray phase contrast image of a multilayer object. This approach allows to study the evolution of the wave front on the path source-object and object-screen (Fig. 2) and to express the complex scalar amplitude of the X-ray field on the screen plane as follows:

$$\psi(x_{scr}, y_{scr}) = \frac{1}{2i\lambda} \iint_{-\infty}^{\infty} \frac{\exp\big(ik(r+s)\big)}{rs} \times \cdot$$

$$\times (\cos(\vec{n}, \vec{r}_i) + \cos(\vec{n}, \vec{s}_i)) \cdot \exp(i\varphi) \cdot dx_i dy_i, \quad (1)$$

Here $\vec{s}$ and $\vec{r}$ are "source − object" and "object − screen" vectors, respectively; λ is wavelength of radiation; $\vec{n}$ is unit normal to the plane of the object; φ is phase shift caused by the impact of the object on the incoming wavefront.

In the case of a monochromatic wave propagating along the z axis, the phase change without taking absorption into account can be written in the following form [1]:

$$\varphi(x, y) = -\frac{2\pi}{\lambda} \int \delta(x, y, z) dz, \quad (2)$$

where the integral is calculated over the thickness of the object in the direction of propagation of the X-ray beam; $\delta(x, y, z)$ is the distribution of decrement of X-ray refraction.

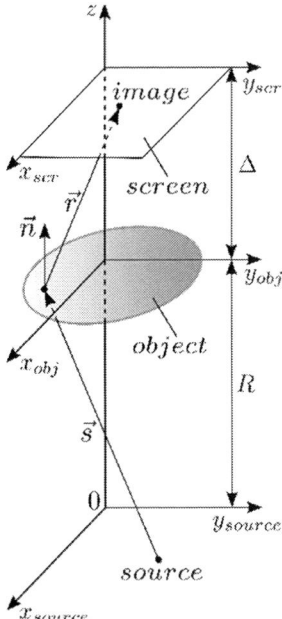

Fig. 2. The scheme of calculating the phase contrast image in the Fresnel-Kirchhoff diffraction theory.

III. RESULTS AND THEIR DISCUSSION

In this study, we developed a method to model the diffraction of X-ray radiation on objects with layered structures, which exhibit anisotropic properties, including optical characteristics. An X-ray phase contrast image was obtained using the free propagation method for a multilayer object with a proportional decrease in the refractive decrement from the center to the edges. This method involves observing the phase shifts of X-rays as they pass through different layers of the object, providing enhanced contrast for weakly absorbing materials.

To eliminate the influence of aperture edges on the diffraction pattern quality at the screen, we subtracted the intensity distribution obtained with the object present from that obtained without the object. This subtraction produced "clean" images of the samples on the screen (see Fig. 3). This approach is feasible for practical applications in experiments, where a detector, instead of a screen, records the intensity distribution and transmits this data to a recording device.

By storing the intensity distribution data from an aperture both with and without an object, we can subtract these signals to calculate the relative intensity signal, effectively removing "noise." The results of these calculations are shown in Fig. 3a, with the relative intensity profile depicted in Fig. 3b. This profile provides quantitative information about the object's internal structure and its multi-layered nature.

Each component of a multilayered object induces a phase shift of varying degrees at the layer boundaries, due to the inhomogeneity in the X-ray refraction index. This phase shift alters the interference pattern. The structural and optical heterogeneity of these objects can enhance X-ray scattering, resulting in the blurring of diffraction maxima and background noise in the diffraction pattern, and may also give rise to additional diffraction maxima.

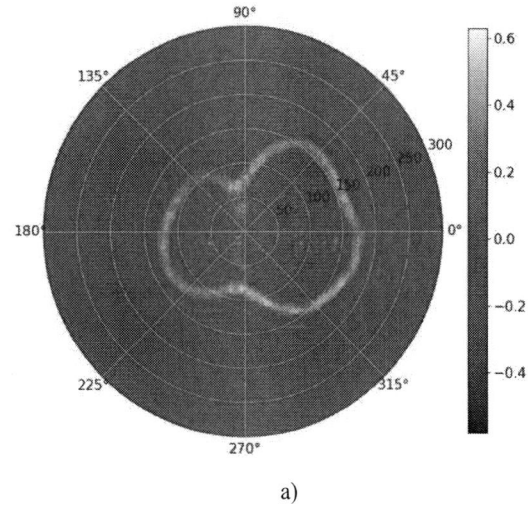

a)

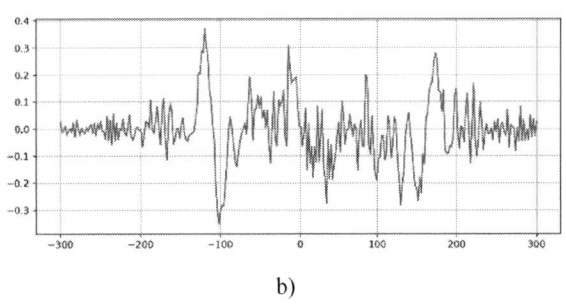

b)

Fig. 3. Aperture diffraction-free X-ray phase contrast image of a six-layer object (a) and the corresponding intensity profile along the x-axis (b).

IV. CONCLUSIONS

It was demonstrated that the modeling results provide quantitative information about the internal structure of the object and its multilayered composition, including critical parameters such as the thickness and refractive indices of individual layers. This detailed data enables a comprehensive understanding of the material's structure at a microscopic level. This capability underscores the method's potential for in-depth analysis of complex materials across a wide range of scientific and medical applications. For instance, in materials science, it could be used to investigate the properties of composite materials, coatings, and biological tissues. Consequently, the method holds promise for advancing research and development in fields like biomedical engineering, materials science, and nanotechnology, where understanding the detailed internal structure of materials is crucial.

ACKNOWLEDGMENT

The research was supported by the Grant of the NAS of Ukraine 2024-2025 for young scientists № 0124U002466.

REFERENCES

[1] D. Paganin, Coherent *X-Ray Optics.* Oxford Univ. Press, 2013.

[2] E. Wolf, M. Born, *Principles of Optics.* Cambridge Univ. Press, 2020.

Narrow Energy Spread Electron Emission from Si-Tip with Ultra-Thin Diamond Like Carbon Coating

Wen Zeng, Yang Chen, WenQi Feng, Yifeng Huang, Runze Zhan, Jun Chen, Shaozhi Deng, Ningsheng Xu, Juncong She*

State Key Laboratory of Optoelectronic Materials and Technologies, Guangdong Province Key Laboratory of Display Material and Technology, School of Electronics and Information Technology,
Sun Yat-sen University, Guangzhou 510275, China

*Corresponding author: shejc@mail.sysu.edu.cn

Abstract—**An ultra-thin Diamond-Like Carbon (DLC) film was deposited onto a clean Si-tip surface for achieving narrow energy spread electron emission. A narrow energy spread electron emission of 0.63 eV @ 44 nA was obtained from a Si-tip with DLC coating, much narrower than that of the Si-tip without deposited DLC (1.5 eV@ 34nA). The removal of natural oxide on Si surface avoids the electron scattering in oxide layer. The low electron affinity of DLC film reduces the electric field needed for emission and decreases the slope of vacuum energy level, which significantly suppresses the emission of electron from the lower energy level and induces a narrow energy spread of electron emission.**

Keywords—*Narrow Energy Spread, field electron emission, Si-Tip, Diamond Like Carbon*

I. INTRODUCTION

Gated Si-tip is one of the promising field electron emitters for modern micro-nano vacuum electronic devices.[1] Narrow energy spread electron emission is highly desired for practical applications such as electron beam lithography/microscopy. However, the energy spread of gated Si-tip is commonly wider than 2 eV @ 10 nA due to the electron scattering in surface oxide of the tip [2, 3]. Although thermal cleaning and ion bombardment are useful for removing the surface oxide, the tip is easy damaged at the same time. In addition, even though the surface is "clean", the oxide usually reform due to the residue oxygen in the working environment. It is still an open issue to obtain narrow energy spread in gated Si-tip field emitter. Numerous studies have showed that surface coated with low-electron-affinity film such as Diamond-Like Carbon (DLC) can effectively improve the field emission performance of Si-tip, i.e., reduced turn-on voltage, improved emission current intensity and stability. However, covering the clean Si-tip surface with an ultra-thin DLC film is also a promising method to achieve narrow energy spread electron emission, which has less been concerned

yet. In this abstract, a clean Si-tip surface deposited with an ultra-thin DLC film was demonstrated to achieve narrow energy spread electron emission.

II. RESULTS AND DISCUSSION

Fig.1 (a) -1 (c) showed the representative scanning electron microscope (SEM), transmission electron microscope (TEM) and high-resolution transmission electron microscope (HRTEM) images (tip apex region) of a typical n-doped Si-tip. The Si-tip was in height of ~1.2 μm (Fig.1 (a)), with apex curvature radius ~5 nm (Fig.1 (b)) and amorphous natural oxide layer with thickness ~2.5 nm on the surface (Fig.1 (c)). The surface oxide was removed through HF etching and immediately followed by H-plasma passivation treatment. This process formed Si-H bonds at the tip surface which can prevent re-oxide. As is showed in Fig. 1 (d), the surface oxide has been completely removed after cleaning process. A layer of DLC with thickness ~5 nm was then deposited onto the clean Si-tip surface (Fig. 1 (e)) by vacuum-arc deposition. The surface DLC can protect the surface Si-H bonds from thermal decomposition during electron emission [4] and thus prevent re-oxide of the tip surface. Both the Si-tips with and without DLC deposited were fabricated and characterized. Experiment demonstrated that narrow energy spread electron emission of 0.63 eV @ 44 nA was obtained (right panel of Fig. 2) from the Si-tip with DLC coated, much narrower than that of the Si-tip without DLC deposited (1.5 eV@34nA). The left panel of Fig. 2 showed the illustration of energy band of Si-tip with deposited DLC film. A low potential barrier forms on the Si/DLC interface due to the narrow band gap of DLC (~2.1 eV). During electron emission, electrons are injected from the conduction band of Si to DLC through tunnelling process, followed by transport to the DLC/vacuum interface by penetrating field and emission to vacuum through quasi-tunnelling. [5] Due to the

979-8-3503-7977-8/24 $31.00 © 2024 IEEE

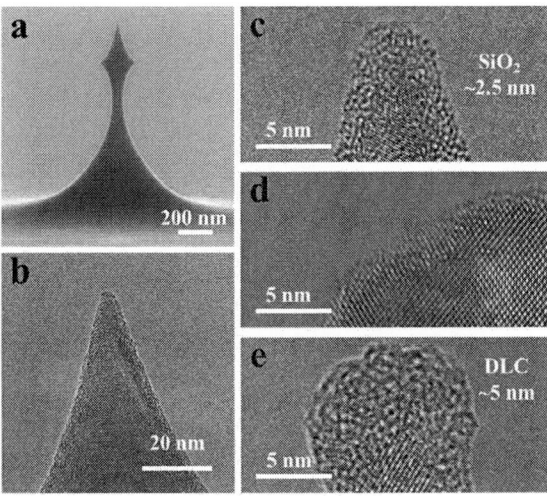

Fig. 1. (a) and (b) SEM and TEM images of typical Si-tip, respectively. (c) - (e) HRTEM images of Si-tip apex before surface cleaning, after surface cleaning and after DLC deposition, respectively.

low electron affinity of DLC (~3.4 eV), a low electric field is needed for electron emission, which induces a low-slope vacuum energy band. Accordingly, the tunnelling of the electron with lower energy is significantly suppressed, leading to a narrow energy spread emission. In contrast, for the Si-tip without DLC film, the electrons are scattered between the defect energy levels in the amorphous natural surface oxide, which causes emission electron energy

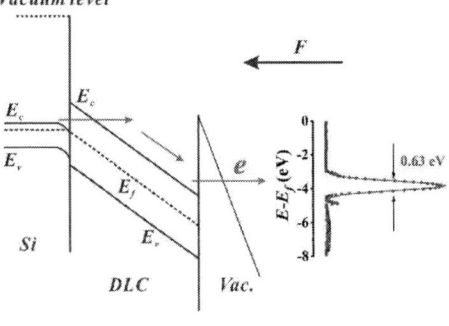

Fig. 2. Left: energy-band illustration of electron emission from Si-tip with deposited DLC film; Right: emission electron energy spectrum obtained at an emission current of 44 nA.

broaden.

III. SUMMARY

A clean Si-tip surface with deposited ultra-thin DLC film was fabricated for achieving narrow energy spread field electron emission. Comparative studies found that narrow energy spread electron emission of 0.63 eV @ 44 nA was obtained from the Si-tip with DLC deposited, much narrower than that of the Si-tip without DLC film (1.5

eV@34 nA). The DLC film prevent the re-oxide of the clean Si-tip surface and avoid scattering of emission electron in oxide layer. Moreover, the low electron affinity of DLC film reduces the electric field needed for emission, which significantly suppresses the emission of electron with low energy and induces a narrow energy spread of electron emission. The DLC coated Si-tip has great potential in electron lithography/microscopy applications.

ACKNOWLEDGEMENT

This work was supported in part by the National Key Research Program of China under Grant 2021YFA1200600, in part by the National Natural Science Foundation of China under Grant U22A2020.

REFERENCES

[1] Y. Huang et al., "P-Type Si-Tips With Integrated Nanochannels for Stable Nonsaturated High Current Density Field Electron Emission," *IEEE Transactions on Electron Devices*, pp. 1-6, 2022.

[2] J. Shaw, "Effects of surface oxides on field emission from silicon," *Journal of Vacuum Science & Technology B: Microelectronics and Nanometer Structures*, vol. 18, p. 1817, 2000.

[3] K. Murakami, T. Igari, K. Mitsuishi, M. Nagao, M. Sasaki, and Y. Yamada, "Highly Monochromatic Electron Emission from Graphene/Hexagonal Boron Nitride/Si Heterostructure," *ACS Appl Mater Interfaces*, vol. 12, pp. 4061-4067, Jan 22 2020.

[4] X. Sun, S. Wang, N. Wong, D. Ma, S. Lee, and B. K. Teo, "FTIR spectroscopic studies of the stabilities and reactivities of hydrogen-terminated surfaces of silicon nanowires," *Inorganic chemistry*, vol. 42, pp. 2398-2404, 2003.

[5] N. S. Xu, J. C. She, S. E. Huq, J. Chen, and S. Z. Deng, "Enhancing electron emission from silicon tip arrays by using thin amorphous diamond coating," *Applied Physics Letters*, vol. 73, pp. 3668-3670, 1998.

Noise Characterization of Graphene Sensors

Patrik Staroň[1], Robert Macků[1], Petr Sedlák[1], Nikola Papež[1], Ramazanov Shihkgasan[1], Farid Orudzhev[1], Mohammed A. Al-Anber[2], Dinara Sobola[1*]

[1]Department of Physics, Faculty of Electrical Engineering and Communication, Brno University of Technology, Brno, Czech Republic
[2]Laboratory of Inorganic Materials and Polymers, Department of Chemistry, Faculty of Sciences, Mutah University, Al-Karak, Jordan

*Corresponding author: sobola@vut.cz

Abstract—**This study investigates the operational dynamics and challenges of graphene-based sensors, focusing on refining data measurement and processing techniques to ensure accurate identification of chemical compounds in different gas atmospheres. The research highlights the importance of sensor cleanliness and the use of analytes with physical properties matching the measurement environment. The cleaning process involves using isopropyl alcohol and heating the sensor to 70°C, with cooling facilitated by the sensor's low thermal mass. This study also explores injecting saturated liquid onto the sensor to eliminate issues related to pressure and gas composition variations. Experimental methods include measuring the graphene sensor in the spectral domain using a setup that minimizes resistive noise and optimizes sensor response. Key variables such as voltage, resistor noise characteristics, signal path to the amplifier, system temperature, and bonding materials are fine-tuned to achieve the lowest parasitic noise. Time-domain and frequency-domain measurement techniques are employed to correlate resistance and impedance changes with gas concentrations, respectively.**

Keywords— *graphene, noise, spectrum, gas, detection*

I. INTRODUCTION

A wide range of sectors, including manufacturing, healthcare, and the automotive industry, require accurate assessment of gas mixtures within certain environments. Workspaces handling hazardous materials must adhere to strict safety standards. The only way to identify and measure volatile compounds in the atmosphere is through the use of gas detection technology. Particularly in places prone to fire risks or with poor ventilation, such as mines, there is a critical demand for gas detection systems. Despite the proliferation of gas sensors in the market, they often fall short in terms of precision, accuracy, sensitivity to specific molecules, affordability, among other issues. Graphene, known for its extraordinary characteristics, exhibits exceptional electrical and thermal conductance, superior mechanical robustness, and offers unique optical and structural benefits unmatched by any other substance[1], [2]. These traits suggest graphene's potential in gas detection technology, aiming for unparalleled sensitivity in identifying various gas molecules. However, practical applications reveal graphene's limitations in detecting certain gases and in processing the data obtained from sensors[3], [4].

As sensor resolution increases, the quality of the signal deteriorates. This study delves into the operational dynamics of graphene-based sensors and their associated challenges. The primary objective is to refine the data measurement and processing techniques for a specific graphene sensor across different gas atmospheres, ensuring the output precisely identifies the chemical compounds detected. The interference of noise, particularly at high resolutions, is a significant concern addressed in this thesis, alongside strategies for its mitigation.

Data can be gathered from sensors through various techniques. Some sensors adjust their capacitance in response to gas levels, while most alter their resistance. Theoretically, changes in the sensor's optical state or variations in acoustic wave propagation could also signal the presence of gases. Typically, sensors are equipped with electrodes that connect to the sensing material, facilitating the measurement of its physical attributes. These assessments range in complexity. The focus of this chapter is on the analysis of chemoreceptive gas sensors, including those based on graphene, which pose unique challenges due to their potential sensitivity. Measurements can be conducted across time and frequency domains.

Time-domain measurements involve recording resistance at a specific moment, requiring data analysis to correlate resistance changes with gas concentrations. Frequency-domain measurements, on the other hand, involve assessing the sensing material's impedance at certain frequencies, exploiting predictable shifts in resonance frequencies for gas detection. Noise significantly affects sensor performance; slow reaction times in the sensing material can allow system noise to dominate, potentially skewing results. This low-frequency noise, or 1/f noise, is a known issue in semiconductor sensors and is relevant to graphene-based sensors due to their inherent noise characteristics. In laboratories, gas sensors are often tested with instruments like the 8753ES, which can evaluate resonant frequencies by measuring a resonator's return loss. Alternatively, resistance measurements with a stable current supply are sufficient. Sensor noise primarily arises from two sources: thermal noise and low-frequency noise, with thermal noise being inherent to all devices. In thermal sensors, this noise results from atomic vibrations leading to inconsistent current flow[5].

II. EXPERIMENTAL

The actual schematic proposed for measuring the graphene sensor in spectral domain is shown in Figure 1. This is a setup that allows to push current through the sensor and a resistor, while measuring the voltage on the resistor. The resistor contributes to the noise, but the tweaking the voltage from the battery should decrease the influence of the resistive noise so that the dominant noise is the one from the sensor.

979-8-3503-7977-8/24 $31.00 © 2024 IEEE

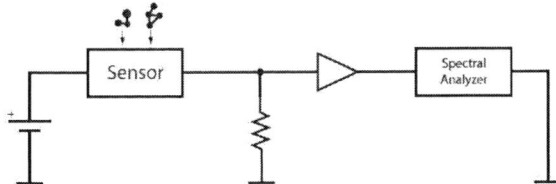

Fig. 1. Diagram of the connections for mesuring the grapene sensor.

By measuring the spectral response of the system while injecting various substances onto the sensor, it is possible to detect the response of the graphene sensing layer to the chemicals it is exposed to. The measurement equipment is very sensitive and it is possible that the resulting information will not be making sense. There are several variables that can be controlled:

- Voltage over the graphene and the resistor.

- The resistor and so its noise characteristics.

- The signal path to the amplifier.

- The temperature of the system.

- The bonding material and metal interfaces.

These variables are need to be fine-tuned in order to achieve lowest parasitic noise for the measurement. Voltage over the graphene will be chosen experimentally. Resistor and its noise should be chosen after setting the voltage threshold[6].

III. CONCLUSIONS

The gas composition, pressure, and temperature have a significant impact on the accuracy of the measurement. To ensure reliable results, it is essential that the sensor be thoroughly cleaned at the beginning of the measurement process. Furthermore, the analyte used should have the same physical properties as the measurement environment for the best possible outcome. The cleaning process involves using isopropyl alcohol to remove any contaminants and raising the sensor's temperature to 70°C. This elevated temperature helps in thoroughly cleaning the sensor. After cleaning, the sensor needs to cool down. However, due to its low thermal mass, this cooling process is relatively quick, minimizing the risk of significant re-contamination. It is important to note that the suggested cleaning temperature of 70°C is not fixed and may be adjusted based on the results obtained from initial measurements.

To address potential issues related to pressure and gas composition, injecting a liquid onto the sensor is an effective method. This approach ensures that the liquid is saturated, thereby eliminating variations in pressure and composition that could otherwise affect the measurement accuracy. This saturation helps maintain consistent conditions, leading to more reliable and accurate measurements.

ACKNOWLEDGMENT

The research described in the paper was financially supported by the Internal Grant Agency of the Brno University of Technology, grant No. FEKT-S-20-6352 and the GACR 23-07384S.

REFERENCES

[1] A. Knápek, J. Sýkora, J. Chlumská, and D. Sobola, "Programmable set-up for electrochemical preparation of STM tips and ultra-sharp field emission cathodes," Microelectron Eng, vol. 173, pp. 42–47, Apr. 2017, doi: 10.1016/J.MEE.2017.04.002.

[2] A. Knápek, D. Sobola, P. Tománek, Z. Pokorná, and M. Urbánek, "Field emission from the surface of highly ordered pyrolytic graphite," Appl Surf Sci, vol. 395, pp. 157–161, Feb. 2017, doi: 10.1016/J.APSUSC.2016.05.002.

[3] D. Sobola, N. Papež, R. Dallaev, S. Ramazanov, D. Hemzal, and V. Holcman, "Characterization of nanoblisters on HOPG surface," Journal of Electrical Engineering, vol. 70, no. 7, pp. 132–136, Dec. 2019, doi: 10.2478/JEE-2019-0055.

[4] A. AlSoud et al., "Electrical properties of epoxy/graphite flakes microcomposite at the percolation threshold concentration," Phys Scr, vol. 99, no. 5, p. 055955, Apr. 2024, doi: 10.1088/1402-4896/AD3B50.

[5] A. Knápek, M. M. Allaham, D. Burda, D. Sobola, M. Drozd, and M. Horáček, "Explanation of the quasi-harmonic field emission behaviour observed on epoxy-coated polymer graphite cathodes," Mater Today Commun, vol. 34, p. 105270, Mar. 2023, doi: 10.1016/J.MTCOMM.2022.105270.

[6] K. Ronoh, S. H. Fawaeer, V. Holcman, A. Knápek, and D. Sobola, "Comprehensive characterization of different metallic thin films on highly oriented pyrolytic graphite substrate," Vacuum, vol. 215, p. 112345, Sep. 2023, doi: 10.1016/J.VACUUM.2023.112345.

Numerial Simulation of a Vacuum Cold Cathode X-Ray Detector Driven by a Dual-Gate Thin Film Transistor

Zhongbin Pu, Zhipeng Zhang*, Jiaquan Kong, Juncong She, Shaozhi Deng, Jun Chen

State Key Laboratory of Optoelectronic Materials and Technologies, Guangdong Province Key Laboratory of Display Material and Technology, School of Electronics and Information Technology, Sun Yat-sen University, Guangzhou 510275, People's Republic of China
*Corresponding author: zhangzhp25@mail.sysu.edu.cn

Abstract—A dual-gate a-IGZO thin film transistor (DG a-IGZO TFT) driven vacuum cold cathode X-ray detector is proposed to achieve a low-dose detection with a high internal gain. The transfer characteristic curves of TFT in the dark and under various X-ray doses are simulated and a high photo/dark current ratio· (10^{10}) is achieved owing to the photocurrent amplification effect operating in the subthreshold region of DG TFTs. The proposed DG a-IGZO TFT driven vacuum cold cathode X-ray detectors are expected to be applied in highly sensitive and low dose X-ray imaging.

Keywords—*X-ray detector, Cold cathode, EBIPC, IGZO TFT, High gain*

I. INTRODUCTION

Large area X-ray detectors have attracted much attention because of their various applications in security, medical and industrial X-ray imaging [1]. In all these applications, high detection sensitivity is paramount because it enables minimizing the detection X-ray dose and thus reduces radiation-related health risks [2]. However, the mainstream a-Se photoconductor based X-ray detectors exhibited limited absorption efficiency and charge collection efficiency for X-ray due to the small atomic number and carrier mobility lifetime product value of the material, limiting their applications in low-dose detection and imaging [1]. Recently, a vacuum flat-panel X-ray detector formed by a photoconductor and a cold cathode was demonstrated to achieve a high detection sensitivity owing to the unique electron bombardment induced photoconductivity (EBIPC) mechanism [3]. However, achieving large area vacuum imaging devices with a low dark current and a wider dynamic range is still challenging.

The amorphous indium–gallium–zinc-oxide thin film transistors (a-IGZO TFTs) are expected to be used in large-area addressable and high-performance vacuum X-ray detectors. On the one hand, a-IGZO TFT exhibited excellent electrical properties such as high mobility, low off-state current and high on-off ratio, making a-IGZO TFT driven vacuum X-ray detectors with high signal-to-noise ratio and fast response speed possible [4]. On the other hand, a-IGZO TFTs driven vacuum cold cathode devices had the significant advantages of a low driving voltage, a uniform emission and a stable emission current [5]. In addition, the reported dual-gate (DG) TFTs achieved a highly sensitive X-ray detection with a high internal gain using the photocurrent amplification effect [6].

In this paper, a DG a-IGZO TFT was proposed to integrate with the vacuum cold cathode X-ray detector. The corresponding photoelectric conversion characteristics were calculated and simulated, demonstrating their promising application in low dose X-ray detection and imaging.

II. THEORY ANALYSIS

Fig.1(a) shows the schematics of a DG a-IGZO TFT driven vacuum cold cathode X-ray detector and Fig.1(b) shows the corresponding equivalent circuit. Electron hole pairs (EHPs) generates in the Ga₂O₃ photoconductor caused by the X-ray irradiation. When a forward bias voltage is applied to the anode, electrons generated in photoconductor are collected by the ITO anode while holes gather in the opposite end. The accumulated hole charges induce the electron emission from ZnO nanowires field emitters and then are neutralized by the emission electrons. It is reported that the generated EHPs can be amplified owing to EBIPC effect [3].

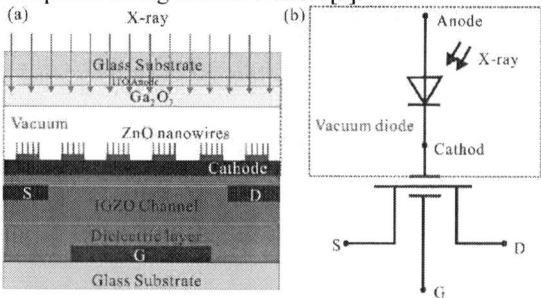

Fig1. (a) Schematics of a DG a-IGZO TFT driven vacuum cold cathode X-ray detector. (b) The corresponding equivalent circuit.

The number of holes accumulated in the top gate electrode is equal to the number of collected holes in Ga₂O₃ photoconductor. The number of collected hole charges of photoconductor can be calculated as [7]:

$$Q_h = \frac{Q_0 \mu_h F \tau_h}{T}\left[(1-e^{-\alpha T}) - \frac{1}{(1/\alpha \mu_h F \tau_h)+1}\left(1-e^{-\alpha L - \frac{T}{\mu_e F \tau_e}}\right)\right] \quad (1)$$

wherein $Q_0 = \frac{5.45\times10^{13}\times eA}{(\alpha_{air}/\rho_{air})W_{\pm}}\left(\frac{\alpha_{en}}{\alpha}\right)$, e is an electronic charge, A is the effective area of X-ray irradiation, α is attenuation coefficient of the photoconductor, α_{air} and ρ_{air} are the energy absorption coefficient and density of air, $W_{\pm}$ is the production energy of an EHP, α_{en} is the energy absorption coefficient of photoconductor, τ_h is the lifetime of hole.

The capacitances of top gate insulation layer and bottom gate insulation layer are labeled as C_{top} and C_{ox}, respectively. The threshold voltage (V_T) of DG TFT can be described as [6]:

$$V_T(t, V_{TG}) = V_{T0} + \gamma V_{TG} = V_{T0} + \gamma \frac{\varepsilon_1 G Q_h}{C_{top}} \quad (2)$$

wherein V_{T0} is initial threshold voltage, V_{TG} is top gate voltage caused by X-ray, γ is sensitivity parameter, ε_1 is the charge collection efficiency of electrodes, G is the internal gain caused by EBIPC effect.

Based on the above analysis, the output current of proposed DG a-IGZO TFT in different operating regions can be obtained [6]. When the TFT operates in the subthreshold region, the output current expression can be described as:

979-8-3503-7977-8/24 $31.00 © 2024 IEEE

$$I_{DS} = I_{D0} \exp\left\{ \frac{q\left[V_{GS} - V_T(t, X, V_{X-ray})\right]}{\eta} \right\} \left[1 - \exp\left(-q\frac{V_{DS}}{kT} \right) \right] \quad (3)$$

wherein $\eta = \frac{qS}{(\ln 10)kT}$, S is subthreshold swing. The output current expression of TFTs operating in the saturation region can be described as:

$$I_{DS} = \frac{W C_i \mu_{FE}}{L} \left[V_{GS} - V_T(t, X, V_{X-ray}) \right]^2 \quad (4)$$

wherein W and L are the width and length of TFT channel, μ_{FE} is the mobility of charge carriers. When the TFT operates in the linear region, the output current expression can be described as:

$$I_{DS} = \frac{W C_i \mu_{FE}}{L} \left[V_{GS} - V_T(t, X, V_{X-ray}) \right] V_{DS} - \frac{1}{2} V_{DS}^2 \quad (5)$$

wherein $V_{DS} \ll V_{GS}$. By measuring the resistance difference of Ga_2O_3 photoconductor under X-ray irradiation, The top gate voltage can be estimated.

III. RESULTS AND DISCUSSION

The electrical characteristics of proposed TFTs are simulated (ATLAS, Silvaco). Fig. 2 shows the simulated electrical transfer characteristic of proposed DG a-IGZO TFTs. The threshold voltage of TFT linearly decreases from 2.2 V to -0.4 V with the increasing X-ray dose rate from 0 to 2 mGy/s (inset of Fig. 2), which is consistent with the equation (2). Owing to the decreasing of threshold voltage, the photocurrent amplification effect occurs in the subthreshold region of DG TFTs, resulting in a high photo/dark current ratio ($\sim 10^{10}$). As shown in Fig. 2, the dark current is $\sim 10^{-16}$A at a bottom gate voltage of -4 V, while the corresponding photocurrent is $\sim 10^{-6}$ A with a X-ray dose rate of the 2 mGy/s. Therefore, the proposed detectors are expected to achieve a low dose X-ray detection and a wider dynamic range.

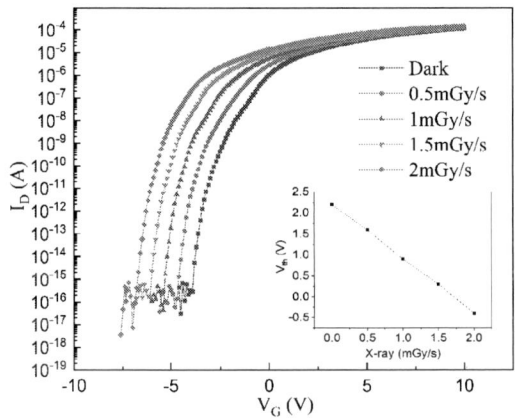

Fig. 2. Simulated transfer characteristic of the DG a-IGZO TFT with the inset showing the relationship between X-ray dose and threshold voltage (V_G represents the bottom gate voltage).

Fig. 3 shows the simulated electrical output characteristic of proposed DG a-IGZO TFTs. It is clearly observed that the a-IGZO TFT exhibits an unsaturation on-current at a large bottom gate voltage (>18V), demonstrating a high voltage resistance of TFTs. In addition, when the bottom gate voltage increases from 2 to 20 V, the saturation current increases from 9.6 µA to 0.5 mA, benefitting for the tuning of internal gain.

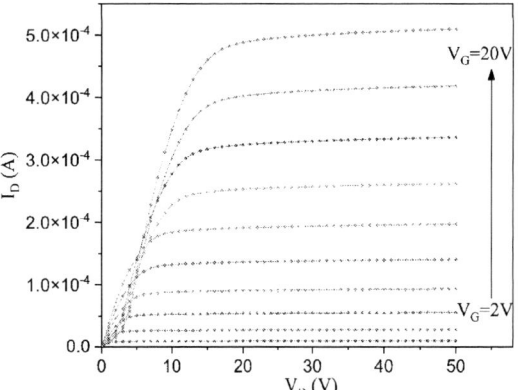

Fig. 3. Simulated output characteristic of the DG a-IGZO TFT (V_G represents the bottom gate voltage).

IV. CONCLUSIONS

In summary, a DG a-IGZO TFT driven vacuum cold cathode X-ray detector was proposed and its photoelectric conversion characteristics were calculated and simulated. The results show that the threshold voltage of DG a-IGZO TFT decreases linearly with the increasing dose rate of irradiated X-ray, resulting in the photocurrent amplification effect operating in the subthreshold region. The proposed detectors are expected to achieve high internal gain and low dose X-ray imaging.

ACKNOWLEDGMENTS

The authors gratefully acknowledge the financial support from the National Key Research and Development Program of China (Grant No. 2022YFA1204200), Key Research and Development Program of Guangdong Province (Grant No. 2023B0101200013), National Natural Science Foundation of China (Grant No. 62271512), Natural Science Foundation of Guangdong Province (Grant No. 2024A1515012852), Guangzhou Municipal Science and Technology Bureau (Grant No. 2023A04J1664), Fundamental Research Funds for the Central Universities.

REFERENCES

[1] J. Pang, et al, "Reconfigurable perovskite X-ray detector for intelligent imaging," Nat. Commun., vol. 15, p. 1769, 2024.

[2] H. Wei, et al, "Sensitive X-ray detectors made of methylammonium lead tribromide perovskite single crystals," Nat. Photonics, vol. 10, no. 5, pp. 333–339, 2016.

[3] Z. Zhang, et al, "Sensitive direct-conversion X-ray detectors formed by ZnO nanowire field emitters and β-Ga₂O₃ photoconductor targets with an electron bombardment induced photoconductivity mechanism," Photonics Res., vol. 9, no. 12, p. 2420, 2021.

[4] Z. Pan, et al, "Approaches to improve mobility and stability of IGZO TFTs: a brief review," Trans. Electr. Electro., 2024, https://doi.org/10.1007/s42341-024-00536-1.

[5] X. Li, et al, "Highly stable field emission from ZnO nanowire field emitters controlled by an amorphous indium–gallium–zinc-oxide thin film transistor," Jpn. J. Appl. Phys., vol. 57, no. 4, p. 045003, 2018.

[6] L. Wang, et al, "A numerical study of an amorphous silicon dual-gate photo thin-film transistor for low-dose X-ray imaging," J. Disp. Technol., vol. 11, no. 8, pp. 646–651, 2015.

[7] V. Richard, et al, "Handbook of medical imaging," SPIE Press, vol. 1, chap. 1, 2000.

On-Chip Integrated Si-tip Field Electron Emission Vacuum Transistor with Saturated Output Characteristics

Zhen Wang[1,†], Yuan Huang[2,†], Yang Chen[1], Yifeng Huang[1], Jun Chen[1], Ningsheng Xu[1], Shaozhi Deng[1], Juncong She[1,*]

[1]State Key Laboratory of Optoelectronic Materials and Technologies, Guangdong Province Key Laboratory of Display Material and Technology, School of Electronics and Information Technology, Sun Yat-sen University, Guangdong Province, People's Republic of China
[2]School of Microelectronics Science and Technology, Sun Yat-Sen University, Guangdong Province, People's Republic of China
*Corresponding author: shejc@mail.sysu.edu.cn
†Z.Wang and Y. Huang contributed equally to this work

Abstract—High performance on-chip integrated field electron emission vacuum transistors (FEVTs) have attracted great attention in recent years. Nevertheless, the reported on-chip integrated FEVTs typically showed non-saturated output characteristic due to the narrow cathode-anode separation (i.e., less than 1 μm), which limits the application of the devices in linear amplifier circuits. Herein, a rational design of vertical on-chip gate-all-around FEVT configured with a Si-tip cathode, a volcano-shape gate electrode and a circular opening anode electrode was proposed for achieving saturated output characteristics. Systematic numerical simulations and experimental investigations were employed for optimizing the gate-height and gate-aperture-radius to shield the cathode surface from the anode field and obtain a saturated high intensity output current.

Keywords—On-chip, field electron emission, vacuum transistor, saturated output characteristic, volcano-shape.

I. INTRODUCTION

Field electron emission vacuum transistors (FEVT) is promising for vacuum electronic applications involving high temperature or radiations [1-3]. In the miniaturized FEVT device, the emission current is significantly influenced by the anode field due to the narrow cathode-anode separation (i.e., less than 1 μm). The anode induced emission results in a non-saturated output characteristic. The non-saturated output characteristics strictly limits the device application in linear amplifier circuits. For obtaining an on-chip FEVT with saturated output characteristics, one approach had been developed by increasing the cathode-anode separation. K. Subramanian et al. has reported a lateral diamond FEVT with saturation characteristics. The anode-cathode separation is up to 500 μm, while the gate-cathode spacing is 2 μm [4]. The lateral diamond FEVT with a large anode-cathode separation leads to a low integration density per unit area. Moreover, it needs a higher anode voltage for ensuring the electron collections. In the present work, we report an on-chip integration of a vertical gate-all-around vacuum transistor. The cathode-anode separation is less than 3 μm. The relationship between device performance and the gate structure was investigated by numerical simulations and experiment. A 40×40 Si-tip array FEVT was obtained, showed well saturated output characteristic, low threshold voltage, and strong gate-to-emitter controllability with low gate leakage current.

II. RESULTS AND DISCUSSION

Fig. 1 showed the schematics illustration of the vertical on-chip gate-all-around FEVT. The FEVT is configured with a Si-tip cathode (source), a volcano-shape gate and a circular opening anode (drain). The FEVT was micro-fabricated by a "self-alignment" process. The separation between the gate, drain and source was determined by the thickness of the SiO_2 layer. The horizontal distance between the tip-apex and the bottom of volcano-gate is 1.2 μm, which is approximately equal to the cathode-anode separation.

Numerical simulation and experimental investigations found that the increase of gate-height (H) and shrink of gate-aperture-radius (R) can significantly shield the cathode surface from the anode electric field, which is beneficial to obtain saturated output characteristics. This is attributed to the reduction of cathode-gate distance and the increase of lateral area of the volcano-shape gate, which enhances the gate-induced electric field at the Si-tip surface. However, this method limits the output current intensity (anode current). The emitted electrons would be captured by gate electrode rather than transported to the anode. Systematic numerical simulations were performed to further optimize the geometric parameters of H and R for achieving high output current. A FEVT with optimized parameters of H=250 nm and R=650 nm was found that showing saturated high intensity output current and high anode collection rate. Accordingly, the FEVT that consist of 40×40 Si-tips (p-type doped) was fabricated, which achieved significant saturated I-V behave, i.e., the anode current was typically saturated at 6 μA at the anode voltage >45 V and gate voltage = 75 V. The device achieved a high I_{on}/I_{off} of 10^6, turn-on voltage of 23 V, a subthreshold slope (SS) of 3 V/decade and a transconductance (g_m) of 0.3 μS. The ratio of anode current to cathode current is 87 %, suggesting that most of the emitted electrons are transported to the anode. The device showed a typical cutoff frequency (f_C, -3dB) of 12 kHz, which is higher than that of the device using n-type Si-tips, i.e., 1 kHz. The gate voltage would induce a depletion-layer capacitance in the p-type FEVT. It is in series with the gate-to-cathode capacitance and thus resulted in a higher f_C.

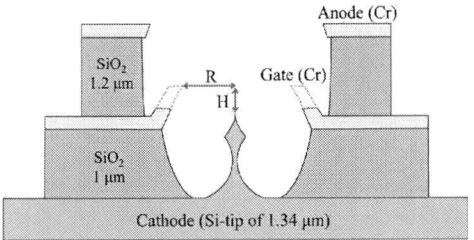

Fig.1. The schematic illustrations of the on-chip integrated Si-tip field electron emission vacuum transistor.

III. CONCLUSION

An on-chip integrated vertical FEVT with a device structure of the Si-tip cathode below the volcano-shape gate electrode was designed and fabricated by using an IC-compatible process. The device structure (the volcano-gate) offers a merit in screening the tip-apex from the anode field. Furthermore, the reduced cathode-gate separation enhanced the gate-to-cathode controllability. As a result, the transistor shows the higher saturated output current and lower threshold voltage. It is interest that the device showed a high output current.

ACKNOWLEDGMENTS

This work was supported in part by the National Key Research Program of China under Grant 2021YFA1200600, in part by the National Natural Science Foundation of China under Grant U22A2020.

REFERENCES

[1] J. W. Han, J. S. Oh, and M. Meyyappan, "Vacuum nanoelectronics: Back to the future?—Gate insulated nanoscale vacuum channel transistor," *Appl. Phys. Lett.*, vol. 100, no. 21, 2012, Art. no. 213505.

[2] J.-W. Han, D.-I. Moon, and M. Meyyappan, "Nanoscale vacuum channel transistor," *Nano Lett.*, vol. 17, pp. 2146–2151, Mar. 2017.

[3] J.-W. Han, M.-L. Seol, D.-I. Moon, G. Hunter, and M. Meyyappan, "Nanoscale vacuum channel transistors fabricated on silicon carbide wafers," *Nature Electronics*, vol. 2, no. 9, pp. 405–411, Aug. 2019, doi: 10.1038/s41928-019-0289-z.

[4] K. Subramanian, W. P. Kang, J. L. Davidson, N. Ghosh, and K. F. Galloway, "A review of recent results on diamond vacuum lateral field emission device operation in radiation environments," *Microelectronic engineering*, vol. 88, no. 9, pp. 2924-2929, Apr. 2011.

Optimization of High-performance Single Island Carbon Nanotube Electron Beam (C-Beam) for Microscopy Application

Ravindra Patil[1], Aniket Karande[1], Ketan Bhotkar[1], Kyu Chang Park[1*]

[1] Kyung Hee University, Information display, Seoul, 02447, Republic of Korea,

*Email: kyupark@khu.ac.kr (Prof. Kyu Chang Park)

Abstract— This study successfully built a carbon nanotube (VACNT) cold cathode (C-beam) as a potential electron source for high-resolution scanning electron microscopes. The C-beam achieved a stable emission current with high transmission and a narrow beam divergence angle. These findings suggest VACNT C-beams hold promise for next-generation microscopy techniques.

Keywords— *C-beam, Carbon nanotube, DC-PECVD, Phosphor screen imaging, Divergence angle.*

I. INTRODUCTION

This study demonstrates the feasibility of carbon nanotubes (CNTs) as high-resolution electron sources for scanning electron microscopy. CNT emitters achieved impressive field emission characteristics, with an anode current exceeding 0.15 mA within a sub-mm² area. Diode and triode configurations further enhance individual tip brightness. Analysis using a commercial tool confirmed minimal beam divergence (1.22°) and a small virtual source size[1][2]. These findings, building upon prior research on high-current CNT electron beams, highlight the potential of CNTs to revolutionize electron microscopy through superior field emission capabilities.

II. MATERIAL AND METHODS

Fabrication process of the carbon nanotube emitter:

This study focuses on the fabrication and characterization of carbon nanotube (CNT) emitters for field emission applications. A meticulous, multi-step process is employed to achieve controlled growth and desired characteristics for the emitters.

The process begins with n-type silicon wafers serving as the substrate for the CNT growth. A thin layer (30 nm) of nickel (Ni) is then deposited onto the silicon using radio frequency magnetron sputtering. Nickel acts as a catalyst for efficient CNT growth. Subsequently, photolithography defines a patterned array of dots on the substrate. Each dot has a diameter of 3 μm and a spacing (pitch) of 15 μm.

To achieve selective growth of CNTs at the designated locations (dots), a triode configuration of direct current plasma-enhanced chemical vapor deposition (DC-PECVD) is utilized. During this process, specific bias voltages are applied to the cathode (-600 V) and mesh electrode (+300 V). The growth conditions are carefully controlled with a gas flow of C2H2: NH3 (65: 453 sccm), a working pressure of 2.2 Torr, and a device temperature of 850°C[3]. This meticulous approach ensures the fabrication of CNT emitters with a controlled and well-defined geometry, a critical factor for achieving optimal field emission properties (Fig.1)[4].

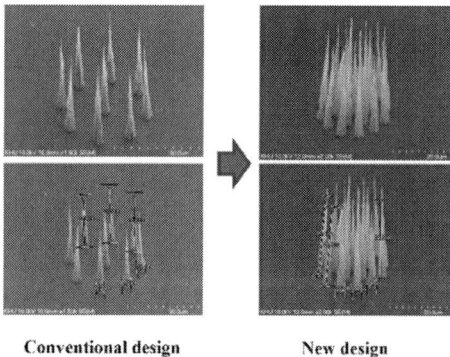

Conventional design   **New design**

Fig. 1. SEM Images of convensional design sample and new design sample.

III. EXPERIMENT RESULT

Our research focused on optimizing and characterizing a carbon nanotube (CNT) electron beam source. We prioritized maximizing beam current while maintaining a minimal emission area. This new design achieved a stable beam current of 58 μA with a high electron transmission rate (Ia/Ic) of 95%, resulting in a significantly higher current density compared to a previous design with a lower emission area (Fig. 2)[5][6].

To characterize the divergence angle of the CNT electron beam, we employed a phosphor screen. Prior to experimentation, electron optical design (EOD) simulations were performed to optimize the measurement setup, particularly the distance between the CNT source and the screen. Following the simulations, the experiment was conducted in a vacuum chamber at a pressure of 10^{-7} Torr. The phosphor screen was positioned 26 mm from the CNT source (Dg-a distance) as determined by the simulations (Fig. 3)[7]. With increasing applied voltage (from 1,000 V to 1,250 V in 50 V increments), the electron beam trajectory exhibited a corresponding increase. Under these optimized conditions, the experiment yielded a focal spot size of 1. 1 mm and a divergence angle of 1. 22⁰ (21. 22 mrad)[3][2]. This data demonstrates the effectiveness of the new design in achieving a high-density electron beam with a well-defined divergence angle, both crucial characteristics for high-resolution microscopy applications.

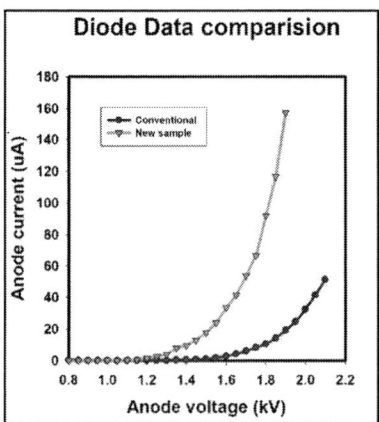

Fig. 2. Comparision of anode current data in diode structure.

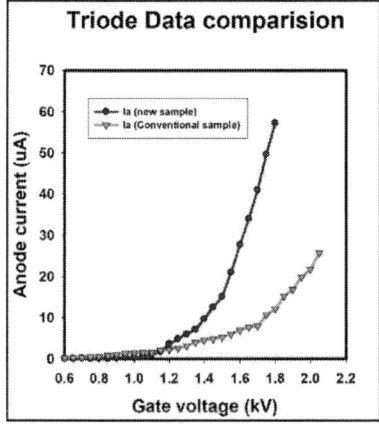

Fig. 3. Comparision of anode current data in triode structure.

IV. CONCLUSION

This study successfully fabricated a vertically aligned carbon nanotube (VACNT) based cold cathode (C-beam) as a potential electron source for scanning electron microscopes. The fabricated C-beam exhibited promising characteristics for high-resolution imaging applications.

The C-beam achieved a stable emission current of 58 μA with a high electron transmission rate (Ia/Ic) of 95%. The beam brightness generated by the VACNT source suggests its potential for high-resolution microscopy. The measured divergence angle of the electron beam was approximately 1. 22 degrees.

These results demonstrate the feasibility and potential of VACNT-based electron sources for next-generation scanning electron microscopes. Further research can explore methods to optimize emission current density and explore the long-term stability of VACNT emitters[8][9]. With continued development, VACNT C-beams have the potential to revolutionize microscopy techniques, opening doors for enhanced imaging capabilities across various scientific and technological fields.

ACKNOWLEDGMENT

This work was supported by the Technology Innovation Program (No. 20013595, Extreme ultraviolet light source using nano electron beam) funded by the Ministry of Trade, Industry and Energy (MOTIE, Korea)

REFERENCES

[1] H. R. Lee, O. J. Hwang, B. Cho, K. C. Park, and others, "Scanning electron imaging with vertically aligned carbon nanotube (CNT) based cold cathode electron beam (C-beam)," *Vacuum*, vol. 182, p. 109696, 2020.

[2] B. C. Adhikari, B. Ketan, J. S. Kim, S. T. Yoo, E. H. Choi, and K. C. Park, "Beam trajectory analysis of vertically aligned carbon nanotube emitters with a microchannel plate," *Nanomaterials*, vol. 12, no. 23, p. 4313, 2022.

[3] K. C. Park, J. H. Ryu, K. S. Kim, Y. Y. Yu, and J. Jang, "Growth of carbon nanotubes with resist-assisted patterning process," *J. Vac. Sci. Technol. B Microelectron. Nanom. Struct. Process. Meas. Phenom.*, vol. 25, no. 4, pp. 1261–1264, 2007.

[4] D. Cai and L. Liu, "The screening effects of carbon nanotube arrays and its field emission optimum density," *AIP Adv.*, vol. 3, no. 12, 2013.

[5] P. Kruit, M. Bezuijen, and J. E. Barth, "Source brightness and useful beam current of carbon nanotubes and other very small emitters," *J. Appl. Phys.*, vol. 99, no. 2, 2006.

[6] Y. Li *et al.*, "Influence of Grid Aperture Ratio on Electron Transmittance and Electron Beam Spot Size in Field Emission Processes of Carbon Nanotubes," *Appl. Sci.*, vol. 14, no. 8, p. 3311, 2024.

[7] H. R. Lee, H. H. Yang, and K. C. Park, "Fabrication of a high-resolution electron beam with a carbon nanotube cold-cathode," *J. Vac. Sci. Technol. B*, vol. 35, no. 6, 2017.

[8] L. Zhu, J. Xu, Y. Xiu, Y. Sun, D. W. Hess, and C. P. Wong, "Growth and electrical characterization of high-aspect-ratio carbon nanotube arrays," *Carbon N. Y.*, vol. 44, no. 2, pp. 253–258, 2006.

[9] K. S. Kim, J. H. Ryu, C. S. Lee, J. Jang, and K. C. Park, "Enhanced and stable electron emission of carbon nanotube emitter arrays by post-growth hydrofluoric acid treatment," *J. Mater. Sci. Mater. Electron.*, vol. 20, pp. 120–124, 2009.

Optimization of Sputtering Condition for TiN-coated Si-FEA

Hiromasa Murata*, Katsuhisa Murakami and Masayoshi Nagao

National Institute of Advanced Industrial Science and Technology, Tsukuba, Japan

*Corresponding author: murata.hiromasa@aist.go.jp

Abstract—TiN coated silicon field emitter arrays (FEA) were fabricated at various sputtering condition. Impact of Ar/N_2 gases flow rate on TiN film properties, such as N fraction and O concentration, was investigated. X-ray photoelectron spectroscopy analysis revealed that N fraction increases and O concentration decreases with the increase of N_2 gas ratio (N_2/(Ar+N_2)). The emission characteristics of FEA coated with TiN using different Ar/N_2 gases flow rate was also investigated. The N-rich and low oxygen TiN emitter showed the lowest operational voltage. Emission start voltage was approximately 20 V and 1 mA was obtained as low as 60 V.

Keywords—volcano structure, field emitter array, titanium nitride, reactive sputtering

I. INTRODUCTION

Volcano-structured double-gate field emitter array (FEA) is promising electron source because of its high beam focusing owing to integration of the precisely arranged focusing electrode on the gate electrode [1] and it has been applied to devices such as image sensor with ultra-high sensitivity [2] and radiation tolerance [3]. These application require relatively small current of the order of microamperes, however, higher current operation is desirable for x-ray sources [4] and high-frequency vacuum tubes [5].

TiN coating technique has achieved low voltage and high current operation owing to its low work function [6,7] and high melting point. In our previous study, we applied TiN coating to volcano-structured single-gate FEA and demonstrated high-current operation of milliamperes from 1000-tip FEA [8]. Emission characteristics are potentially improved by the film property of TiN coating. In this study, we deposited TiN coating layer on Si-FEA at a different sputtering condition and compared their emission characteristics.

II. TiN FILM ANALYSIS

Prior to FEA fabrication, composition and O concentration of TiN were evaluated by x-ray photoelectron spectroscopy (XPS). TiN was deposited on a flat Si substrate using direct current (DC) magnetron sputtering (sputtering pressure: 0.1 Pa, DC power: 300 W) using Ti target and Ar/N_2 gases. Here, the Ar/N_2 gases flow rates were set to 20/1, 20/2, 20/3, 20/4, 20/5, and 0/20 sccm. The background pressure just before gas introduction was ranged from 2×10^{-5} to 5×10^{-5} Pa. XPS analysis was performed after removing TiN top surface by Ar^+ ion beam. The N fraction x and O concentration of TiN was estimated based on the peak intensities in XPS narrow spectra. Fig. 1 shows dependence of x and O concentration on N_2 gas ratio (N_2/(Ar+N_2)). As shown in this figure, x rapidly increased for N_2/(Ar+N_2) < 0.2 and slightly increased for N_2/(Ar+N_2) $\geq$ 0.2. Conversely, O concentration rapidly decrease for N_2/(Ar+N_2) < 0.2 and slightly decreased for N_2/(Ar+N_2) $\geq$ 0.2. TiN film deposited by 100% N_2 showed over-stoichiometry and lowest O contaminations. Thus, we deposited TiN on Si-FEA using three conditions, Ar/N_2 = 20/3, 20/5, and 0/20 sccm, that allowed us to form near-stoichiometric TiN.

III. FABRICATION AND CHARACTERIZATION

Fig. 2 shows the fabrication process of the FEA (1000 tips). SiO_2 was deposited by plasma-enhanced chemical vapor deposition and patterned by reactive ion etching (RIE) (Fig. 2(a)). Si surface was thermally oxidized and etched by buffered hydrofluoric acid (BHF) (Fig. 2(b),(c)). TiN, insulating layer, and gate electrode were deposited by sputtering (Fig. 2(d)). After gate holes were formed (Fig. 2(e)), tip open was performed by BHF (Fig. 2(f)). Fig. 3 shows the cross-sectional scanning electron microscopy (SEM) images of fabricated TiN-coated Si emitter.

To evaluate emission characteristics, we used the electrical circuit as shown in the inset of Fig. 4(a). Fig. 4(a) shows I-V characteristics of the FEA coated by TiN deposited at different N_2 gas flow rate. Electron emission started at 20–30 V and achieved almost 1 mA. The operational voltage depends on the N_2 gas flow rate. The TiN emitter deposited at high N_2-gas-rate showed lower operational voltage. This is probably due to a lower work function of N-rich TiN owing to the low oxygen contamination as shown in Fig. 1. The ratio of work function among the FEA of Ar/N_2 = 0/20, 20/5, and 20/3 is approximately 1:1.4:1.5, which is estimated from the slope of F-N plot shown in Fig. 4(b) by assuming the emission area and beta factor is the same among the FEA.

IV. CONCLUSION

We deposited TiN coating layers on Si-FEA by DC magnetron sputtering at different Ar/N_2 gases flow rates and compared their electron emission characteristics. XPS analysis indicate that N fraction of TiN increased and O concentration in TiN decreased with increasing N_2 gas flow rates. Electron emission began at lower gate voltages for FEAs coated with N-rich TiN, likely reflecting the lower work function owing to reduction of O concentration. The findings of this study promote the development of FEA that achieve high-current operation with focusing electron beam.

ACKNOWLEDGMENT

A part of this work was financially supported by JSPS KAKENHI (No. 23K13371) and supported by "Advanced Research Infrastructure for Materials and Nanotechnology in Japan (ARIM)" of the Ministry of Education, Culture, Sports, Science and Technology (MEXT).

REFERENCES

[1] M. Nagao, Y. Gotoh, Y. Neo, H. Mimura, "Beam profile measurement of volcano-structured double-gate Spindt-type field emitter arrays" J. Vac. Sci. Technol. B, vol. 34, pp. 02G108-1-6, 2016.

[2] M. Nanba, Y. Hirano, Y. Honda, K. Miyakawa, T. Ookawa, T. Watanabe, S. Okazaki, N. Egami, K. Miya, K. Nakamura, M. Taniguchi, S. Itoh, and A. Kobayashi, "640x480 Pixel HaRP Image

Sensor with Active-matrix Spindt-type FEA", Proc. 13th IDW '06, pp. 1817-1820, Dec. 2006.

[3] Y. Gotoh, M. Nagao, T. Masuzawa, Y. Neo, H. Mimura, T. Okamoto, M. Akiyoshi, I. Takagi, "Research Project on Development of Radiation Tolerant Compact Image Sensor with a Field Emitter Array", Technical Digest of International Vacuum Nanoelectronics Conference, pp. 240-241, Jul. 2015.

[4] Y. Y. Yu, and K. C. Park, "Fabrication of high quality X-ray source by gated vertically aligned carbon nanotube field emitters", J. Vac. Sci. Technol. B, vol. 41, pp. 023203-1-7 (2023).

[5] H. Makishima, S. Miyano, H. Imura, J. Matsuoka, H. Takemura, A. Okamoto, "Design and performance of traveling-wave tubes using field emitter array cathodes," Applied Surface Science, vol.146, pp.230-233, 1999.

[6] S. Y. Kang, J. H. Lee, Y. H. Song, Y. T. Kim, K. I. Cho, H. J. Yoo, "Emission characteristics of TiN-coated silicon field emitter arrays", J. Vac. Sci. Technol. B, Vol. 16, pp. 871-874, 1998.

[7] M. Nakamoto and J. Moo, "Suitability of low-work-function titanium nitride coated transfer mold field-emitter arrays for harsh environment applications", J. Vac. Sci. Technol. B, vol. 29, pp. 02B112-1-5 (2011).

[8] H. Muarta, K. Murakami, M. Nagao, "Electron emission properties of titanium nitride coated volcano-structured silicon emitters", J. Vac. Sci. Technol. B, Vol. 42, pp. 013203-1-4, 2024.

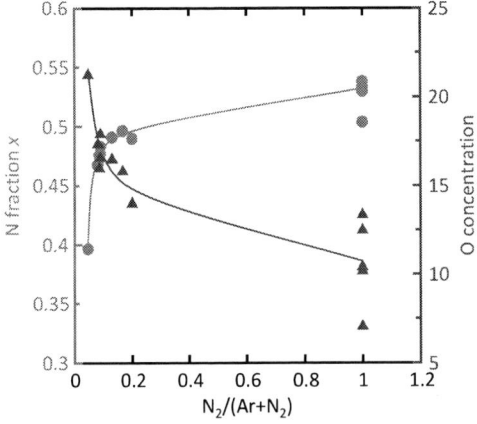

Fig. 1 N fraction x of $Ti_{1-x}N_x$ and O concentration in TiN as a function of N_2 gas ratio ($N_2/(Ar+N_2)$).

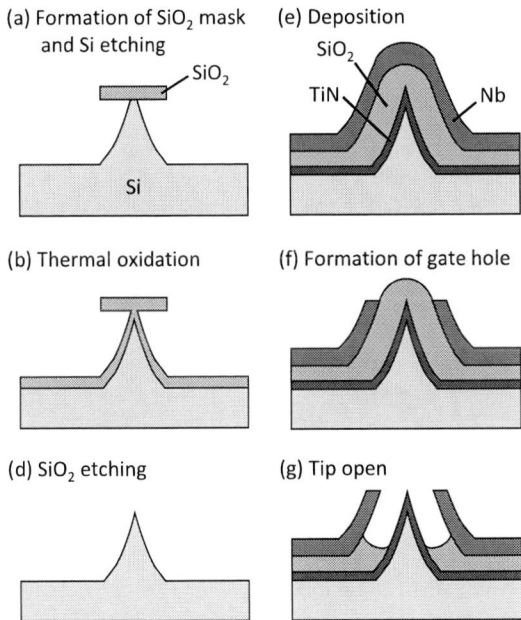

Fig. 2 Fabrication process of TiN-coated Si-FEA.

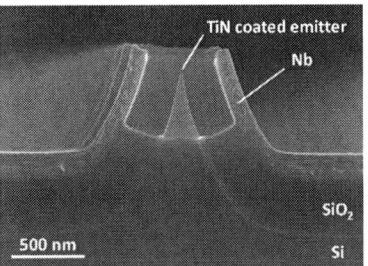

Fig. 3 SEM image of the TiN-coated Si-FEA.

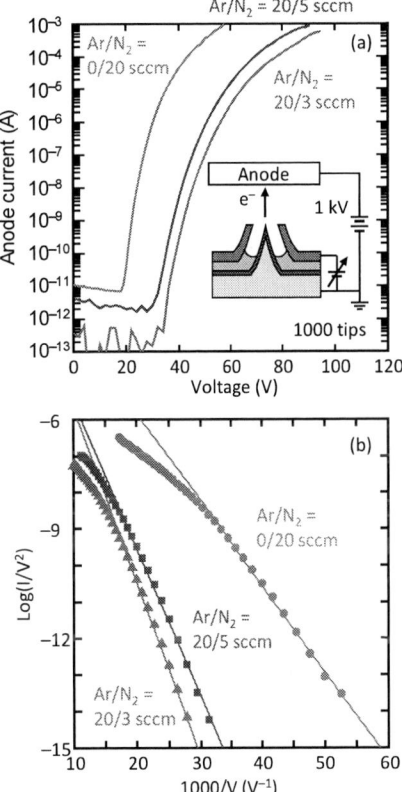

Fig. 4 Emission characteristics of TiN-coated Si-FEA. (a) I-V curves. (b) FN plots calculated from (a).

979-8-3503-7977-8/24 $31.00 © 2024 IEEE

Origin of the Slope-Intercept Linear Relationship in Field Emission

Anthony Ayari, Pascal Vincent, Sorin Perisanu, Philippe Poncharal, Stephen T. Purcell

Institut Lumière Matière, UMR5306
Universite Claude Bernard Lyon 1, CNRS
F-69100, Villeurbanne, France
anthony.ayari@univ-lyon1.fr

Abstract—Field emitters are often analyzed by linear fitting of a Fowler-Nordheim plot. The fitted slopes and intercepts can show a strong correlation, but its origin remains unclear. We propose a simple model showing that this correlation is due to fluctuations in the slope and the fact that the data in the abscissa of the Fowler-Nordheim plot are far from the origin. We show that this explanation correctly predicts other results in the field emission literature.

Index Terms—field emission, SK plot, linear regression

I. Introduction

A simple and widely used method of analysing the properties of a field emitter is to use Fowler-Nordheim (FN) theory although this approach is not very reliable theoretically [1]–[4] and experimentally [5]. In a standard FN plot field emission I-V data are fitted to a straight line to extract the slope A_{FN} and intercept B_{FN} such that :

$$\log \frac{I}{V^2} = \frac{A_{FN}}{V} + B_{FN} \qquad (1)$$

where I is the emission current, V is the applied voltage difference between the anode and cathode and $\log$ is the natural logarithm.

Consecutive measurements of the current in a given voltage range, yield different I-V curves and thus different A_{FN} and B_{FN}. A plot of the slopes against the corresponding intercepts for different voltage sweeps often shows an approximate linear relationship [6]. This representation is commonly referred to as an "SK plot" and this relationship has been observed in a several types of field emitters. Its origin has never been convincingly explained.

II. Experiments

An FN plot is shown in Fig. 1. It was obtained from a standard <111> tungsten tip measured in an ultra-high vacuum chamber with a base pressure of 3×10^{-10} Torr. The tip was fabricated by electrochemical etching of a tungsten wire with a diameter of 125 μm. The tip was degassed, several times, for 30 seconds, with a resisting loop at a temperature of 1700 K. A quadrupole was placed in front of the tip. The current was measured with a coupled microchannel plate MCP/phosphor screen system connected to a homemade current amplifier. The MCP/phosphor allows visualization of the emission area and amplify the very low currents. The FN plot was reproduced 10 successive times on the same tip.

The corresponding SK plot is shown in Fig. 2 and can be fitted linearly. It has a very good coefficient of determination $R^2 = 0.9855$.

III. Analysis

We propose that this correlation has a simple geometrical interpretation illustrated graphically in Fig. 3. In an field emission experiment, the linear fit of the FN plot is tangent to the actual slighlty non-linear data at a point $(1/V_m, \log \frac{I_m}{V_m^2})$ approximatly at the middle of the data range in the F-N plot. Usually, the intercept is far from the tangent point, thus a cantilever effect amplifies the contribution from slope fluctuations. This contribution from slope fluctuations dominates the observed correlation in Fig. 2.

Furthermore, from linear regression theory, the correlation of the slope and intercept estimates is given by the non-diagonal terms of the variance-covariance matrix :

$$cov(A_{FN}(t), B_{FN}(t)) = -var(A_{FN}(t)) < \frac{1}{V} > \qquad (2)$$

where $< \frac{1}{V} >$ is the average over the different values of the applied voltages. As the different applied voltages are equally spaced this term is almost equal to $1/V_m$ introduced above. The correlation increases with distance from the origin. When the data are shifted by $-1/V_m$ in order to be centered around zero, no correlation can be observed in our data. It can be shown [7] that the linear relationship of the SK plot is given by :

$$A_{FN}(t) = V_m \log \frac{I_m}{V_m^2} - V_m B_{FN}(t) \qquad (3)$$

where $A_{FN}(t)$ and $B_{FN}(t)$ are the FN slope and intercept measured after each I-V curve. From (3), it can be predicted that slope in an SK plot should be equal to $-V_m$ and the intercept to $V_m \log \frac{I_m}{V_m^2}$.

For our data on a single tungsten tip, the predicted intercept is -64,650 V corresponding to a discrepancy of 6 % compared to the intercept fitted above of -68,578 V (± 2 % uncertainty). The predicted slope is -1,459 V which corresponds to a deviation of 14 % from the slope fitted above of -1,693 V (± 4 % uncertainty). This simple model can therefore reasonably predict more than 85 % of the slope and intercept values.

A meta-analysis of the field emission literature on SK plots indicates that this explanation correctly predicts the results of

979-8-3503-7977-8/24 $31.00 © 2024 IEEE

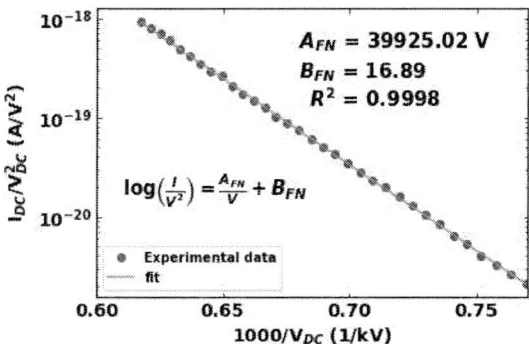

Fig. 1. FN plot for a W field emitter. The slope is equal to 39,925.02V, the intercept is equal to 16.89 and the coeficent of determination is equal to 0.9998.

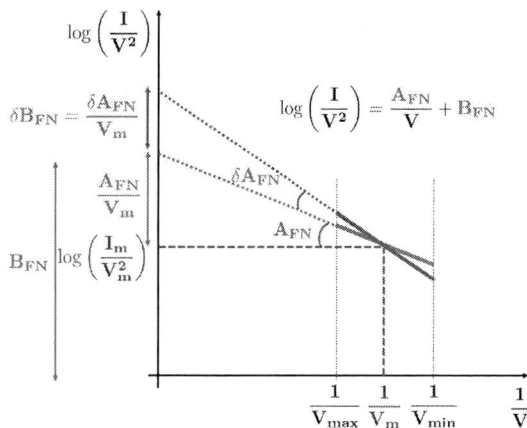

Fig. 3. Schematic illustration of the origin of the correlation between the FN slope and FN intercept.

other groups [7]. The different slopes and inteceps extracted over two orders of magnitude from the SK plots in the literature are close to the predicted theoretical values.

[5] A. Ayari, P. Vincent, S. Perisanu, P. Poncharal, and S. T. Purcell, "All field emission experiments are noisy,. . . are any meaningful?," Journ. Vac. Sci. Tech. B, vol. 41, pp. 024001, 2023.

[6] J. Ishikawa, et al. , "Influence of cathode material on emission characteristics of field emitters for microelectronics devices," Journ. Vac. Sci. Tech. B, vol. 11, pp. 403–406, April 1993.

[7] A. Ayari, P. Vincent, S. Perisanu, P. Poncharal, and S. T. Purcell, "Is the linear relationship between the slope and intercept observed in field emission S-K plots an artifact?," unpublished.

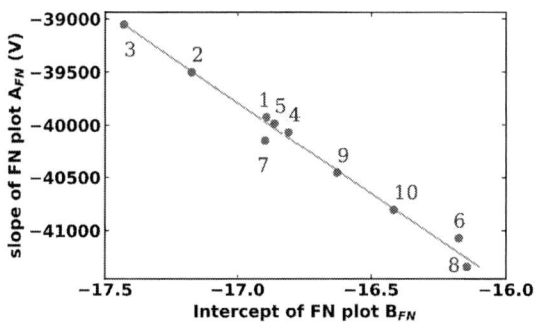

Fig. 2. Evolution of the FN slope as a function of the FN intercept in a standard S-K plot representation for a W field emitter. The number close to each data point corresponds to the chronological order of the measurements. The solid line is a fit of the data point.

IV. CONCLUSION

In conclusion, we have proposed an explantion of the origin of an SK plot. We have showed that the linear relationship between the FN slope and intercept is probably an artifact due to some purely statistical correlations. Our experiments on a W field emitter and our analysis of the data in the litterature are in agreement with this explanation.

REFERENCES

[1] L. Murphy and R. Good Jr, "Thermionic emission, field emission, and the transition region," Physical review, vol. 102, pp. 1464, 1956.

[2] R. G. Forbes, "Call for experimental test of a revised mathematical form for empirical field emission current-voltage characteristics," Applied Physics Letters, vol. 92, pp. 193105, 2008.

[3] A. Kyritsakis and J. Xanthakis, "Derivation of a generalized fowler–nordheim equation for nanoscopic field-emitters," Proceedings of the Royal Society A: Mathematical, Physical and Engineering Sciences, vol. 471, pp. 20140811, 2015.

[4] A. Ayari, P. Vincent, S. Perisanu, P. Poncharal, and S. T. Purcell, "All field emission models are wrong, ... but are any of them useful ?," Journ. Vac. Sci. Tech. B, vol. 40, pp. 024001, 2022.

Outgassing and Leak Rates of Bonding Technologies for Quantum Systems

Verena Velthaus and Jakob Buchheim*

Institute of Quantum Technologies, German Aerospace Center, Ulm, Germany
*Corresponding author: jakob.buchheim@dlr.de

Abstract—**Miniaturization is the key parameter to bring quantum systems to commercialization. For sealing and characterization of miniaturized ultra-high vacuum cells for quantum systems, a new ultra-high vacuum system was developed. With the new setup we investigated different chip bonding methods and measured their outgassing and leak rates.**

Keywords—*Anodic Bonding, Outgassing, Leak rate, UHV, Quantum systems*

I. INTRODUCTION

To bring quantum systems into broad application, miniaturization is one of the most important necessities [1]. Usually quantum phenomena can only be observed and exploited in atoms well insulated from the surrounding environment like in ultra-pure crystals or ultra-high vacuum. Widely used systems to insulate atoms are magneto-optical traps (MOTs) which need ultra-high vacuum (UHV) for their operation. Therefore, the development of miniaturized vacuum systems is required. Chip bonding techniques for the enclosure of MEMS are well established [2; 3], but MOTs require sealings with ultra-low leak rates [4] and use rubidium atoms which have high vapor pressure even at low temperatures [5]. Therefore, bonding techniques have to be optimized for low process temperatures.

II. EXPERIMENTAL SETUP

For testing bonding technologies for their outgassing and leak rates, a new UHV setup was build up (Fig. 1). To perform the bonding process in vacuum, it includes a heated sample stage and the possibility to apply load and high voltage to the bonding sample. A quadrupole mass spectrometer and a Bayard-Alpert ionization vacuum gauge are used to determine the outgassing rate during the bonding process.

For measuring the leak rate of the bonds, a special chip design was developed consisting of a 22x22 mm substrate chip with a 5x5 mm hole in the center. The hole is sealed by an 8x8 mm lid using the bonding technique under test. The bonded chip is then mounted to a special flange for connecting one site of the chip to the vacuum chamber and the other site to a helium reservoir. Helium leaking through the bond is detected by measuring the helium partial pressure in the chamber with the mass spectrometer. An attached load lock system allows to mount a bonded chip without venting the main chamber.

III. RESULTS

Fig. 2 shows an exemplary pressure evolution during anodic bonding of a borosilicate chip (8x10 mm) to a silicon chip (20x20 mm). The bonding was conducted at approx. 350 °C and 1 kV applied voltage. The total pressure was measured with the Bayard-Alpert ionization vacuum gauge, while the partial pressure data were determined with the quadrupole mass spectrometer. The highest pressure increases due to the bonding process showed the masses 2 (H_2), 32 (O_2) and 44 (CO_2). The pressure increase was $1.0 \cdot 10^{-9}$ mbar, which corresponds to an outgassing rate of approx. $7 \cdot 10^{-8}$ mbar·l/s. Outgassing due to sample heating is not considered in this analysis.

For different bonded chips and glass cells, the helium leak rates were measured using the setup and method described in Section II. Again, bonding of borosilicate glass to silicon was tested. The sealings were made by anodic bonding at 400 °C and 1 kV in atmosphere. For all bonded chips, the leak rate was below the minimal measurable leak rate of the setup. The minimal measurable leak rate is $<7.5 \cdot 10^{-13}$ mbar·l/s.

IV. CONCLUSION AND OUTLOOK

First tests on anodic bonding demonstrate that our new UHV bonding setup shows good performance. Future work will focus on the optimization of the bonding process for lower working temperatures to make it applicable also for thermal sensitive components. Additionally, a lower working temperature reduces the outgassing of the surrounding components during the process and ensure to maintain UHV conditions.

ACKNOWLEDGMENT

The authors thank S. Jenisch and J. Appiah for their valuable support during the installation of the setup and the sample preparation. Furthermore, the authors want to express their thanks to the team of the cleanroom at Ulm University.

REFERENCES

[1] K. Bongs, S. Bennett, A. Lohmann, "Quantum sensors will start a revolution - if we deploy them right", Nature, vol. 617, pp. 672–675, 2023.

[2] S. Ke, D. Li, S. Chen, "A review: wafer bonding of Si-based semiconductors", J. Phys. D: Appl. Phys., vol. 53, pp. 323001, 2020.

[3] J. Xu, Y. Du, Y. Tian, C. Wang, "Progress in wafer bonding technology towards MEMS, high-power electronics, optoelectronics, and optofluidics", Int. J. Optomechatronics, vol. 14, pp. 94–118, 2020.

[4] J. A. Rushton, M. Aldous, M. D. Himsworth, "Contributed Review: The feasibility of a fully miniaturized magneto-optical trap for portable ultracold quantum technology", Rev. Sci. Instrum., vol. 85, pp. 121501, 2014.

[5] A. M. van der Spek, J. J. L. Mulders, L. W. G. Steenhuysen, "Vapor pressure of rubidium between 250 and 298 K determined by combined fluorescence and absorption measurements", J. Opt. Soc. Am. B, vol. 5, pp. 1478–1483, 1988.

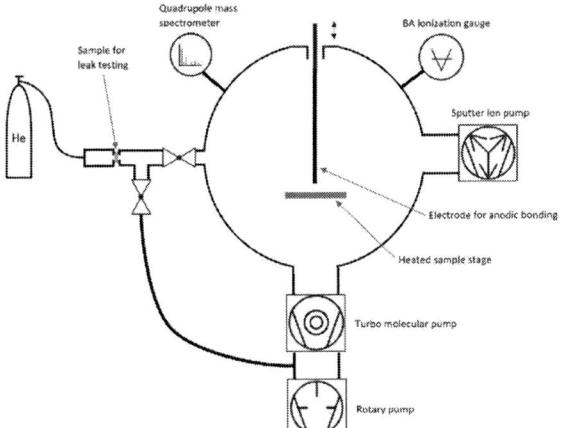

Fig. 1. Schematic view of the experimental setup.

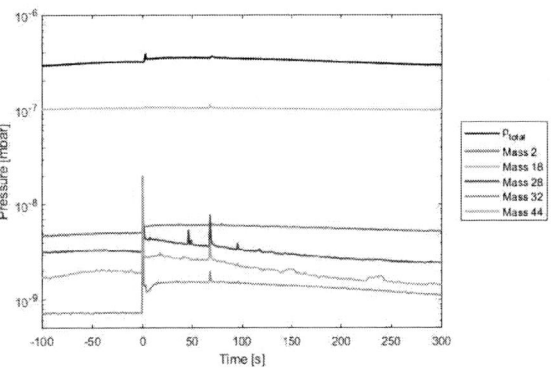

Fig. 2. Evolution of the total pressure and selected partial pressures during anodic bonding at 350 °C. The high voltage (1 kV) for the bonding was switched on at 0 seconds.

Photoelectron Emission from Molybdenum Disulfide/Hexagonal Boron Nitride/Graphene Heterostructure

Guichen Song, Shaozhi Deng and Jun Chen*

State Key Laboratory of Optoelectronic Materials and Technologies, Guangdong Province Key Laboratory of Display Material and Technology, School of Electronics and Information Technology, Sun Yat-sen University, Guangdong Province, People's Republic of China
*Corresponding author: stscjun@mail.sysu.edu.cn

Abstract—**The planar metal-insulator-semiconductor (MIS) emitter using van der Waals heterostructure of two-dimensional (2D) materials have potential applications in vacuum microelectronic and optoelectronic devices. In this work, the MoS₂/h-BN/FLG heterostructure was prepared by all-dry transfer technique. The photoelectron emission characteristics of the device were studied under white laser irradiation. Maximum responsivity of 7.1×10^{-5} A/W for anode current was recorded under the gate voltage of 7 V. The time-resolved photo-response of device exhibits long rise and fall time. The mechanisms for observed phenomena were discussed.**

Keywords—Photoelectron Emission, Metal-insulator-semiconductor heterostructure, 2D Materials.

I. INTRODUCTION

The photoelectron cathode finds wide applications in optoelectronic detector and energy conversion devices. Conventional photoelectron cathodes usually use complex surface processing and is susceptible to environmental influence. Typical photoelectron cathodes also suffer from limited lifetime and low quantum efficiency (10^{-3}–10^{-2}%) [1]. Recent studies have shown that the planar emitter using metal-insulator-semiconductor (MIS) structure composed of 2D materials is more inert to environmental influence and shows unique photoelectron performance. Furthermore, by applying a bias voltage, the photo-responsivity of MIS structure can be effectively tuned to be sensitive to incident light of different wavelengths [2].

MoS₂, a typical two-dimensional semiconductor material, demonstrates outstanding performance such as direct band gap (1.8 eV) and high carrier mobility (200 cm² V⁻¹ s⁻¹). It is widely used in the fields of transistors, photodetectors, and electrocatalysis devices [3], making it a promising candidate as the semiconductor layer in the MIS structure for achieving high-performance photoemission.

In this work, the MoS₂/Hexagonal Boron Nitride (h-BN)/Few-layer Graphene (FLG) heterostructure was prepared and the photoelectron emission characteristics of the heterostructure was examined and the underlying mechanisms were discussed.

II. EXPERIMENTAL

The MIS heterostructure was prepared on a Si substrate with SiO₂ layer by all-dry transfer technique with polydimethylsiloxane (PDMS) assisting. The FLG, h-BN and MoS₂ were delaminated by bule tape from their bulk crystal, respectively, onto the elastomeric PDMS layer. A homemade micromanipulator is used to manipulate the stamp to the target location and release the flakes on the substrate. The Cr/Au electrodes are deposited using thermal evaporation and patterned by lithography and lift-off process.

The photoelectron emission properties were measured using a tungsten probe under base pressure of 1.6×10^{-8} Pa. The photoelectron emission was excited by white-light laser (Fianium WhiteLase SC400) and the wavelength range from 210 to 2400 nm. The vacuum electrons were collected by the tungsten probe anode with a bias voltage of 20 V and anode-to-cathode distance is about 5 μm. The emission currents were collected by using semiconductor parameter analyzer (Keithley 4200) with and without laser irradiation.

III. RESULT AND DISCUSSION

The optical image of the prepared MoS₂/h-BN/FLG heterostructure is shown in Fig. 1. The FLG, h-BN and MoS₂ were identified by orange, bule and purple lines, respectively. The FLG is applied with gate voltage (V_g), and gate current (I_g) and anode current (I_a) are recorded at same time. All the flakes lie smoothly on the substrate. However, bubbles and wrinkles could be observed on the interface.

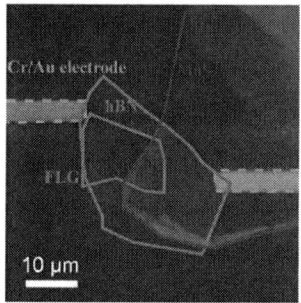

Fig. 1. Optical image of prepared MoS₂/h-BN/FLG heterostructure

Fig. 2 shows the I-V curves of the heterostructure obtained in dark and under light illumination. The corresponding F-N plots were plotted in insert of Fig. 2. As shown in Fig. 2, the I_g and I_a increase with the increasing laser power density. The gate current shows a 347% and 145% increase under laser power density of 6.37 and 2.76 W/cm² compared to the dark current, whereas the anode current experiences a 240% and 47% rise, respectively.

Fig. 3 shows the band diagram of device in working state. The electrons of MoS₂ in the valence band are excited by laser to the conduction band of MoS₂. And the excited electrons tunnel into h-BN and accelerated in h-BN under electric field induced by gate bias, then inject into the FLG. The electrons, possessing an energy exceeding the work function of FLG, are emitted into vacuum, and collected by anode. While those electrons with energy less than the work function fall into FLG and become gate current. The responsivity of I_g and I_a is calculated, and the responsivity of I_g and I_a is found to increase with increasing gate voltage. The maximum values are 9.4×10^{-4} for I_g and 7.1×10^{-5} A/W for I_a under the gate voltage of 7 V, respectively, which is limited by the breakdown field of the h-BN. The low responsivity of I_a might be due to low absorption of the MoS_2 and the scattering in h-BN. The low responsivity of I_g is due to large gate leakage current induced by hole injection from FLG to MoS_2 [2].

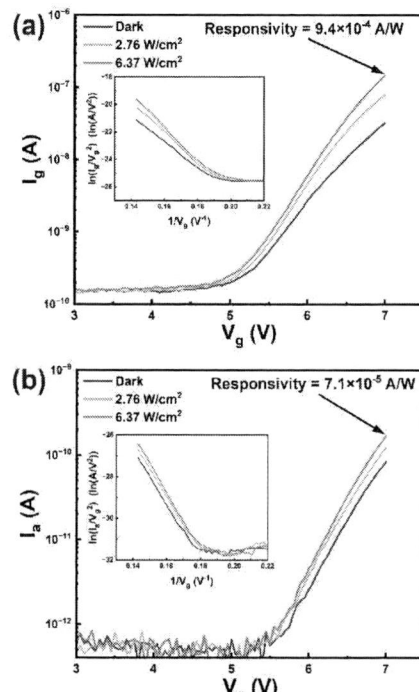

Fig. 2. The result of (a) I_g-V_g and (b) I_a-V_g for photoelectric emission characteristics under different laser power density. Insert: the corresponding F-N plots of I_g and I_a for V_g.

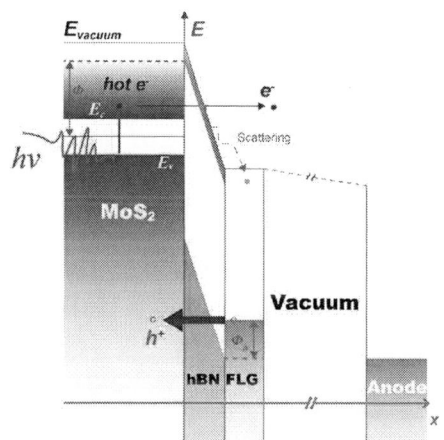

Fig. 3. Band diagram and photoelectric emission process of MoS_2/h-BN/FLG heterostructure.

To eliminate the semiconductor thermal effect caused by strong light irradiation (2.76 W/cm^2), the photo-response of the device was measured under pulsed light irradiation. The laser is switched on manually using a shutter and the duration of the pulsed light is about 5 seconds. The laser power density is 2.76 W/cm^2 and the applied bias of 6.8 V. The raise time and fall time of the single light cycle is shown in Fig. 4. When the sample is exposed to laser irradiation, there is an increase in both I_a and I_g. The device has a rise time of 0.7 s and a slower decay time of 4.1 s. The slow decay time could be due to the trap states in MoS_2 and h-BN layer. In addition, the decay time edge has two stages. One is characterized by a rapid decrease, which can be attributed to the defect recombination in MoS_2 [4]. The other exhibits a slower decline and can be attributed to the deep defects in the bandgap of hBN [5].

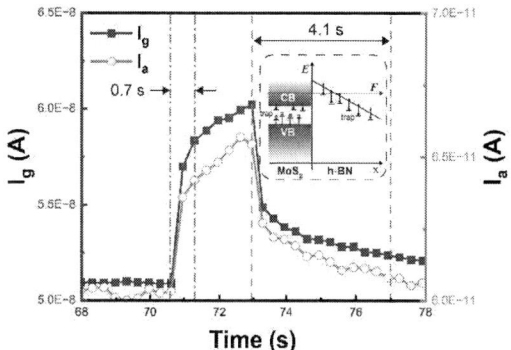

Fig. 4. Time-resolved photo-response of the device under laser irritation.

IV. CONCLUSIONS

In this work, we have constructed a photocathode composing of the MoS_2/h-BN/FLG heterostructure. Photoelectron emission was observed from the device under white laser irradiation. The responsivity of 7.1×10^{-5} A/W and rise time of 0.4 s for anode current were obtained. The results can be explained by the electron injection process in the heterostructure. The device performance can be improved by optimizing the preparation process and tuning the energy barrier.[6]

ACKNOWLEDGMENTS

The authors gratefully acknowledge the financial support from the National Key Research and Development Program of China (Grant No. 2022YFA1204200), Key Research and Development Program of Guangdong Province (Grant No. 2023B0101200013), National Natural Science Foundation of China (Grant Nos. 82272131 & 62001527), the Science and Technology Department of Guangdong Province (Grant No. 2023B1212060025), Fundamental Research Funds for the Central Universities.

REFERENCES

[1] F. Rezaeifar, R. Ahsan, Q. Lin, H. U. Chae, and R. Kapadia, "Hot-electron emission processes in waveguide-integrated graphene," Nature Photonics, vol. 13, no. 12, pp. 843-848, 2019.

[2] Y. Chen, S. Deng, N. Xu, and J. Chen, "Energy-tunable photon-enhanced thermal tunneling electrons for intrinsic adaptive full spectrum solar energy conversion," Applied Physics Letters, vol. 116, no. 6, p. 063902, 2020.

[3] J. Jiang et al., "Flexo-photovoltaic effect in MoS_2," Nature Nanotechnology, vol. 16, no. 8, pp. 894-901, 2021.

[4] Q. A. Vu et al., "Tuning Carrier Tunneling in van der Waals Heterostructures for Ultrahigh Detectivity," Nano Letters, vol. 17, no. 1, pp. 453-459, 2017.

[5] D. Wang and R. Sundararaman, "Layer dependence of defect charge transition levels in two-dimensional materials," Physical Review B, vol. 101, no. 5, p. 054103, 2020.

[6] Y. Chen, D.-K. Ki, Z. Li, and J. Chen, "Concept for a fractional energy barrier tunneling junction," Applied Physics Letters, vol. 123, no. 24, p. 243503, 2023.

Photoresponse of Field Emission Current from FAPbI₃ Perovskite Film

Bin Wen, Zhuoran Ou, Guofu Zhang, Manni Chen, Juncong She, Shaozhi Deng, and Jun Chen*

State Key Laboratory of Optoelectronic Materials and Technologies, Guangdong Province Key Laboratory of Display Material and Technology, School of Electronics and Information Technology, Sun Yat-sen University, Guangzhou 510275, People's Republic of China

* E-mail: stscjun@mail.sysu.edu.cn

Abstract—Perovskite materials have been extensively studied for applications in light emitting diode, detector, photovoltaic devices due to their excellent optoelectronic properties. Study of photoresponse of field emission from perovskite materials may obtain interesting physics and novel optoelectronic devices based on field emission can be developed, which has not been explored before. In this work, we prepared FAPbI₃ films on ITO glass by anti-solvent one-step spin coating and the photoresponse of field emissions properties from the FAPbI₃ films were measured. Distinct photoresponse of field emission current and persistent photoconductivity were observed. The effect of defects in the polycrystalline perovskite film on the photoresponse of field emission was discussed.

Keywords—FAPbI₃, field emission, photoresponse, persistent photoconductivity

I. INTRODUCTION

Cold cathode using field emission phenomenon exhibits superiority in terms of lifetime, response speed, and device volume, having wide applications in field emission displays, X-ray sources and so on. Furthermore, novel optoelectronic detector can be realized using photoresponse of field emission current. With excellent and tunable photoelectric characteristic properties, perovskite materials show excellent performance in solid-state photovoltaics and photodetectors devices. It is of great significance to explore the photoresponse of field emissions properties of the perovskite materials, which have interesting physics and potential applications due to their extraordinary high optical absorption coefficient and large carrier mobility-lifetime product ($\mu \cdot \tau$). However, no reports on the photoresponse of field emissions properties of the perovskite have been reported [1-2].

FAPbI₃ is one of the important metal halide perovskite materials, which was widely used in the solar cells due to its direct bandgap of about 1.48 eV, great UV-vis absorption and good carrier transport properties [3]. In previous studies, perovskite films were prepared by blade-coating and vapor co-deposition, but these methods exist issues such as uneven nucleation and poor crystallinity [4-5]. In this work, we prepared the FAPbI₃ film on ITO glass by anti-solvent one-step spin coating method. The structure of FAPbI₃ was characterized and their field emission properties and photoresponse of the field emission current was studied. The underlying mechanism for the observed phenomena was also discussed.

II. EXPERIMENT

The FAPbI₃ films were prepared on ITO glass by solution spin coating method, with a sample area of 1.5 cm × 1.5 cm. The 1.4 M FAPbI₃ perovskite precursor solutions were prepared by mixing FAI, PbI₂ and 35 mol% MACl in DMF/DMSO binary solvents at a volume of 8:1. Before spin-coating, the ITO glass was treated with UV ozone for 15mins. The precursor solution was spin-coated onto the substrate at speeds of 1000 rpm for 10s and 5000 rpm for 15s, then the anti-solvent dropwise was added at the countdown of 5s. Afterwards, the sample was quickly transferred to a hot plate and heated at 150 °C for 15 minutes. The SEM images of FAPbI₃ films were obtained by a scanning electron microscope (SUPRA TM60). The PL characteristic was measured by Raman spectroscopy (InVia Reflex) using 325 nm excitation light. The XRD pattern was obtained by an X-ray diffraction system (Empyrean3).

The field emission measurement was carried out using a diode structure with the FAPbI₃ films as cathode and the ITO glass as the anode in the vacuum chamber at 2×10^{-5} Pa. Fig. 1(a) and (b) shows the schematic diagram of field emission measurement and device structure. The sample underwent an aging process before photo response measurement. The photoresponse of field emission were measured by illuminating the sample using a white light source.

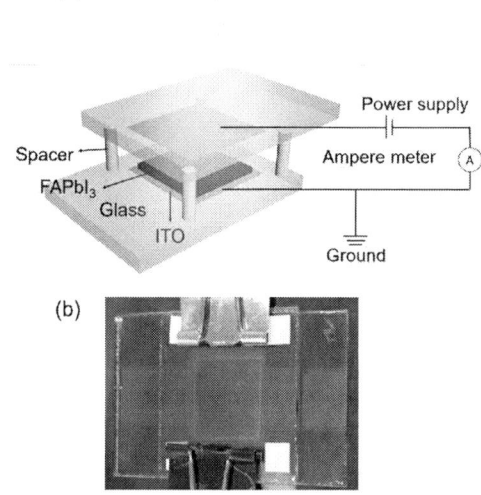

Fig. 1(a) Schematic diagrams of field emission measurement; (b) The diode device of FAPbI₃ film.

III. RESULT AND DISCUSSION

SEM images of samples show that the grain size of FAPbI₃ is relatively uniform, with an average size of 1~2 μm and no obvious voids exist. The XRD of the FAPbI₃ film exhibits sharp peaks at 14° and 28.1° which are related to (001) and (002) diffraction planes corresponding to the cubic structure of α-FAPbI₃ [6]. No peak of the δ- FAPbI₃ phase was observed in the XRD pattern, indicating that the prepared FAPbI₃ films were almost pure α-FAPbI₃ phase structure.

979-8-3503-7977-8/24 $31.00 © 2024 IEEE

The J-E curves with and without light illumination were presented in Fig. 2(a). Upon illumination, the turn-on electric field ($J = 10 \mu A/cm^2$) changed from 23.4 V/μm to 22.3 V/μm. The enhanced field emission current is related to the photo-generated carriers, which increased the supply of electrons in the conduction band for emission. The stability of field emission current was continuously measured for 60 mins and no decay in the emission current was observed. It can be seen in Fig. 2(b) that after being illuminated, the field emission current gradually increased from 9.7 μA to a maximum current of 12.5 μA (an increase of 28%) in about 50 s. We also observed that the emission current decreased from 12.5 μA to 10.5 μA after the illumination was turned off. However, the current still exceeded the initial dark emission current. We attributed this to the persistent photoconductivity (PPC) phenomenon. The observed persisting photocurrent may be related to the defects in the polycrystalline perovskite film. The defects trap the photo-generated carriers under illumination and the carriers are continuously released in the subsequent dark state measurement [7].

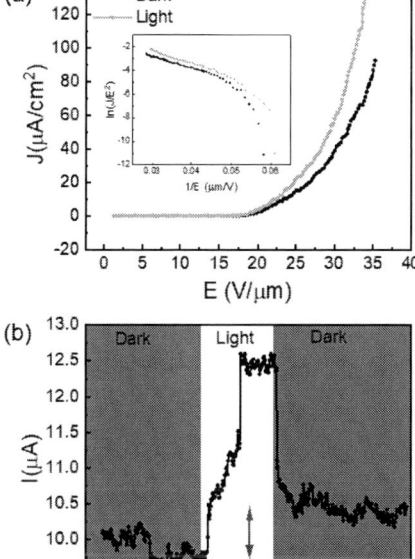

Fig. 2 (a) J-E characteristics of FAPbI$_3$ film (Insert shows the corresponding F-N plots); (b)Response of field emission current to the light illumination

To characterize the defects in the sample, we measured its steady-state PL spectrum [8]. The fluorescence peak position of perovskite located at 800 nm. and the relationship between PL intensity under illumination was studied. It is found that the PL intensity gradually increased under continuous illumination, which indicates that the defects were filled by photogenerated carriers. The PL results confirmed that the defects in the FAPbI$_3$ films is the reason for the observed PPC phenomenon.

IV. CONCLUSION

Photoresponse of field emission current was studied from FAPbI$_3$ film. The photoresponse in the field emission current were observed. In addition, we found that existence of persistent photoconductivity phenomenon which is attributed to the defects in the FAPbI$_3$ film.

ACKNOWLEDGMENTS

The authors gratefully acknowledge the financial support from the National Key Research and Development Program of China (Grant No. 2022YFA1204200), Key Research and Development Program of Guangdong Province (Grant No. 2023B0101200013), National Natural Science Foundation of China (Grant Nos. 82272131 & 62301620), the Science and Technology Department of Guangdong Province (Grant No. 2023B1212060025), Fundamental Research Funds for the Central Universities.

REFERENCES

[1] N. S. Xu, and S. E. Huq, "Novel cold cathode materials and applications," *Materials Science and Engineering: R: Reports,* vol. 48, no. 2-5, pp. 47-189, 2005.

[2] M. Ahmadi, T. Wu, and B. Hu, "A Review on Organic–Inorganic Halide Perovskite Photodetectors: Device Engineering and Fundamental Physics," *Advanced Materials,* vol. 29, no. 41, pp.1605242, 2017.

[3] F. F. Targhi, Y. S. Jalili, and F. Kanjouri, "MAPbI$_3$ and FAPbI$_3$ perovskites as solar cells: Case study on structural, electrical and optical properties," *Results in Physics,* vol. 10, pp. 616-627, 2018.

[4] T. Moser, K. Artuk, Y. Jiang, T. Feurer, E. Gilshtein, A. N. Tiwari, and F. Fu, "Revealing the perovskite formation kinetics during chemical vapour deposition," *Journal of Materials Chemistry A,* vol. 8, no. 42, pp. 21973-21982, 2020.

[5] W.-J. Hsu, E. C. Pettit, R. Swartwout, T. Z. Kadosh, S. Srinivasan, E. L. Wassweiler, G. Haugstad, V. Bulović, and R. J. Holmes, "Efficient Metal-Halide Perovskite Photovoltaic Cells Deposited via Vapor Transport Deposition," *Solar RRL,* vol. 8, no. 1, p. 2300758, 2024.

[6] G. Murugadoss, P. Arunachalam, S. K. Panda, M. Rajesh Kumar, J. R. Rajabathar, H. Al-Lohedan, and M. D. Wasmiah, "Crystal stabilization of α-FAPbI$_3$ perovskite by rapid annealing method in industrial scale," *Journal of Materials Research and Technology,* vol. 12, pp. 1924-1930, 2021.

[7] A. Sumanth, K. Lakshmi Ganapathi, M. S. Ramachandra Rao, and T. Dixit, "A review on realizing the modern optoelectronic applications through persistent photoconductivity," *Journal of Physics D: Applied Physics,* vol. 55, no. 39, p. 393001, 2022.

[8] Y. Yuan, G. Yan, C. Dreessen, T. Rudolph, M. Hulsbeck, B. Klingebiel, J. Ye, U. Rau, and T. Kirchartz, "Shallow defects and variable photoluminescence decay times up to 280 micros in triple-cation perovskites," *Nat Mater,* vol. 23, no. 3, pp. 391-397, 2024.

Planar On-chip Auto-ponderomotive Devices for Electron Beam Control

Franz Schmidt-Kaler, Michael Seidling, Robert Zimmermann, Nils Bode, Fabian Bammes, Lars Radtke and Peter Hommelhoff
Department Physik, Friedrich-Alexander-Universität Erlangen-Nürnberg (FAU),
91058 Erlangen, Germany

Corresponding author: franz.schmidt-kaler@fau.de

Abstract—**New means to manipulate electron beams are highly sought after as they might enable new electron optical functionalities and devices. We recently demonstrated guiding of electron beams with energies of up to 4 keV with the help of a transversely confining potential akin to that of a Paul trap. It resulted from a periodic arrangement of electrodes with alternating polarity, leading to an oscillatory field of the moving electrons. We call this scheme auto-ponderomotive electron guiding. With a more refined electrode pattern, we could also demonstrate splitting of the guided electron beam. The focus of the presentation will be on a resonator for guided electrons, consisting of a linear guide and two switchable electron mirrors. With these elements demonstrated, we aim for quantum electron microscope operation.**

I. INTRODUCTION

Cryo-electron microscopy is the method of choice employed to resolve organic and beam-sensitive samples. Despite its achievements concerning high resolution down to 0.4 nm, its employment necessitates the imaging of thousands of the same sample specially in a vitreously frozen state [1]. On top, small samples cannot yet be efficiently imaged [2].

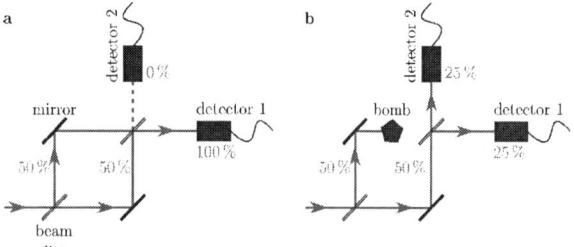

Fig. 1: *Interaction-free measurement with Mach-Zehnder setup [11]. Interference prohibits detector 2 from clicking with both arms open.*

Thus, true single-particle detection with atomic resolution has not yet been made feasible with electrons. A new approach employing "wave mechanical properties of electrons" should enable imaging of macromolecules with little to no damage" [3, 4]. In an interferometric setup see Fig 1, with no absorbing sample only detector 1 clicks as for the other detector. With introduction of a beam-blocking sample, the interference at the second beam splitter ceases to exist and with 25% probability, detector 2 clicks. The efficiency of the interaction-free measurement can be arbitrarily increased by coherent probability amplitude build-up in an electron resonator. Here, like Rabi oscillations, the probability amplitude oscillates between the two resonator cavities separated by the electron beam splitter. The beamsplitter separating the two cavity arms is now tuned in reflectivity depending on the desired number of roundtrips. Even with low numbers of repeated interrogations, an advantage in efficiency can be achieved compared to a single IFM realized with electrons at a IFM probability of 14% [5]. A particular

challenge is the demonstration of the feasibility of individual electron optical elements such as guides, splitters and resonators. Only with these three, the repeated electron sample interrogation with the same electron can achieve an efficient interaction free measurement.

II. AUTO-PONDEROMOTIVE ELECTRON BEAM MANIPULATION

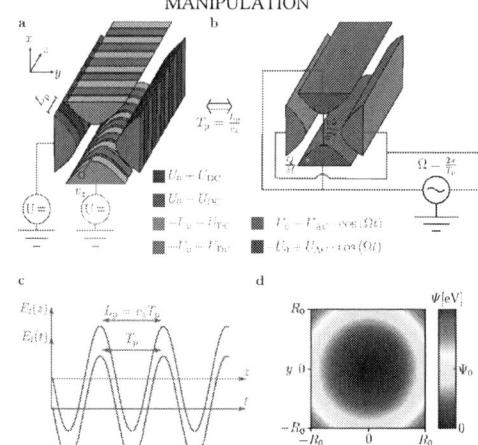

Charged particles can be confined within a conventional Paul trap [6], but higher particle velocities seen with lighter particles such as electrons in the keV range pose a significant challenge as for these higher microwave frequencies and

Fig. 2: *From conventional Paul traps to auto-ponderomotive potentials [11]. Auto-ponderomotive equivalent of the conventional Paul trap. Charged particles experience similar motion and pseudopotential*

powers are required. The electrostatic auto-ponderomotive approach expands the useable trap frequencies by micro-segmenting the electrodes. Here, moving electrons experience

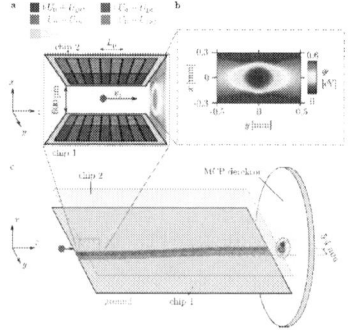

Fig. 3: *Auto-ponderomotive beam guiding around S-curved electrostatic electrode layout [11]. Guiding and transmission separated spatially on micro-channel plate detector.*

an alternating field as the hyperbolic electrodes are now separated into small polarities each on constant but alternating potential, see Fig 2. The resulting pseudo-potential is equivalent to that of a conventional Paul trap and thus similar manipulation of charged particles can be performed. In a further step, the hyperbolic geometry is projected onto planar electrode layouts, we realize these with printed-circuit boards.

III. ELECTRON BEAM GUIDE

We aim for seamless integration of our electron optical elements into conventional scanning electron microscopes (SEM) operating at electron energies from 200eV up to 30keV. A demonstration of our electron beam guiding capabilities is shown in Fig. 3, where a double layered electrode layout leads to an offset of the fed-in electron on the micro-channel plate detector with applied electric guiding potential. Electron energies from 20eV up to 4keV have been guided [7, 8].

IV. ELECTRON BEAM SPLITTER

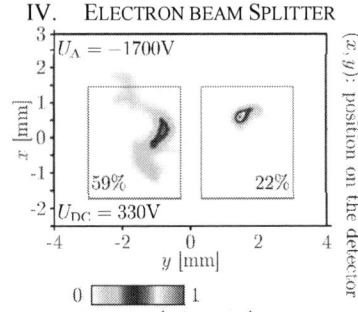

Fig. 4: Electrostatic beam splitting demonstrated up to primary energies of 1.7keV [11].

Electron beam splitting has been demonstrated in SEM and with our home-built tungsten nanotip, see Fig. 4. The double-layered electrode approach guaranteed beam splitting of up to 1.7keV electrons [7]. The two inputs in the auto-ponderomotive beam splitter slowly diverge realizing and output angle of ~ 3 degrees.

V. ELECTRON BEAM RESONATOR

Our electron beam resonator consists of two electron mirrors and a linear auto-ponderomotive guide in between for reduced electron losses [9]. The time synchronization for the erbium fiber laser, as well as the electron mirror pulsers [10] and the delay-line detector (DLD) is achieved by a Keysight 81150A arbitrary pulse form generator. Electron bunches with a length of about 1ns were laser-triggered off a tungsten nano-tip biased at -50V and steered into the resonator with home-built electron optics, see Fig. 5(a). Mirror operation for a requested number of round trips is shown in Fig. 5(b). After an arbitrary mirror 2 delay, the count rate is measured at the delay-line detector. This first electron optical resonator demonstrated up to 7 roundtrips with a steady round trip time of about 42ns [9].

VI. OUTLOOK

The demonstrators of electrostatic electron-optical elements for beam guiding, splitting and resonating have been

shown. One of the next steps is the miniaturization and tests for coherence after beam manipulation.

ACKNOWLEDGMENT

The authors acknowledge funding from the Gordon and

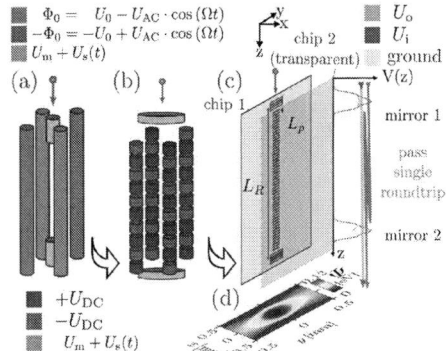

Fig. 5: Electrostatic resonator operated at 50eV for up to 7 round trips [11]. From left to right: Paul trap with end caps, segmented auto-ponderomotive equivalent and planar projection with indicated switching potentials.

Betty Moore Foundation via Grant 5723 (Quantum Electron Microscope Project) and 11473 (Imaging quantum coherence with shaped electrons) and ERC Grants 616823 (NearFieldAtto) and 884217 (AccelOnChip).

REFERENCES

[1] D. Kuster, C. Liu, Z. Fang, J. Ponder, and G. Marshall, High-Resolution Crystal Structures of Protein Helices Reconciled with Three-Centered Hydrogen Bonds and Multipole Electrostatics, PLoS One, 2015, 10: 4.

[2] M. Herzik, M. Wu, and G. Lander, High-resolution structure determination of sub-100 kDa complexes using conventional cryo-EM, Nature Communications, 2019, 10: 1032

[3] R. Glaeser, How good can cryo-EM become? Nature Methods, 2016, 13: 28

[4] Kruit et al. Designs for a quantum electron microscope. Ultramicroscopy, 2016, 164: 31-45

[5] Turner et al. Interaction-free measurement with electrons. Physics Review Letters, 2021, 127: 110401

[6] Wolfgang Paul, Electromagnetic traps for charged and neutral particles. Reviews of Modern Physics, 1990, 62, 531

[7] R. Zimmermann, M. Seidling and P. Hommelhoff, Nature Communications, 2021, 390: 12

[8] M. Seidling, R. Zimmermann and P.Hommelhoff, Applied Physics Letters, 2021, 118: 034101

[9] in press: Seidling and Schmidt-Kaler et al. Resonating electrostatically guided electrons. Phys. Rev. Lett., 2024

[10] Simonaitis JW, Slayton B, Yang-Keathley Y, Keathley PD, Berggren KK, Precise, sub nanosecond, and high-voltage switching enabled by gallium nitride electronics integrated into complex loads, Rev. Sci. Instrum., 2021, 92: 074704

[11] Dissertation Michael Seidling, Friedrich-Alexander-Universität Erlangen-Nürnberg, 2021

Planar-Gate Zinc Oxide Nanowire Cold Cathode for Line-Coded Flat Panel X-ray Source

Junhang Xie, Qi Liu, Guofu Zhang, Song Kang, Shaozhi Deng, Ningsheng Xu and Jun Chen*

State Key Laboratory of Optoelectronic Materials and Technologies, Guangdong Province Key Laboratory of Display Material and Technology, School of Electronics and Information Technology, Sun Yat-sen University, Guangzhou, China

*Corresponding author: stscjun@mail.sysu.edu.cn

Abstract—**Line-coded flat panel X-ray source can be applied in recently proposed coded array beam X-ray imaging scheme for static CT. In this study, planar-gate zinc oxide nanowire cold cathode has been developed for a line-coded flat panel X-ray source application. Field emission measurement was carried out and line addressability with gate modulation was achieved.**

Keywords—X-ray sources, planar-gate, ZnO nanowire

I. INTRODUCTION

X-ray imaging has a wide range of applications in medical diagnosis, security screening, industrial non-destruction testing and scientific instruments. [1] The cold cathode flat panel X-ray source (FPXS) can achieve large-area and addressed X-ray emission, having the advantages of low power consumption, compact structure, and low anode heat load. [2] By using FPXS, one can realize static computed tomography (CT) without mechanical scanning, avoiding the problems of slow imaging speed, large volume, high dose, and motion artifacts caused by mechanical rotation systems in traditional CT.[3]

The coded array beam X-ray imaging scheme is proposed recently which can realize static CT imaging using coded cold cathode FPXS. [3] The selective emission of X-ray beam can be realized by applying voltage to the gate of FPXS. By applying voltage to a gate column can produce line-coded beams, from which CT images can be reconstructed. Line-coded cold cathode FPXS is one of the devices which is needed for realizing such an imaging technique.

Field emitter arrays (FEAs) are the most important components of cold cathode FPXS. ZnO nanowires are ideal candidates for cold cathode material for large area FEAs, by virtues of easy preparation on large area and uniform emission. [4] Addressable ZnO nanowire FEAs and its applications in the cold cathode FPXS have been demonstrated. [5, 6]

In this study, planar-gate ZnO nanowire field emitters with line-addressing capability were fabricated for potential application in a line-coded cold cathode FPXS. The emission characteristics under gate voltage driving were studied. The results verified the line-addressing capability of the device.

II. EXPERIMENTAL

Fig. 1(a) shows schematic diagram of planar-gate ZnO nanowire field emitters. The cathode electrodes and gate electrodes are parallelly arranged, with a spacing of 2 μm between them. The patterned ZnO nanowires are prepared on the indium-tin-oxide (ITO) cathode electrodes. The size of each ZnO nanowire pattern is 15 × 15 μm, with a spacing of 70 μm between two patterns.

The glass substrate was cleaned and an ITO layer with a thickness of 540 nm was prepared on it using direct-current (DC) magnetron sputtering. The gate and cathode electrodes are formed by wet etch after lithography. Zn thin film layer was deposited on the ITO layer through e-beam evaporation, followed by the formation of Zn pattern arrays using a lift-off method. Subsequently, patterned ZnO nanowires were prepared by thermal oxidation by placing the substrate, achieved by placing the substrate in a horizontal quartz tube furnace. The fabricated device was inspected by a scanning electron microscopy (SEM; Zeiss SUPRA 55, Jena, Germany).

Fig. 1(b) shows schematic diagram of the measurement set-up. For the measurement of gated ZnO nanowire field emitters, a phosphor screen is used as the anode and was connected to the high-voltage power supply (V_a) while the cathode electrodes were grounded. By applying voltages to different gates through a home-made gate driving circuit, line-coded addressing of the gated ZnO nanowire field emitters can be achieved.

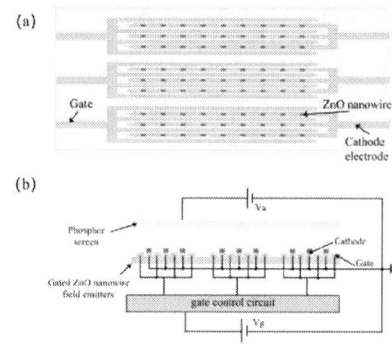

Fig. 1. Schematic diagram of (a)planar-gate ZnO nanowire field emitters and (b) the measurement set-up.

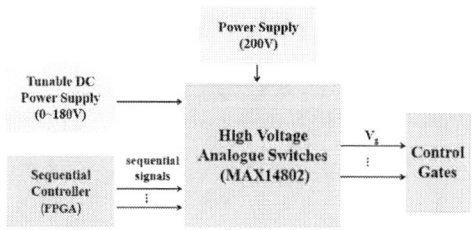

Fig. 2. Schematic diagram of the gate control circuit.

The schematic diagram pf the gate control circuit is shown in Fig. 2. MAX14802 high-voltage analogue switches are used, which can provide separately 16 output of high-voltage SPST switches. Inside MAX14802, a digital interface to transfer data into an internal 16-bit shift register and hold the data through a programmable latch with enable and reset control functions. The gate voltage is supplied by an tunable DC power supply, which is connected to the driving voltage input of the MAX14802 chip, and the output voltage (V_g) pins are connected to the control gates of the device. The MAX14802 chip is powered by another 200 V voltage source. A Field-Programmable Gate Array (FPGA) is programmed to supply sequential signals for the high-voltage switches. Decoupling capacitors are used to lower the fluctuation of the output voltage.

III. RESULTS AND DISCUSSION

Fig. 3 shows the morphology of the fabricated planar-gate ZnO nanowire FEAs. Fig. 2(a) shows the parallel arrangement of strip ITO electrodes and arrays of ZnO nanowire patterns. Fig. 2(b) shows one ZnO nanowire pattern located in the central region of the cathode substrate. The average length of the ZnO nanowires is about 3 μm and the population density is about ~6 μm^{-2}. The ZnO nanowires in each region basically maintain similar population density, length and tip diameter, which indicates that the growth of ZnO nanowires in the device structure is relatively uniform.

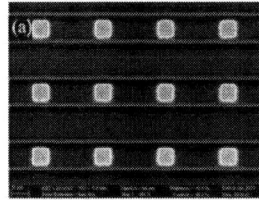 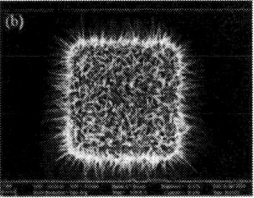

Fig. 3. SEM images of (a) the fabricated planar-gate ZnO nanowire FEAs and (b) a single ZnO nanowire pattern.

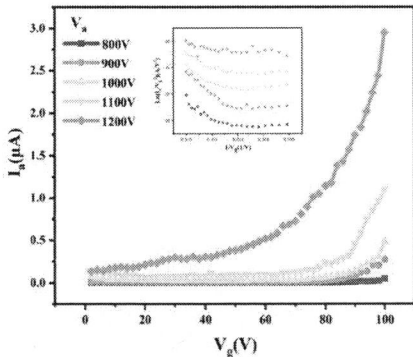

Fig. 4. The I_a-V_g curves of fabricated planar-gate ZnO nanowire field emitters measured under different anode voltages. The inset shows the corresponding F-N plots.

Fig. 4 shows the emission current vs gate voltage (I_a-V_g) characteristics of the planar-gate ZnO nanowire field emitters. The current was measured by applying voltage to a single column of gate electrode, while simultaneously connecting all the cathode electrodes to the ground. The emission characteristics were obtained by varying the gate voltage at anode voltages of 800 to 1200 V at step of 100 V. When V_g=100V and V_a=1200 V, the obtained anode current is ~2.95 μA. The F-N plots are presented in the inset. Non-linearity is observed in the F-N plots. Previous has indicated the

phenomenon is induced by the diode emission effect caused by the anode voltage [5].

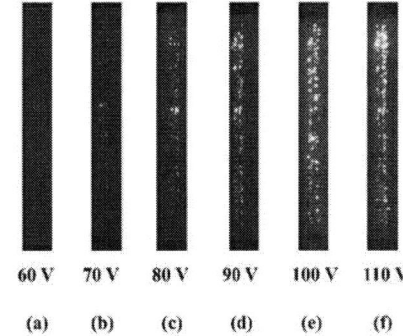

Fig. 5. The variation of emission images with gate voltage.

Fig. 5 shows the emission images recorded from one line under various gate voltage. When the anode was biased at 800V and all the cathode electrodes were connected to the ground, the gate voltage changes from 60 V to 110 V. With increasing gate voltage, more emission spots are observed from the phosphor screen. The results verified the line-addressing capability of the fabricated gate-structure ZnO nanowire FEAs.

IV. CONCLUSIONS

In summary, we have developed a planar-gate zinc oxide nanowire cold cathode for a line-coded flat panel X-ray source application. Gate control circuit is developed to realize line-addressing function. The emission characteristics under gate voltage driving were studied. The line-addressing capability with gate modulation was realized.

ACKNOWLEDGMENT

The authors gratefully acknowledge the financial support from the National Key R & D Program of China (Grant No. 2022YFA1204200), Key Research and Development Program of Guangdong (Grant No. 2023B0101200013), NSFC (Grant Nos. 82272131 & 62001527), the Science & Technology Department of Guangdong (Grant No. 2023B1212060025).

REFERENCES

[1] S. Carmignato, W. Dewulf, and R. Leach, Industrial X-ray Computed Tomography. Switzerland: Springer International Publishing, 2018.

[2] C. Wang, G. Zhang, Y. Xu, et al, Nanomaterials. vol. 11, p. 3115, 2021.

[3] J. Duan, Y. Li, X. Mou, et al, 17th Virtual international meeting on fully 3d image reconstruction in radiology and nuclear medicine, pp. 86-89, 2023.

[4] Y. Chen, S. Deng, N. Xu, et al, Nanomaterials. vol. 11, p. 2150, 2021.

[5] X. Cao, J. Yin, L. Wang, et al, IEEE Transactions on Electron Device. vol. 67, pp. 677-683, 2020.

[6] X. Cao, G. Zhang, Y. Zhao, et al, Applied Physics Letters. vol. 5, p. 119, 2021.

Portable Scanning and Acquisition System for miniature SEM

Marcin Białas
Department of Electronics, Photonics and Microsystems
Wroclaw University of Science and Technology
Wrocław, Poland
marcin.bialas@pwr.edu.pl

Abstract—**This paper presents a device that constitutes a nearly complete control and measurement system intended for conducting research on a miniature scanning electron microscope (μSEM)[1]. The subject of this publication is also sample data recorded using this equipment.**

Keywords—octupole, high voltage electronics, electron imaging

I. INTRODUCTION

In research conducted on electron beam microsystems, it is necessary to develop appropriate electronic systems responsible for power supply, control and data collection. When the subject of research is a scanning electron microscopy, the required systems are: an electron gun power supply, a deflection systems controller and electron detectors. The system discussed in this article is designed to work with a MEMS microscope with a single, octupole stage of deflection and is a complete solution for imaging tests - it controls the beam and records signals from the detector. Only the electron gun requires power from a separate source, depending on its type and design.

II. OVERVIEW

The presented system (fig. 1) mostly fits into a 19" housing. It consists of an octupole deflector controller, a detected signal preamplifier (separate unit) and a data acquisition system.

Fig. 1. *19" RACK housing, containing octupole deflector controller, detected signal converter, communication circuits and power supplies.*

The input and output circuits are isolated (floating), which allows working with μSEM structures of any configuration (detector or cathode at ground potential). A diagram of an example μSEM setup is shown in fig. 2.

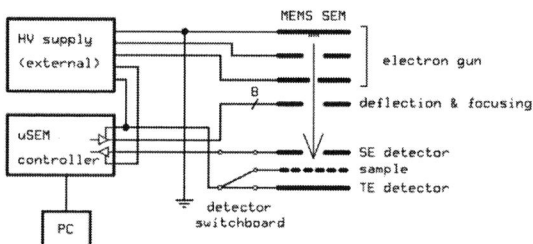

Fig. 2. *Example of miniature SEM test setup (electrical). External, multi-channel HV supply is used to power the electron source and to bias other electrodes in electron column of the instrument. There are three possible sources of detected signal – TE (transmitted electrons), SE (secondary e.) and AE (absorbed e., conductive sample itself) detector. While one of them was used, the rest was connected to anode supply.*

The control software (fig. 3) allows working in three modes: line scan, raster scan and detector test mode. The deflection parameters are set separately in each mode via text input or graphical interface ("drag and drop" method). The data is saved in CSV files from which entire images can be restored and further processed with another specialized program, or data can be processed in any other way.

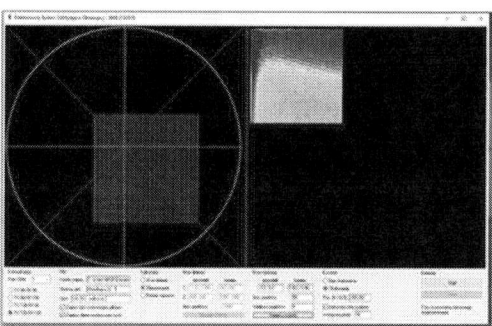

Fig. 3. *Window of the control software, raster scan mode. Preview of the scanning area (shown as a grid of green dots, each representing point to be scanned) is on the left, while the actual acquired image is displayed on the right.*

III. DETAILS

Communication between the computer and the device takes place via a virtual serial port at a speed of 115.2 kb/s. This allows for the analysis of up to 1280 points per second. The program sends the settings of 8 output channels that control the deflection system, and then receives the current measurement result from the detector. A block diagram of the system is shown in fig. 4.

This work has been financed by statutory grant of department of Electronics, Photonics and Microsystems, WUST

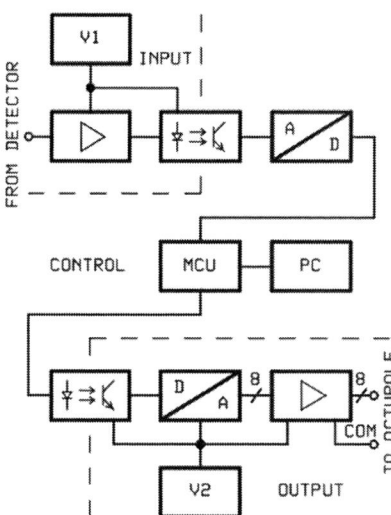

Fig. 4. A block diagram of the system. V1 and V2 represents floating power supplies for input and output section. The control section is powered via USB cable from PC.

Eight 12-bit D/A converters (MCP4922) are responsible for setting the output voltages. Data is sent to them via optocouplers, thanks to which the entire circuit is floating and can be polarized with additional voltage - up to 4 kV. This allows for experiments in which the octupole is also used as a focusing electrode.

Each output circuit consists of a differential amplifier (NE5534) and a voltage amplifier made of discrete parts. Its design was modeled on [2], with slight changes. The amplitude of the output voltages can reach 120 V. Each channel has a bandwidth of 8 kHz.

The input circuit (detector signal preamplifier) has the form of a separate module (fig. 5), so it can be connected to the detection electrode with the shortest possible, coaxial cable.

Fig. 5. Photo of the external current preamplifier. Upper left part is the actual preamplifier (with 4 selectable measurement ranges). Upper right part is the analog optocoupler circuitry, which allows biasing of the input with high voltages.

It consists of a nanoampere current amplifier with a JFET input (TL072) and an analog opto-isolation system (IL300), allowing easy polarization of the detector with any voltage up to 2 kV. Floating (insulated) part of the circuit is battery powered. Output voltage is digitalized by 10-bit A/D converter built in MCU (AVR ATmega328).

IV. IMAGES

An example image of a calibration grid with 30 μm wide crossbeams obtained with this system is shown in fig. 6. An SE detector in the form of a silicon plate biased with a voltage of +30 V relative to the sample (anode) was used to record it. Original resolution was 300×300 px. Contrast (difference between currents in most dark and most bright part of the picture) was 0.9 nA at 4 kV acceleration voltage and approx. 100 nA probe current. An electron gun of JEOL JSM-IT100 scanning microscope was used as an electron beam emitter instead of the MEMS structure, in order to achieve as small as possible spot diameter and to overcome related limitations of resolution.

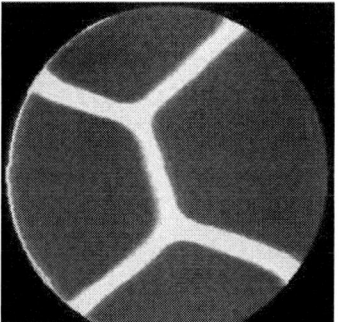

Fig. 6. 30 μm wide crossbeams of hexagonal calibration grid – an image obtained with presented electronic system and MEMS octupole scanning device.

Fig. 6 can be compared to similar image of the same specimen, fig. 7, acquired with JSM-IT100 microscope on its own.

Fig. 7. 30 μm wide crossbeams of hexagonal calibration grid – an image obtained with JEOL JSM-IT100 scanning electron microscope.

V. SUMMARY

Presented system, made of popular and easily accessible semiconductors, is noticeably slower than factory-made SEM, however it allows to experiment with custom software and hardware (electronics as well as SEM structure) and gives images of comparably quality.

REFERENCES

[1] Michał Krysztof, Marcin S. Białas, Piotr Szyszka, Tomasz P. Grzebyk, Anna Górecka Drzazga, "Fabrication and characterization of a miniaturized octupole deflection system for the MEMS electron microscope", Ultramicroscopy, 2021, vol. 225, art. 113288, p. 1-8

[2] Analog Devices LT1055 datasheet LT0815 rev. D, p. 2

Potential and Charge Density at the Surface of a Field Emitter

Chris Edgcombe[1,*] and Janis Huns[1,2]

[1]Department of Physics, University of Cambridge, Cambridge, UK;
[2]now at Department of Physics, University of Latvia, Riga, Latvia.
*Corresponding author: cje1@cam.ac.uk

Abstract— **Density functional analysis of a hemispherically capped (5,5) carbon nanotube in zero field by ONETEP shows that both charge density and the *difference* between self-consistent potential and vacuum level decrease exponentially as distance from the Fermi equipotential surface increases into the barrier region.**

Keywords---field emitter, density functional theory, carbon nanotube, many-state interaction, exchange effect, screening.

I. INTRODUCTION

To investigate tunnelling at a field emitter, we explored data generated by earlier analysis [1,2] with the order-N density functional code ONETEP [3] for the ground state of a capped (5,5) carbon nanotube (CNT). We aimed to identify effects of many-state interaction (including exchange and screening) that are not described by known single-state theory (such as [4]). Here we describe the results found with zero applied field, for potential $V(z)$ on the axis of the CNT and for charge density $\rho(r, \theta)$ on the axis and on a radius from the centre of the cap through a carbon nucleus in the penultimate ring of nuclei in the cap.

II. RESULTS

In Fig. 1, the variation of electronic charge density near the carbon nuclei is visible. The line plot of Fig. 2 shows that the density falls off as $\exp(-k_\rho r)$ with radius r from the centre of the hemispherical cap, both on the axis and at the polar angle of 60°, with a decay rate k_ρ of approximately 29 nm^{-1}. The exponential changes extend over a radial range of at least 0.5 nm from their maxima near the hemisphere of carbon nuclei at radius 0.315 nm.

The effective potential is computed by ONETEP self-consistently with the charge density distributions in the occupied states of the structure. On comparing the computed potential with the classical potential of a point source on the axis at the maximum of electron density and of magnitude to produce the Fermi level E_F at z_F, the computed potential rises towards the vacuum potential (V_{vac}) over $z > z_F$ appreciably faster than the potential from the point source.

Using the computed axial $V(z)$, we then plotted axial values of $dV(z)/dz$, averaged over intervals of 2 pm outside the Fermi equipotential, against $V(z)$. This produced a sequence of straight lines with near-equal slopes as shown in Fig. 3, which also shows the relation between $V(z)$ and z. On approximating to $dV(z)/dz$ by a straight line between the intercepts on the axes, we find

$$V(z) - E_F \approx (\Phi + 0.23 \text{ eV}) \{1 - \exp[-k_v (z - z_F)]\}$$

$$(0 < z - z_F < 0.42 \text{ nm})$$

where Φ, the work function, ≈ 4.43 eV; z_F is the location on axis of the equipotential at the Fermi level E_F, and $k_v = 11.6$ nm^{-1} is a decay rate (less than half that for charge density).

ONETEP also identified separately the contributions to potential from (i) Coulombic interactions of electrons, (ii) 'ion-ion', and (iii) exchange-and-correlation effects (XC). They show that the valley in the Coulombic potential (for this single-walled CNT) is confined in the axial direction almost entirely within the Fermi equipotential, while the potential change due to XC adds a valley of similar magnitude but greater axial extent. The greater extent implies that for this CNT cap, over a range of 0.5 nm outward from the Fermi equipotential, the change of total axial potential from E_F to V_{vac} that forms the work function is due almost completely to XC effects.

The exponential variation of axial $V(z)$ in the barrier may suggest a screening effect as discussed in chapters 17 and 26 of [5]. However, the results in Fig. 3 are for the barrier region outside the emitter, where we expect that there are no positive charges.

III. CONCLUSION

Analysis with ONETEP of the hemispherically-capped (5,5) CNT in zero field finds that both ρ and ($V_{vac} - V(z)$) decrease exponentially as distance from the surface of the Fermi equipotential increases into the barrier region. Other potential distributions also produce an exponential decrease of ρ in the barrier, but our self-consistent calculation shows further that XC effects cause the total potential to rise towards the vacuum level exponentially

It would be interesting to compare the behaviour with an external field applied, and to analyse an emitter that is more than a monolayer thick.

ACKNOWLEDGMENTS

The authors thank Dr. S.M. Masur for permission to extract the results shown here from files computed in the course of her work for the PhD degree. They also thank Prof. M. Payne for helpful comments.

REFERENCES

[1] S.M. Masur, C.J. Edgcombe, and C.H.W. Barnes, "On modeling the induced charge in density-functional calculations for field emitters", J. Vac. Sci. Technol. B, 40, 042802; 2022. doi: 10.1116/6.0001886.

[2] S.M. Masur, E.A. Linscott, and C.J. Edgcombe, "Modelling a capped carbon nanotube by linear-scaling density-functional theory", J. Electr. Spectr. 241, 146896; 2020. doi: 10.1016/j.elspec.2019.146896

[3] https://onetep.org .

[4] E.L. Murphy and R.H. Good, "Thermionic emission, Field Emission and the Transition Region", Physical Review, 102 (6) 1464-1473; 1956.

[5] N.W. Ashcroft and N.D. Mermin, Solid State Physics, International Edition, W B Saunders Company, Philadelphia, 1976.

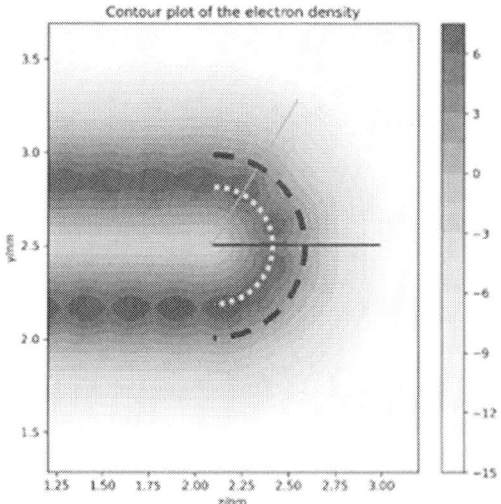

Fig. 1. Electron density on a longitudinal section through the CNT axis and end cap. The semi-circles show approximate locations of sections through the frame of carbon nuclei (dotted) and the Fermi equipotential in zero field (dashed).

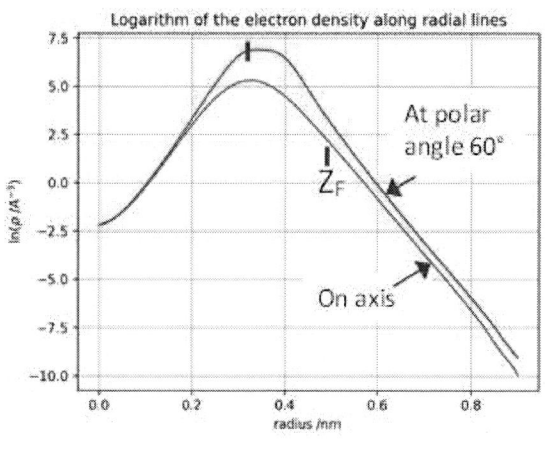

Fig. 2. Log$_e$ (electron density) as a function of radius from the centre of the end cap, for the 2 directions shown in Fig. 1. The radius of the hemisphere of carbon nuclei is marked on the red curve. The location of the Fermi equipotential on the axis is marked as Z$_F$.

Computed axial potential and gradient

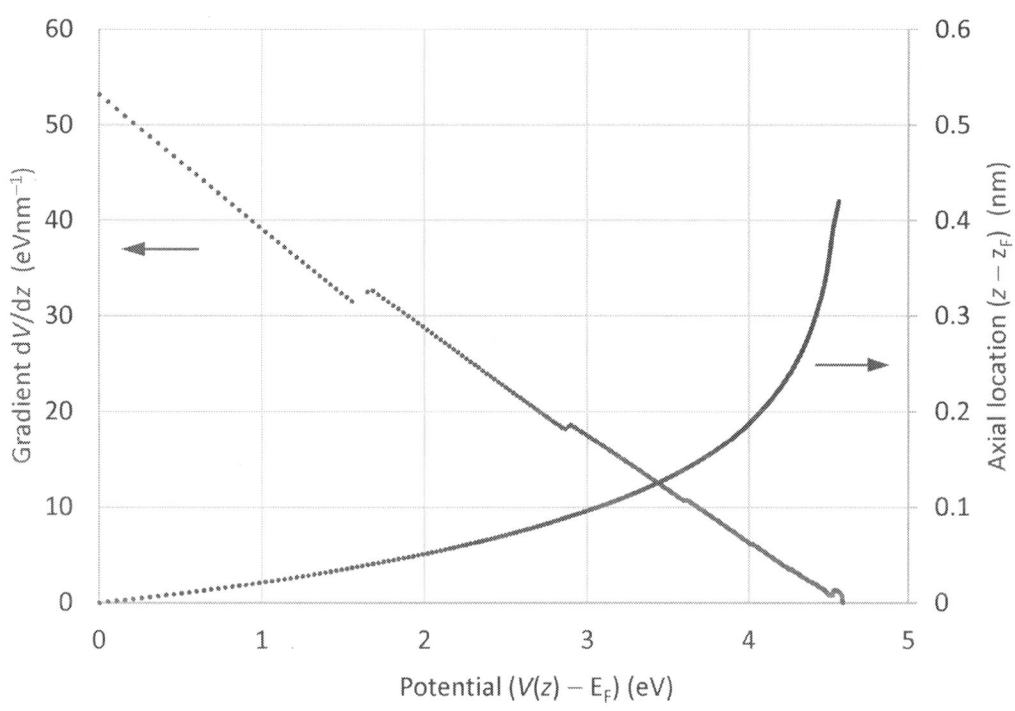

Fig. 3. Computed gradient of the effective axial potential $V(z)$ and axial location z (measured outward from the Fermi equipotential), shown with $V(z)$ (relative to E$_F$) as independent variable. Here z$_F$ is the axial location of the Fermi equipotential, E$_F$.

Scanning Enhancement of STM-tungsten Probes by Applying Colloidal Graphite Coatings

Mohammad M. Allaham[1,2,*], Zuzana Košelová[1,3], Daniel Burda[1], and Alexandr Knápek[1,3]

[1] *Electron and plasma technologies, Institute of Scientific Instruments of CAS, Královopolská 147, 61264 Brno, the Czech Republic.*
[2] *Central European Institute of Technology, Brno University of Technology, Purkyňova 123, 1200 Brno, the Czech Republic.*
[3] *Department of Microelectronics, Faculty of Electrical Engineering and Communication, Brno University of Technology Technická 3058/10, 616 00 Brno, the Czech Republic.*
*Corresponding author: allaham@isibrno.cz

Abstract—**Tungsten cold field emission nanotips are known for their excellent performance as scanning tunneling microscopy (STM) probes. However, their performance is limited when the STM scan is performed in ambient gas pressure, or without in-situ cleaning of the tip. As a reasonable solution for this problem, tungsten nano tip of 70 nm in diameter was coated by a thin layer of conductive graphite paint. The graphite paint was diluted by isopropanol to ensure the creation of a thin graphenic film on the nanotip apex. This process produced a tungsten-graphite composite probe of enhanced conductivity when compared to the oxidized tungsten tip, performing STM scans outside ultra-high vacuum environments.**

Index Terms—**colloidal graphite, tungsten-graphite STM probes, cold field emission cathodes in STM.**

I. INTRODUCTION

Scanning tunneling microscopy (STM) is a high-resolution surface analytical technique that became a pivotal innovation in nanoscience and its applications such as atomic level imaging and the studies of surface structures, topography, and reactions. STM is performed in ultra-high vacuum, ambient gas pressure, and even liquids after a proper choice of the STM-probe [1, 2].

To perform an STM scan, a needle-shaped probe of good conductive material and nano-sized apex (such as tungsten field emission cathodes) is brought to a sample-probe separation distance of 1 nm from the surface of a conductive sample. At this vacuum gap distance, the tunneling of electrons between the two electrodes (probe and sample) is possible at exceptionally low applied voltages through a single atom [3, 4]. According to Murphy-Good cold field emission model, this is possible because of the reduced Schottky-Nordheim potential energy barrier as shown in Fig.1 for a tungsten probe of local work function $\phi = 4.5$ eV, when biased at 1 V at electrodes separation distance of 1 nm [5].

STM-probes are usually prepared as metallic tips made from tungsten, gold, or platinum/iridium thermocouples. Tungsten STM probes require in-situ cleaning of any oxide or contamination layers that limit the conductivity of the used probe. Moreover, tungsten probes are limited to operate in ultra-high vacuum conditions.

In this work, we investigate the performance of tungsten STM probes (W-STMP) before and after applying a thin

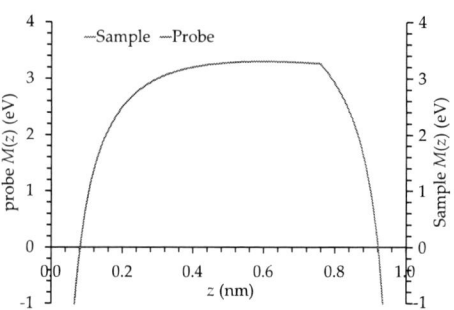

Fig. 1. STM electrodes Schottky-Nordheim merged potential energy barrier.

graphenic layer of conductive colloidal graphite (WG-STMP). The results showed outstanding enhancement in the scanning performance of the WG-STMP compared to the W-STMP when operated outside the ultra-high vacuum environment.

II. MATERIALS AND METHODOLOGY

The tungsten probes were fabricated from 99.9% high-purity polycrystalline tungsten wires of 1 cm in length and 0.3 mm in diameter. The tungsten field emission nanotips were fabricated and cleaned following the procedure described in [6], with one extra submerging step in isopropanol (IPA). To prepare the WG-STMPs, clean W-STMPs were dipped inside colloidal conductive graphite liquid, which is prepared by diluting the liquid KONTAKT CHEMIE graphite 33 conductive paint (CRC Industries Europe, Zele, Belgium) in ultrasonic IPA bath.

The prepared probes were tested using the NenoVision litescope multi-functional atomic force microscope (Nenovision, Brno, the Czech Republic). Moreover, the experiments were operated inside the chamber of TESCAN scanning electron microscope at pressure of <0.01 Pa.

III. RESULTS AND DISCUSSION

To test the performance of the prepared W-STMP and WG-STMP, TESCAN calibration sample was used to scan its pattern, which is shown in Fig.2(a). The first step was to scan the pattern using a clean W-STMP (Fig.2(b)). As

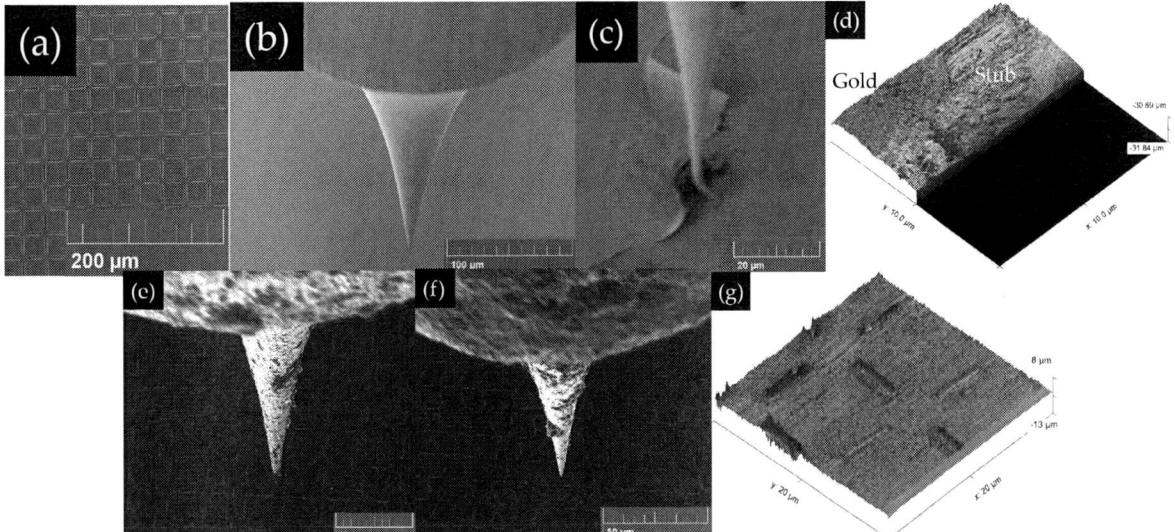

Fig. 2. sequence of SEM and STM images showing the (a) scanned pattern, (b) clean W-STMP before starting the STM scan, (c) clean W-STMP during the STM scan, (d) scanned gold-stub edge with a clean W-STMP, (e) WG-STMP before the STM scan, WG-STMP during the STM scan, and (g) the scanned pattern with WG-STMP.

expected, the scan was not successful from the beginning because of the limited conductivity of the oxidized nanotip, and relatively limited conductivity of the calibration sample. However, another experiment was performed by scanning the height of a gold layer when dposited on SEM stub. The W-STMP probe scanned part of the tested region. The tip of the W-STMP was slowly bending during this experiment, and finally it was destroyed after scanning $10 \times 20 \, \mu m^2$ as seen in Fig.2(c and d).

The second step was to test the performance of WG-STMP. The prepared probe is presented in Fig.2(e) before performing the scan of the calibration sample (which the W-STMP failed to scan). the probe survived three full scans ($20 \times 20 \, \mu m^2$) before the tip was slightly bent, and Fig.2(f) shows the status of the tip during the STM scan. Note that the zigzag shape of the scanned pattern is formed because of the motion of the stage during the scan. All the three scans were successful and the results from the first scan is presented in Fig.2(g).

The enhancement of the scanning performance is related to the better conductivity of graphite, its be According to the structure of graphite, its higher resistance to the environmental factors inside the chamber, and the unique hexagonal honeycomb structure of the graphenic layers.

IV. CONCLUSIONS

Tungsten is one of the most recommended metals for the fabrication of STM probes. However, the performance of W-STMP is limited to ultra-high vacuum environments and the need for in-situ cleaning of the tip before the scanning process.

Coating the scanning nanotip of tungsten field emission cathodes by graphite successfully enhanced the scanning performance of W-STMP in ambient pressure conditions. The results of this study showed a higher scanning performance, a higher resistance for environmental conditions, and a higher resistance to the adsorption of the scanned sample surface contamination.

ACKNOWLEDGMENT

The research described in the paper was financially supported by the Internal Grant Agency of the Brno University of Technology, grant number CEITEC VUT/FEKT-J-23-8307 and CEITEC VUT/FEKT-J-24-8567.

The described research was supported by the Czech Technology Agency (FW03010504). We also acknowledge the Czech Academy of Sciences (RVO:68081731).

CzechNanoLab project LM2018110 funded by MEYS CR is gratefully acknowledged for the financial support of the measurements/sample fabrication at CEITEC Nano Research Infrastructure.

REFERENCES

[1] G. Binning, and H. Rohrer, "Scanning tunneling microscopy". Surf. Sci. 126, 1983, 236–244.
[2] H. Wu, G. Li, J. Hou, and K. Sotthewes, "Probing surface properties of organic molecular layers by scanning tunneling microscopy". Adv. Colloid Interface Sci. 318, 2023, 102956.
[3] L. Bi, K. Liang, G. Czap, H. Wang, K. Yang, and S. Li, "Recent progress in probing atomic and molecular quantum coherence with scanning tunneling microscopy". Prog. Surf. Sci. 98, 2023, 100696.
[4] T. Tiedje, J. Varon, H. Deckman, J. Stokes, "Tip contamination effects in ambient pressure scanning tunneling microscopy imaging of graphite". J. Vac. Sci. Technol. A 6(2), 1987, 372-375.
[5] M. M. Allaham, R. G. Forbes, A. Knápek, D. Sobola, D. Burda, P. Sedlak and M. S. Mousa, "Interpretation of field emission current–voltage data: Background theory and detailed simulation testing of a user-friendly webtool". Mater. Today Commun. 31, 2022, 103654.
[6] Z. Košelová, L. Horáková, D. Burda, M. M. Allaham, A. Knápek, Z. Fohlerová, "Cleaning of tungsten tips for subsequent use as cold field emitters or stm probes". J. Elect. Eng. 75 (1), 2024, 41–46.

Scattering Matrix Approach to Electron Transmission Probability Through a Flat Semiconductor / Vacuum Interface

Nathaniel Hernandez [1,2], Marc Cahay [1,2,*], James Hart [1,3], Jonathan O'Mara [3,4], Jonathan Ludwick [2,5],
Dennis E. Walker Jr. [3], Tyson Back [5], and Harris Hall [3]

[1] Spintronics and Vacuum Nanoelectronics Laboratory, University of Cincinnati, Cincinnati, Ohio, USA
[2] UES, a BlueHalo company, Dayton, Ohio, USA
[3] Air Force Research Laboratory, Sensors Directorate, Wright-Patterson Air Force Base, USA
[4] KBR, Beavercreek, Ohio, USA
[5] Air Force Research Laboratory, Materials and Manufacturing Directorate, Wright-Patterson Air Force Base, Ohio, USA

*Corresponding author: cahaymm@ucmail.uc.edu

Abstract—**A scattering matrix technique is used to study the transmission probability through a flat semiconductor / vacuum interface considering the effective mass difference between the two materials. A comparison is made between the longitudinal and transverse energy dependence of the transmission probability calculated using the BenDaniel-Duke (BD) and Zhu-Kroemer (ZK) boundary conditions at the semiconductor / vacuum interface. For longitudinal energy between the minimum of the conduction band and a few $k_B T$ above the Fermi level in the semiconductor, the transmission probability can be an order of magnitude smaller when using the Zhu-Kroemer boundary conditions compared to the BenDaniel-Duke boundary conditions. The Zhu-Kroemer boundary conditions should be used to calculate the field emission current through a semiconductor/vacuum interface.**

Keywords—Scattering Matrix, Semiconductor Field Emission

I. Introduction

The scattering matrix technique [1] is used to calculate the tunneling probability through the sharp barrier present at the semiconductor / vacuum interface, such as shown in Fig. 1 (where the effects of Schottky barrier lowering were neglected and band bending in the semiconductor were neglected). Fig. 2 is a plot of the transmission probability through the potential barrier as a function of the longitudinal kinetic energy normalized by the barrier height V_0 (i.e., energy in the direction perpendicular to the semiconductor / vacuum interface) for the transverse kinetic energy (parallel to the interface) equal to 0 and 100 millielectronvolts (full and dashed lines, respectively). The green and purple curves in Fig. 2 show the transmission probability calculated using the Zhu-Kroemer boundary conditions [2] at the interface. This set of boundary conditions is more rigorous [3] than the commonly used BenDaniel-Duke boundary conditions [4] to treat tunneling problems between materials with different effective masses.

The transmission probability calculated using the Zhu-Kroemer boundary conditions is found to be smaller at low incident longitudinal energy. There is a crossover at higher energy where the BenDaniel-Duke boundary conditions predict a smaller transmission probability. However, this range of energy is far into the tail of the Fermi-Dirac distribution function, even at elevated temperature and high value of the external electric field and will have no influence on the calculation of the field emission current.

Fig. 3 is a contour plot of $\log_{10}(T_{BD}/T_{ZK})$ versus both normal and transverse energies to stress the difference between the transmission probabilities calculated using the BenDaniel-Duke versus Zhu-Kroemer boundary conditions. The potential energy barrier is shown in Fig.1.The effective mass in the semiconductor is set equal to $0.067 \times m_0$ and the barrier height is set to 4.07 electronvolts. The externally applied electric field is 4 volts per nanometer. Fig. 4 shows that the transmission probability calculated using the BenDaniel-Duke boundary conditions can be one order of magnitude larger than the one calculated using the Zhu-Kroemer boundary conditions.

II. Conclusions

A scattering matrix technique is used to calculate the transmission probability through a flat semiconductor / vacuum interface. The importance of the effective mass difference between the two materials was assessed using both the BenDaniel-Duke and Zhu-Kroemer boundary conditions at the semiconductor / vacuum interface. In the past, the Zhu-Kroemer boundary conditions have been shown to be more suitable boundary conditions to include the effective mass variation. We have shown that the latter can give transmission probability one order of magnitude smaller than the BenDaniel-Duke boundary conditions and should be used when calculating the FE current through a semiconductor / vacuum interface, which will be the subject of future investigation. The effects of Schottky barrier lowering and band bending in the semiconductor, which were neglected in this preliminary report, can easily be included in the scattering matrix technique. These will also be addressed in a future publication.

Acknowledgement

This material is based upon work supported by the Air Force Office of Scientific Research (Program Manager, Dr. John Luginsland) under Award No. FA9550-20RXCOR027. M. Cahay and N. Hernandez acknowledge support from the Air Force Summer Research Fellowship Program (SFFP is FA9550-20-F-0005) during the Summers of 2021, 2022, and 2023 and the support by the Air Force Research Labs under contract No. FA8650-22-F-5815. J. Ludwick was supported by the Air Force Research Labs under contract No. FA8650-16-D-5408.

979-8-3503-7977-8/24 $31.00 © 2024 IEEE

REFERENCES

[1] M. Cahay, M. McLennan, and S. Datta, "Conductance of an array of elastic scatterers: A scattering-matrix approach," *Phys. Rev. B*, vol. 37, no. 17, pp. 10125–10136, Jun. 1988.

[2] Q.-G. Zhu and H. Kroemer, "Interface connection rules for effective-mass wave functions at an abrupt heterojunction between two different semiconductors," *Phys. Rev. B*, vol. 27, no. 6, pp. 3519–3527, Mar. 1983.

[3] W. A. Harrison, "Effects of matching conditions in effective-mass theory: Quantum wells, transmission, and metal-induced gap states," *J. Appl. Phys.*, vol. 110, no. 11, p. 113715, Dec. 2011.

[4] D. J. BenDaniel and C. B. Duke, "Space-charge effects on electron tunneling," *Phys. Rev.*, vol. 152, no. 2, pp. 683–692, Dec. 1966.

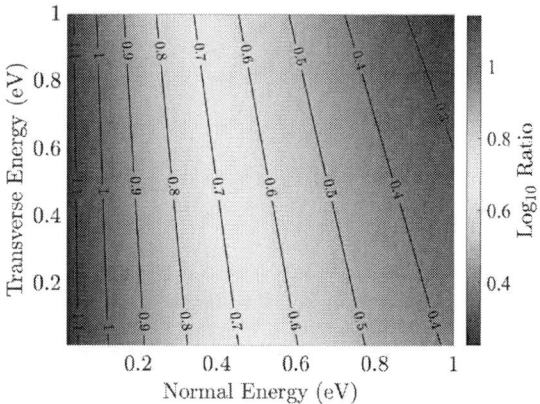

Fig. 3: Contour plot of $\log_{10}(T_{BD}/T_{ZK})$ versus both normal and transverse energies to stress the difference between the transmission probabilities calculated using the BenDaniel-Duke versus Zhu-Kroemer boundary conditions. The potential energy barrier is shown in Fig. 1.

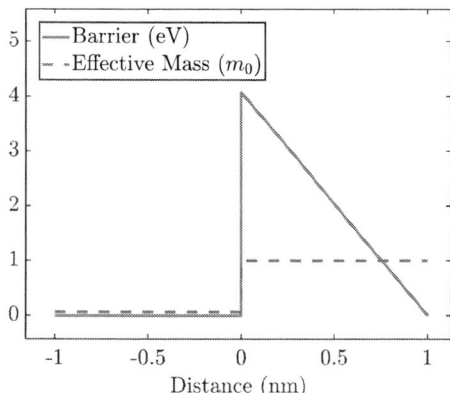

Fig. 1. Potential energy barrier at an GaAs /vacuum interface. The effective mass in the semiconductor is set equal to $0.067 \times m_0$ and the barrier height is set to 4.07 electronvolts. The externally applied electric field is 4 volts per nanometer. The red dashed line shows the difference in the electron mass in the semiconductor and vacuum (in units of m_0)

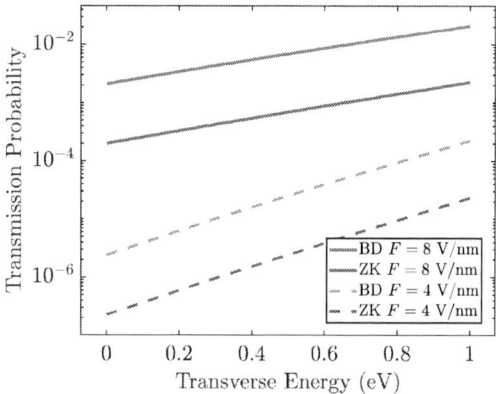

Fig. 4: Transmission probability through a potential energy barrier similar to Fig. 1 as a function of the transverse kinetic energy of an electron incident on the semiconductor interface. The normal kinetic energy was set equal to 100 millielectronvolts. The full and dashed lines are the transmission probability calculated for an externally applied electric field, F, of 8 volts per nanometer (two top curves) and 4 volts per nanometer (two bottom curves), respectively.

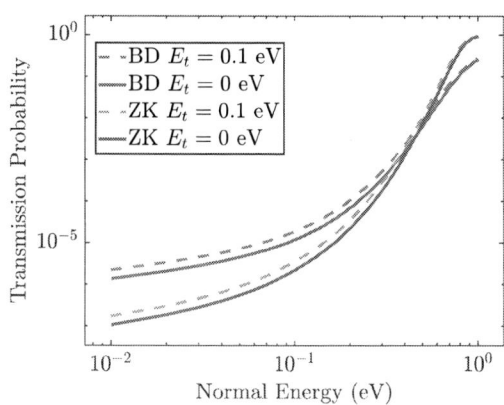

Fig. 2: Transmission probability through the potential energy barrier shown in Fig. 1 as a function of the normalized kinetic energy (i.e., in the direction perpendicular to the semiconductor interface). The transverse kinetic energy was set equal to 0 and 100 millielectronvolts for the full and dashed lines, respectively. The longitudinal energy is normalized by the electron affinity at the semiconductor/vacuum interface ($V_0 = 4.07$ eV).

Silicon Field Emitter Arrays for Vacuum Integrated Circuits

Nedeljko Karaulac[*] and Akintunde I. Akinwande

Microsystems Technology Laboratories, Massachusetts Institute of Technology, Cambridge, MA 02139, USA
[*]Corresponding author: karaulac@mit.edu

Abstract—**We present a proof-of-concept inverter based on silicon field emitter arrays (Si FEAs) that could be fabricated as a vacuum integrated circuit (IC). A circuit model for Si FEAs is developed, and the voltage transfer characteristics of the FEA inverter are simulated. In addition, a sample of 30 Si FEAs is characterized to determine the variations in a_{FN} and b_{FN} caused by the fabrication process. A Monte Carlo analysis is used to test the impact of variations on the FEA inverter performance.**

Keywords—**silicon, field emitter arrays, inverter, IC**

I. INTRODUCTION

Nanoscale vacuum-channel transistors (NVCTs) are expected to show better performance in a wide variety of high-frequency, high-power, and harsh environment applications due to their ballistic transport and higher breakdown field [1]. Silicon field emitter arrays (Si FEAs) are a proven and mature technology that can be implemented as vacuum transistors, and they could also be used in vacuum integrated circuits (ICs). Many of the challenges regarding uniformity, reliability, and lifetime have been addressed in this technology [2], [3]. Most recently, we addressed the scalability of the emission current by designing a layout-independent fabrication process for Si FEAs [4]. However, several questions regarding the feasibility of vacuum ICs still remain. For instance, in digital and analog circuits, the transistor parameters (e.g., K, V_T, λ) must be closely matched, and the impact of variations in FEA parameters, a_{FN} and b_{FN}, on future vacuum ICs is unknown. In this work, we simulate a proof-of-concept inverter based on Si FEAs. We first develop a FEA circuit model and test the model on experimental data. Next, the transfer characteristics of the Si FEA inverter circuit are simulated. Lastly, we characterize and model the statistical variation resulting from our recent fabrication process [4], and we perform a Monte Carlo analysis to determine if it is feasible to build a FEA inverter using our Si FEA fabrication process.

II. RESULTS AND DISCUSSION

The Fowler-Nordheim (FN) equation is widely used to model the emission current, I_E, from FEAs as a function of gate-to-emitter voltage, V_{GE}:

$$I_E = a_{FN} V_{GE}^2 \exp\left(\frac{-b_{FN}}{V_{GE}}\right). \tag{1}$$

However, the FN equation fails to predict the partition of the emission current between the anode and gate at different anode-to-emitter voltages, V_{AE}. For example, Fig. 1 shows the

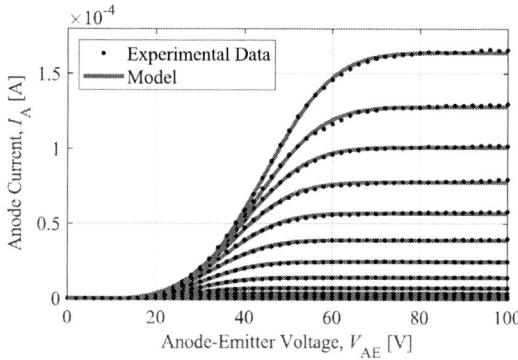

Fig. 1. Experimental output characteristics of a $100{\times}100$ Si FEA. The anode was placed approximately $100\ \mu$m above the FEA. Each curve is biased with a different value of V_{GE}. More precisely, from top to bottom, $V_{GE} = 50$ V, 48 V, 46 V, etc.

experimental output characteristics of a $100{\times}100$ Si FEA. This plot shows the variation of the anode current, I_A, as a function of V_{AE}. The data shows a characteristic s-shape, which resembles a sigmoid function. We propose this s-shape can be modeled by an "anode transport factor," α, which takes a value between 0 and 1, and is given below:

$$\alpha = \left[\frac{1}{2} + \frac{1}{2}\mathrm{erf}\left(\frac{V_{AE}-\mu}{\sigma\sqrt{2}}\right)\right] e^{-be^{-c\left(\frac{V_{AE}-\mu}{\sigma\sqrt{2}}\right)}} \tag{2}$$

where erf is the error function, σ is the slope parameter, μ is the location of the inflection point, and b and c are fitting parameters. The proportion of emission current that is collected by the anode is then $I_A = \alpha I_E$, and the remaining emission current is collected by the gate, $I_G = (1-\alpha)I_E$. Fig. 2(a) shows a graphical representation of the proposed FEA circuit model. To test our model, we fitted α for the data in Fig 1. We found the following empirical relationship for σ and μ as a function of V_{GE}:

$$\sigma = (0.2678)V_{GE} - 0.8179 \tag{3}$$

$$\mu = (0.9949)V_{GE} - 6.1923 \tag{4}$$

In addition, $b = 0.0009213$, $c = 5.45$, $a_{FN} = 4.41 \times 10^{-5}$ (A/V²), and $b_{FN} = 309$ V. The model fits the data very well in Fig 1.

Using the FEA circuit model, we simulated the inverter circuit shown in Fig. 2(b). The inverter consists of a $1000{\times}1000$ Si FEA with a pull-up resistor. This circuit can also be used

979-8-3503-7977-8/24 $31.00 © 2024 IEEE

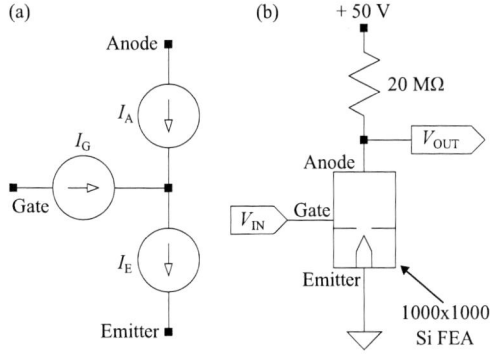

Fig. 2. Diagram of (a) FEA circuit model where $I_A = \alpha I_E$, $I_G = (1 - \alpha)I_E$, and $I_A + I_G = I_E$; (b) inverter circuit consisting of a Si FEA and pull-up resistor.

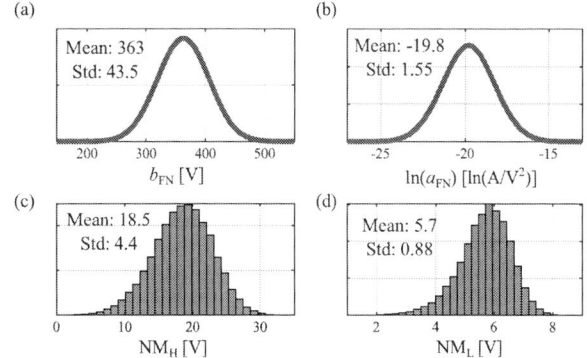

Fig. 4. Probability distribution of (a) b_{FN} and (b) $\ln(a_{FN})$ from 30 Si FEAs. Monte Carlo analysis results showing variation in (c) NM_H and (d) NM_L.

as an amplifier, but we focus on its use as a logic inverter. For the design of the FEA, we used $a_{FN} = 2.52 \times 10^{-3}$ (A/V^2) and $b_{FN} = 363$ V. The supply voltage was 50 V, and the resistor was 20 MΩ; this results in a transition between logic "high" and logic "low" at 25 V, thus ensuring adequate noise margin.

The simulated voltage transfer characteristics (VTC) of the FEA inverter are plotted in Fig. 3, which shows the output voltage, V_{OUT}, as a function of input voltage, V_{IN}. An important figure of merit is the noise margin, which is the voltage range that the input gate can tolerate while still maintaining the correct output logic level. The terms V_{OH}, V_{OL}, V_{IH}, and V_{IL} are defined as the voltages where the slope of the VTC is equal to -1. Thus, the noise margin "high" is $NM_H = V_{OH} - V_{IH}$, and noise margin "low" is $NM_L = V_{IL} - V_{OL}$. For the VTC in Fig. 3, we obtained $NM_H = 18.7$ V and $NM_L = 5.8$ V.

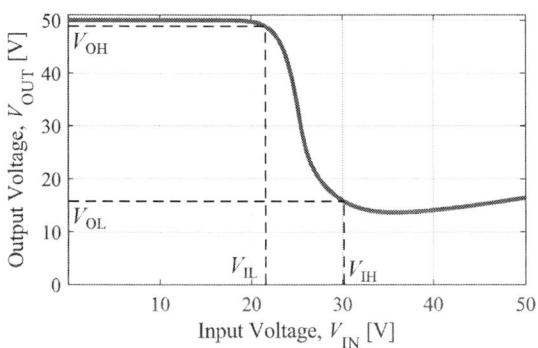

Fig. 3. Voltage transfer characteristics of the FEA inverter, including definitions of V_{OH}, V_{OL}, V_{IH}, and V_{IL} required for calculating noise margin.

The actual noise margin will vary due to variations in the emission current caused by our fabrication process [4]. To test the effects of variation, we first measured 30 Si FEAs and used the FN plot to extract the slopes and y-intercepts, b_{FN} and $\ln(a_{FN})$, respectively. Fig. 4(a)-(b) shows the distribution we modeled using the mean and standard deviation (std) from the data. Note that $\ln(a_{FN})$ is per field emitter, and hence

a_{FN} can be scaled by array size, as we did for the circuit in Fig. 2(b). Next, we performed a Monte Carlo analysis by randomly sampling 1000 pairs of b_{FN} and $\ln(a_{FN})$ values from the distributions we modeled. The FEA inverter was simulated for each pair, and the noise margin was calculated. Fig. 4(c)-(d) shows the distribution of NM_H and NM_L resulting from the Monte Carlo analysis. For 95% of the inverters, the NM_H is 9–27 V and NM_L is 4–7 V. These ranges are greater than 0 V, so we anticipate the majority of FEA inverters will operate properly.

III. CONCLUSION

We developed a circuit model for FEAs that predicts the current at all three terminals as a function of voltage. This model allowed us to simulate an inverter consisting of a Si FEA and pull-up resistor. The VTC of the FEA inverter predicts successful operation as a logic inverter with adequate noise margin. Lastly, we simulated the variation in noise margin that could occur from process variation and found that vacuum ICs are feasible given our current process in [4].

ACKNOWLEDGMENT

This material is based upon work supported by AFOSR Grant No. FA9550-18-1-0436. This work was carried out in part through the use of MIT's Microsystems Technology Laboratories and MIT.nano facilites.

REFERENCES

[1] J.-W. Han, M.-L. Seol, D.-I. Moon, G. Hunter, and M. Meyyappan, "Nanoscale vacuum channel transistors fabricated on silicon carbide wafers," *Nature Electronics*, vol. 2, no. 9, p. 405–411, 2019.

[2] S. A. Guerrera and A. I. Akinwande, "Nanofabrication of arrays of silicon field emitters with vertical silicon nanowire current limiters and self-aligned gates," *Nanotechnology*, vol. 27, no. 29, p. 295302, 2016.

[3] N. Karaulac, G. Rughoobur, and A. I. Akinwande, "Highly uniform silicon field emitter arrays fabricated using a trilevel resist process," *Journal of Vacuum Science & Technology B*, vol. 38, no. 2, p. 023201, 2020.

[4] N. Karaulac, W. Chern, G. Rughoobur, and A. I. Akinwande, "Silicon field emitter arrays fabricated using a layout-independent process," in *2021 34th International Vacuum Nanoelectronics Conference (IVNC)*, 2021, pp. 1–2.

Silicon Nanowire Field Emitters with Integrated Extraction Gates Using Benzocyclobutene as an Insulator

Philipp Buchner[1]*, Alexander Kaiser[1], Matthias Hausladen[1], Mathias Bartl[1], Michael Bachmann[2] and Rupert Schreiner[1]

[1] Faculty of Applied Natural Sciences and Cultural Studies, OTH Regensburg, D-93053 Regensburg, Germany

[2] Ketek GmbH, D-82737 Munich, Germany

*Corresponding author: philipp.buchner@oth-regensburg.de

Abstract— **We are continuously improving the performance of our field emission electron sources. In this work a geometrically optimized design of electron sources with silicon nanowire field emitters on pillars was fabricated. This new design increased the packing density of the emitters by using a hexagonal arrangement of the pillars and a pillar spacing of 40 µm. Benzocyclobutene was used as the insulator material for an integrated (Cr/Ni) extraction gate. A modified fabrication process for the field emitters further improved reproducibility and reliability. An emission current of about 0.4 mA was measured for 30 minutes at an extraction voltage of 250 V and an anode voltage of 500 V. Electron transmission through the gate reached almost 100%.**

Keywords— *field emission, field emitter array, silicon nanowire emitters, integrated gate*

Introduction

Field emission electron sources based on silicon nanowires show an excellent performance in terms of achievable emission current [1]. The electron sources consist of two silicon chips: The silicon nanowire cathode array and the extraction grid. However, the alignment of these two silicon chips has to be done manually source by source under an optical microscope in order to adjust the grid openings to the emitter pillars. Alternatives such as integrated gate electrodes mitigate this problem [2]. In addition, these gates reduce the extraction voltages required due to the smaller tip to gate distances possible and increase the electron transmission due to the perfect alignment, while allowing a tighter packing of emitters, which increases the utilisation of the available chip space. Benzocyclobutene (BCB) [3] is suitable as an insulator material for integrated gate electrodes, as it can be easily spun onto the already etched emitter pillars and then thermally cured.

Design and Fabrication

The spatial density of the emission array has been increased by switching from a square to a hexagonal packing of the emitters and by reducing the emitter spacing from 50 µm to 40 µm (Fig. 1a). The basic fabrication process for the emission pillars is based on the process described in [1] also using low doped (1-10 Ωcm) n-type silicon. However, the process has been modified to further improve its reliability and repeatability. This was achieved by not relying solely on the anisotropic plasma etching process, which gradually etches the photoresist as the pillars grow in height. Instead, the anisotropic etching process was interrupted multiple

times (at pillar hights of roughly 10, 15, 20 and 25 µm pillar hight), and the photoresist was etched separately from the silicon with an oxygen plasma (90 s each) (Fig. 1b). The BCB was subsequently spin-coated onto the emission pillar arrays in multiple layers until the pillars were completely covered.

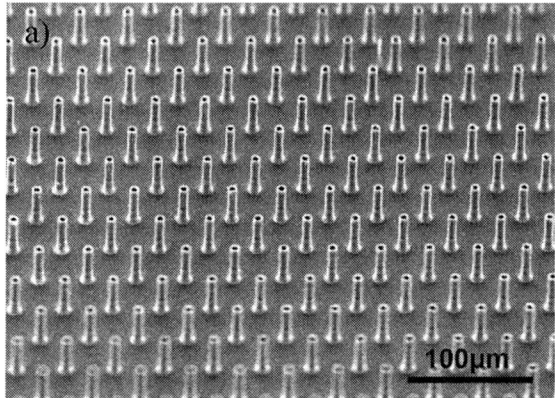

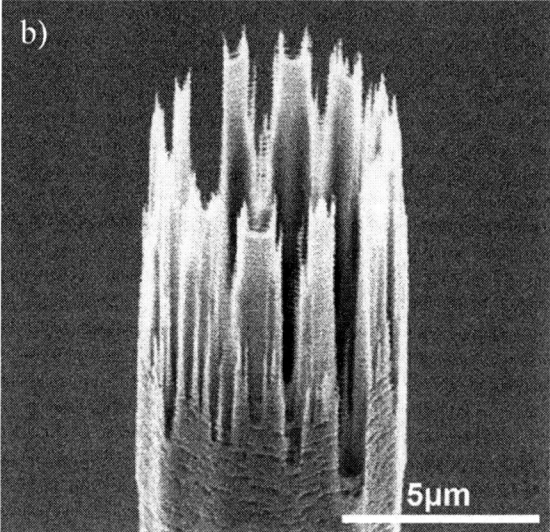

Fig. 1. a) SEM of a large area of the hexagonal array of pillars. One can see how the individual emission pillars are homogenious, identical and show no defects. b) Detailed view of a single emission pillar after etching before BCB coating. Its diameter is 5 µm and its hight is about 30 µm. One can clearly see twelve distinct prongs with nanowires on their tops.

The research work was funded by the Bavarian Research Foundation under project-number AZ-1583.

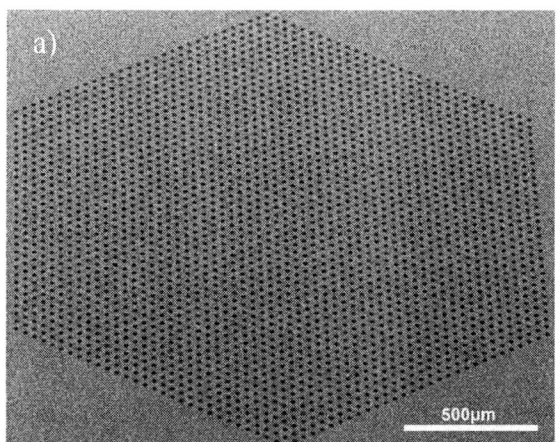

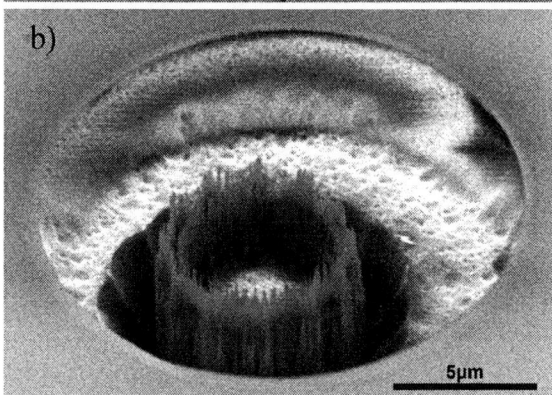

Fig. 2. a) SEM of the silicon nanowire electron source. One can see the extraction gate with the circular apertures in which the emission pillars are centered. (Hexagonal packing, 40 μm spacing, 4219 emitter pillars, area: 5.5 mm²). b) Detail view of one of the silicon nanowire field emitter pillars covered in BCB. The emitter is centered in its gate opening, which has a diameter of 17 μm. The BCB was selectively removed in the vicinity of the emission pillar by plasma etching, after the gate electrode was structured by a lift-off process.

The BCB was then fully thermally cured in a nitrogen atmosphere (300 °C for 2 hours). The extraction gates were fabricated and aligned on wafer level by photolithography, followed by metallisation with Cr (10 nm) and Ni (190 nm) with consecutive lift-off. Afterwards the BCB around the emitters was removed by plasma etching until the emission pillars were released and a long isolation path was achieved (Fig. 2).

Experimental

The electron sources were characterised under ultra-high vacuum (10^{-10} hPa), following the same measurement procedure as described in [1]. A tungsten needle was used as an anode. The needle was placed 10 mm centred above the gate. A fixed anode voltage (U_A) of 500 V was supplied and the anode current (I_A) measured. The gate electrode was kept at 0 V (gate voltage U_G and gate current I_G). The varying extraction voltage (U_C) was applied to the cathode by a second tungsten needle and the cathode current (I_C) was measured, too.

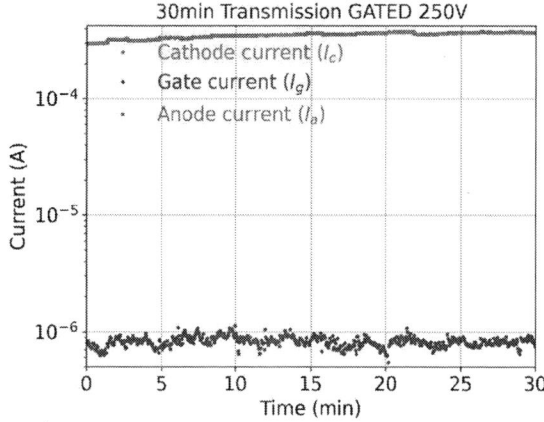

Fig. 3. Plot of a 30 minute measurement. The anode voltage is 500 V, the gate voltage is held at 0 V and the extraction voltage of 250 V is applied to the cathode. One can see that I_c and I_A are almost identical. An average transmission of 99.61% is achieved.

A 10 kΩ resistor was added into the cathode path to protect the measurement circuit in the case of a short circuit. The characterisation cycle was started at 50 V (U_C) and increased in 50 V steps for subsequent runs until 250 V. For each cycle, U_C was ramped up and applied to the cathode for one hour with current measurements taken every 2 seconds. Then U_C was ramped down and the next cycle started by ramping up to the next voltage level. The first 30 min of each cycle were discarded because the electron sources showed instabilities due to activation and burn-in processes. For the other data points the transmission was calculated as the ratio of I_A and I_C. Subsequently the average transmission was calculated as the average of the individual transmissions. An emission current of about 0.4 mA was measured for 30 minutes at an extraction voltage of 250 V. Electron transmission through the gate averaged 99.61% (Fig. 3).

Conclusion

We have demonstrated the use of BCB as an insulator material for gated field emission electron sources on wafer level. We achieved almost 100% electron transmission in comparison to 85% for silicon nanowire electron sources with manually aligned. Further improvement would be a self-aligning fabrication process for the extraction gate mitigating even the second photolithography step.

References

[1] Buchner, P.; Hausladen, M.; Bartl, M.; Bachmann M.; Schreiner, R. High current field emission from Si nanowires on pillar structures. *J. Vac. Sci. Technol. B*, 2024; 42 (2): 022208.

[2] Prommesberger, C.; Langer, C.; Ławrowski, R.; Schreiner, R. Investigations on the long-term performance of gated p-type silicon tip arrays with reproducible and stable field emission behavior. *J. Vac. Sci. Technol. B*, 2017; 35 (1): 012201.

[3] Kirchhoff, R.; Carriere, C.; Bruza, K.; Rondan, N.; Sammler, R. Benzocyclobutenes: A New Class of High Performance Polymers. *J. Mac. Sci.: A – Chemistry*; 1991; 28:11-12, 1079-1113.

Simulating Vacuum Arc Initiation by Coupling Emission, Heating, and Plasma Processes

Roni Aleksi Koitermaa*[†], Tauno Tiirats*, Veronika Zadin*, Flyura Djurabekova[†] and Andreas Kyritsakis*[†‡]

*Institute of Technology, University of Tartu, Nooruse 1, 50411 Tartu, Estonia

[†]Helsinki institute of Physics and Department of Physics, University of Helsinki, PO Box 43, 00014 Helsinki, Finland

[‡]Email: andreas.kyritsakis@ut.ee

Abstract—**Although vacuum arcs have been extensively studied, the physical initiation processes remain unclear. Multiscale and multi-physics models that concurrently simulate these complex interconnected processes are necessary to investigate these phenomena. This paper presents further developments in the modeling of vacuum arc plasma initiation from intensively emitting field emitters. Our model concurrently couples particle-in-cell with Monte Carlo collisions (PIC-MCC) simulations of plasma processes with electron emission and thermal effects calculations of the tip. It includes various processes such as evaporation, impact, and field ionization, offering insights into the dynamics of plasma buildup from an initially cold field emitting tip surface.**

Index Terms—**FEMOCS, vacuum breakdown, plasma onset, coupled multiphysics models, particle-in-cell**

I. INTRODUCTION

Vacuum arcing hinders various devices, such as electron sources, vacuum switches, X-ray sources, fusion reactors, and particle accelerators. Nevertheless, the physical mechanisms leading from field emission to plasma in a vacuum are not fully understood. Recent computational models [1], [2], [3], [4] have given a rigorous quantitative understanding of the processes by coupling the most significant processes of vacuum arc initiation: tip shape evolution, space-charge suppressed thermal-field electron emission, tip heating effects, thermal evaporation, plasma collisions, sputtering, and electromagnetic coupling with the macroscopic system.

Here we present our latest static model that focuses on the interplay between the thermal processes on an intensively emitting nanotip and plasma initiation. This model concurrently couples particle-in-cell with Monte Carlo Collisions (PIC-MCC) simulations with finite element-based calculations of electron emission and the associated thermal effects. Our model includes and assesses the significance of the essential particle and heat exchange processes, i.e., neutral evaporation, impact ionization, direct field ionization, backward ion bombardment, Nottingham, and evaporating heating (or cooling). Our simulation results give deep insights into the dynamics of plasma buildup from an intensively emitting metal nanotip.

This work is funded by the EU's H2020 program ERA Chair "MATTER" (856705) and by the Estonian Research Council, projects RVTT3 and SJD61

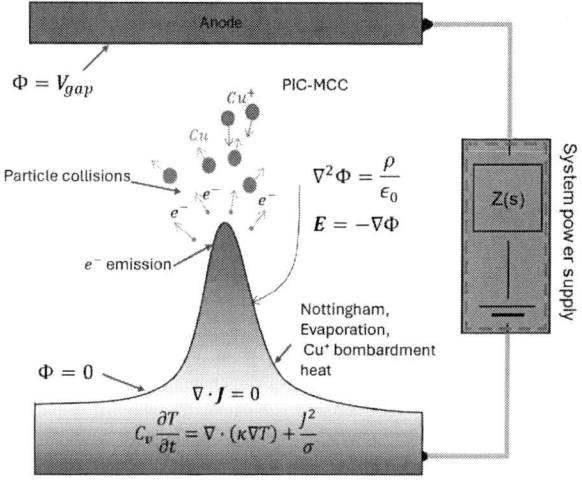

Fig. 1. Model summary.

II. METHOD

Our methodology is summarized in figure 1. We concurrently couple a Particle Cell with a Monte Carlo collision (PIC-MCC) model with a solution to the thermal problem in the tip. The most important element of our method is the concurrent coupling of the plasma processes occurring in the vacuum region (white in the schematic) and the tip interior (colored). This coupling starts by calculating the interface surface's thermal-field electron emission using the GETELEC code [5]. This is then used to calculate the local injection rate of electrons passed into the PIC code. GETELEC also calculates the Nottingham heat, which is passed as a Neumann boundary condition into the heat equation. Similarly, the current density is used as a Neumann boundary condition for the continuity equation that calculates the distribution of the current in the heat, which is used to calculate the Joule heat component. The temperature distribution resulting from solving the heat equation is fed back to GETELEC, which is also used to calculate the local evaporation rate on the interface and inject neutrals in the PIC system. Many neutrals get ionized due to collisions with electrons or direct field ionization, which are accelerated back onto the cathode surface. The resulting

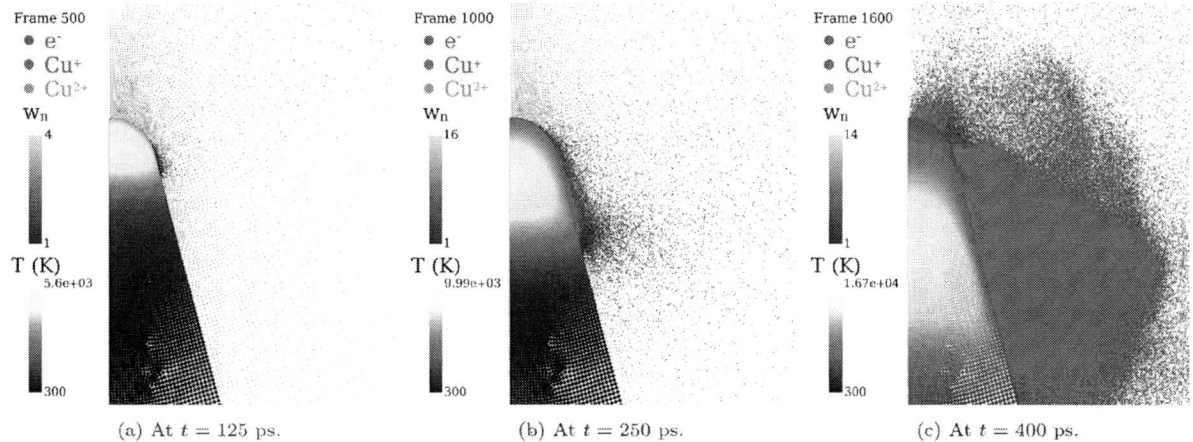

Fig. 2. Evolution of the temperature distribution in the bulk mesh points ("hot" color code), the electric field in the vacuum (blue arrows), and the positions of PIC particles: e^- (blue dots), ions (red and orange dots), and neutral vapor ("viridis" color coding representing the particle weight).

bombardment heat is also included in the surface Neumann boundary condition for the heat equation.

III. RESULTS AND DISCUSSION

We simulated a 50nm apex radius tip, at an applied apex field (without space charge) of 15 GV/m. Figure 2 summarizes the evolution process. The tip starts reaching high temperatures at the apex, which leads to vapor emission. The vapor ionizes, leading to backward bombardment in a thermal runaway plasma initiation process. We find that even if we neglect the nanotip kinetics of an intensively field-emitting nanotip (static approximation), the tip goes into thermal runaway and forms plasma if a sufficiently high electric field is applied. Tip heating increases the evaporation rate from its surface, which ionizes by electron impacts and field ionization, leading to plasma formation and vacuum breakdown. Within the static tip approximation, we identified the important processes for vacuum arcing: direct field ionization dominates at the beginning of plasma formation. In contrast, impact ionization takes over when the plasma density increases. Furthermore, the injection of neutral vapor is mostly by thermal evaporation, with sputtering having a secondary role. Nottingham heating, evaporative cooling, and bombardment heating all significantly contribute to the tip's heating.

IV. CONCLUSIONS

We demonstrate that even a static tip can go into thermal runaway and form vacuum arc plasma when a strong enough electric field is applied. Tip heating leads to evaporation that provides the vapor to be ionized and form plasma, which is maintained by further evaporation enhanced by heating caused by the backward bombardment of the plasma ions. Still, capturing the dynamic evolution of the tip (e.g., by molecular dynamics) is necessary to understand the process fully. Our main future direction is addressing the challenges of a full coupling between the current plasma simulation and molecular dynamics.

REFERENCES

[1] R. Koitermaa, A. Kyritsakis, T. Tiirats, V. Zadin, and F. Djurabekova, "Simulating vacuum arc initiation by coupling emission, heating and plasma processes," *Vacuum*, vol. 224, p. 113176, 2024.

[2] A. Kyritsakis, M. Veske, K. Eimre, V. Zadin, and F. Djurabekova, "Thermal runaway of metal nano-tips during intense electron emission," *Journal of Physics D: Applied Physics*, vol. 51, no. 22, p. 225203, 2018.

[3] M. Veske, A. Kyritsakis, F. Djurabekova, K. N. Sjobak, A. Aabloo, and V. Zadin, "Dynamic coupling between particle-in-cell and atomistic simulations," *Phys. Rev. E*, vol. 101, p. 053307, May 2020.

[4] H. Timko, K. Ness Sjobak, L. Mether, S. Calatroni, F. Djurabekova, K. Matyash, K. Nordlund, R. Schneider, and W. Wuensch, "From field emission to vacuum arc ignition: A new tool for simulating copper vacuum arcs," *Contributions to Plasma Physics*, vol. 55, no. 4, pp. 299–314, 2015.

[5] A. Kyritsakis and F. Djurabekova, "A general computational method for electron emission and thermal effects in field emitting nanotips," *Computational Materials Science*, vol. 128, pp. 15–21, 2017.

Simulation of the Electrical Properties of a Graphene Monolayer Field Effect Transistor

Ammar Al Soud[1,2*], Ahmad M D (Assa'd) Jaber[3], Vladimír Holcman[2,] Petr Sedlak[2], Dinara Sobola[1,2]

[1] Central European Institute of Technology Brno University of Technology Purkyňova 656/123,61200 Brno, Czech Republic.
[2] Department of Physics, Faculty of Electrical Engineering and Communication, Brno University of Technology, Technická 2848/8, 61600 Brno, Czech Republic.
[3] Department of Basic Medical Sciences, Faculty of Medicine, Aqaba Medical Sciences University, Aqaba, 12 Jordan.

*Corresponding author: Ammar.al.soud@vutbr.cz

Abstract— Field-effect graphene transistors are finding increasing commercial and research applications. Simulation is an important step in facilitating this transition. It contributes to understanding the work process, identifying potential issues and minimizing the cost of production. In this work, the electrical characteristics of back-gated graphene field-effect transistor were simulated using ANSYS electronics software. The output current was studied by applying a voltage difference between the source and drain ranging from -5 mV to 5 mV. The back-gate applied voltage was -50 to 50 V. The results show that the gate voltage induced a similar change in both the contact and channel resistance but did not change the density of mobility of positive and negative carriers

Keywords— field effect transistors, graphene, mobility.

I. INTRODUCTION

Since its isolation in 2004, graphene has attracted the materials science community's attention firmly into the world of two-dimensional (2D) materials [1]. The main focus of research into graphene-based electronic devices has been the realisation of field-effect transistors (FETs), due to their potential to in overcoming the limitations of conventional CMOS devices. One of the main attractions of GRMs is that they overcome short-channel effects [2]. In addition, graphene is able to change its electronic properties by the effect of the voltage applied to the back gate. The field effect electric changes the concentration of the carrier in the graphite channel, thereby changing the current through it depending on the value of the applied voltage on the back gate [1]. Graphene field-effect transistors (GFETs) can have a channel made of a monolayer graphene, rather than a semiconductor materials [3]. Unlike conventional semiconductors, the conduction and valence domains in graphene do not overlap. Instead, they converge at a single point known as the Dirac point. At this point the electrons are effectively massless and therefore have very high electron mobility [2]. Therefore, FETs are characterised by V-shaped diodes, the p-branch dominating at the negative conduction and the n-type at the positive gate vol tage [4].The aim of this work is to use simulation to study the effect of positive and negative back-gate voltage on contact, channel resistance, and both Hall effect and electron mobility in a monolayer GFET with Au contacts.

II. MATERIALS AND METHODS

Figure 1 shows an schematic diagram of the monolayer GFET structure. A monolayer GFET consists of a 675 nm thick silicon substrate. It is topped with silicon oxide (100 nm). The electrodes and contacts are made of gold with a thickness of 140 nm. designations. The dimensions of the graphene channel was 100×100

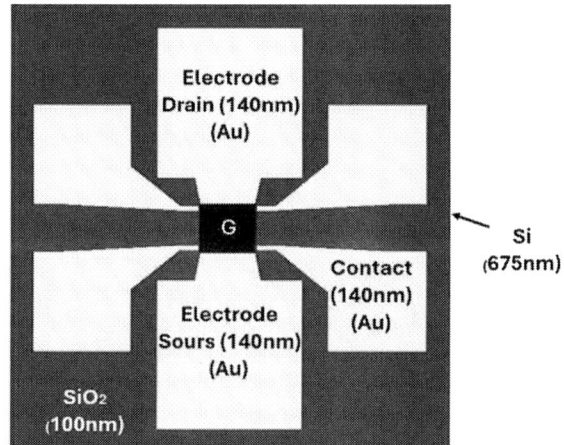

Fig. 1. schematic diagram of the monolayer GFET structure.

III. RESULTS AND DISCUSSIONS

Figure 2 shows a symmetrical behaviour of the transfer characteristic for both a hole carrier branch and an electron carrier branch where the slope was equal for both. The lowest current value (Dirac point) was directly above V = 0 V The symmetry between the hole and electron carriers can be explained by the efficiency of the metal contacts and the interaction of graphene with SiO_2 under ideal conditions [5]. Figure 3 shows the linear behaviour of the current and voltage bias source-drain (Ids-Vds) which confirms the ohmic character of the Au contacts /graphene channel in the applied voltage range was -5 to 5 mV between source and drain while the applied gate voltage was between -50 to 50 V.

979-8-3503-7977-8/24 $31.00 © 2024 IEEE

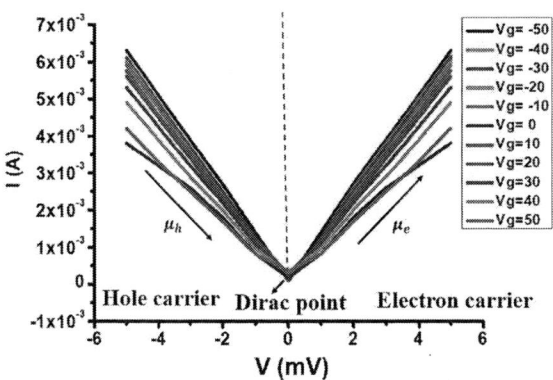

Fig. 2. The a transfer characteristic between hole and electrons carriers.

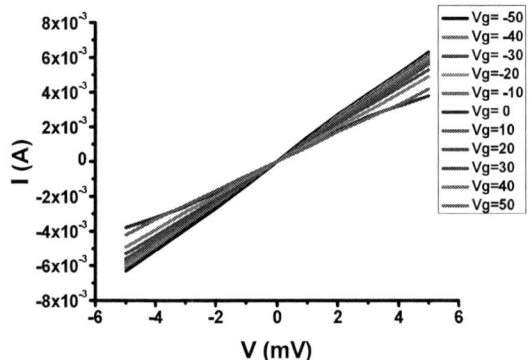

Fig. 3. I_{ds}-V_{ds} output.

The values of the hole and electron carrier mobility (4490 cm^2 V^{-1} s^{-1}), calculated by the equation [3], [6].

$$\mu_{\text{hole (electron)}} = \mp \frac{gl}{C V_{ds} W} \qquad (1)$$

Where $\mu_{\text{hole (electron)}}$ is the electron or hole carrier mobility. While g is the transconductance of the hole or electron carrier at GFET, C is the capacitance of the silicon dioxide nano layer (C = 33 nFcm^{-2} for 100 nm SiO$_2$) [3]. In addition, l and W are the length and width of the channel, here equal to 100 μm each.

In real measurements, a small shift of the Dirac point towards positive or negative is expected depending on the polarity of the gate voltage, this shift can be illustrated as shown in Figure 5. Initially, fermi level of pristine graphene is centered on the Dirac point when the gate voltage is 0 V. Furthermore, the Fermi level of graphene can move up within the conduction band or down within the valence band, depending on the polarity of the gate voltage relative to the Dirac point. Thus, the charge carriers converge at the graphene/SiO2 interface due to the capacitive charging effect. The positive gate voltage induces electrons, while the negative voltage induces holes. Studies have shown that the concentration of charge carriers induced by the electric field can reach levels as high as 10^{13} cm^2 V^{-1} s^{-1} [6].

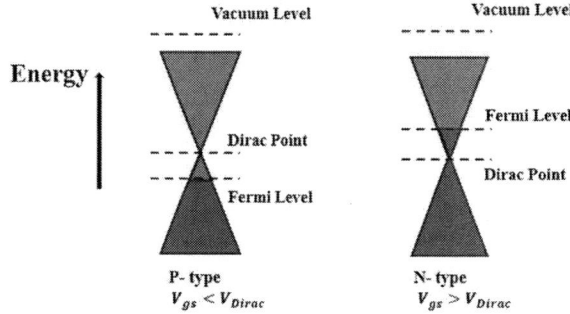

Fig.4. Energy band plots of graphene impacted by Back-gate voltage.

IV. CONCLUSION

In this work, the monolayer GFET was simulated using Ansys software. There was a symmetric behaviour of the transport property for both the hole and the electron carrier branch as the slope was equal for both. The value of the hole and electron carriers was (4490 cm^2 V^{-1} s^{-1}). The Dirac point was above 0 V. The I_{ds}-V_{ds} behaviour was linear, indicating the ohmic nature of the Au/graphene contacts. These results can be compared with experimental results that are expected to be obtained in the future where the Dirac point is expected to be shifted towards positive or negative depending on the polarity of the gate voltage.

ACKNOWLEDGMENT

The Czech Nano Lab project LM2023051 funded by MEYS CR is gratefully acknowledged for the financial support of the measurements/sample fabrication at CEITEC Nano Research Infrastructure. The research described in this paper was financed by the grant 23-07384S of the Czech Science Foundation (GACR) as well as. Internal Grant Agency of Brno University of Technology, grant No. FEKT-S-23-8228.

REFERENCES

[1] K. S. Novoselov *et al.*, "Electric Field Effect in Atomically Thin Carbon Films," *Science (1979)*, vol. 306, no. 5696, pp. 666–669, Oct. 2004.

[2] J.-H. Chen, C. Jang, S. Xiao, M. Ishigami, and M. S. Fuhrer, "Intrinsic and extrinsic performance limits of graphene devices on SiO2," *Nat Nanotechnol*, vol. 3, no. 4, pp. 206–209, Apr. 2008.

[3] F. Urban, G. Lupina, A. Grillo, N. Martucciello, and A. Di Bartolomeo, "Contact resistance and mobility in back-gate graphene transistors," *Nano Express*, vol. 1, no. 1, p. 010001, Jun. 2020.

[4] A. N. Mina, A. A. Awadallah, A. H. Phillips, and R. R. Ahmed, "Simulation of the Band Structure of Graphene and Carbon Nanotube," *J Phys Conf Ser*, vol. 343, p. 012076, Feb. 2012.

[5] A. Toral-Lopez *et al.*, "GFET Asymmetric Transfer Response Analysis through Access Region Resistances," *Nanomaterials*, vol. 9, no. 7, p. 1027, Jul. 2019.

[6] J. Wei, B. Liang, Q. Cao, H. Ren, Y. Zheng, and X. Ye, "Understanding asymmetric transfer characteristics and hysteresis behaviors in graphene devices under different chemical atmospheres," *Carbon N Y*, vol. 156, pp. 67–76, Jan. 2020.

Simulations and Investigations of Silicon Nanowire Field Emitters

Mathias Bartl[1*], Philipp Buchner[1], Matthias Hausladen[1], Ali Asgharzadehkhorasani[1], Michael Bachmann[2] and Rupert Schreiner[1]

[1]Ostbayerische Technische Hochschule Regensburg, Germany, [2]Ketek GmbH, Munich, Germany
*Corresponding author: mathias.bartl@oth-regensburg.de

Abstract— The emission behavior of field emission electron sources consisting of a silicon nanowire cathode, an extraction grid electrode and a planar anode was investigated based on a particle tracing simulation using FEM. The focus was on the influence of the grid geometry as well as the positioning of the grid relative to the emitters on the electron transmission. The highest transmission can be achieved with the emitter tips protruding 10 μm trough the extraction grid openings. The transmission decreases more rapidly with increasing distance between the tip and the grid the thicker the grid is.

Keywords— *field emission, field emission simulation, silicon nanowire emitter*

I. INTRODUCTION

Recently we reported on silicon nanowires on pillar structures for high current field emission [1]. The focus of the previous work lied upon the fabrication process and a first performance evaluation of these structures applicated in a field emission electron source. Besides the design of the emitter structure itself, the geometry of the grid and their relative positioning to one another have a great influence on the emission behavior, especially on the electron transmission [2]. Therefore, the aim of this work is to better understand the influence of the grid geometry on the electron transmission. We developed a simplified model of the emission structure in combination with a grid and an anode using FEM. Trough the variation of grid parameters such as the grid thickness and the distance between the emission tips and the grid, we tried to find an optimal parameter selection.

II. MODELING AND SIMULATION SETTINGS

A simplified 2-dimensional geometry was used for the simulation of the silicon nanowire structure. It comprises of two emitter pillars with a height of 50 μm and a tip radius of 15 nm. These are arranged at a distance of 5 μm. The basic grid was modeled with a thickness of 10 μm, 30 μm openings with an opening angle of 5° and a grid spacing of 50 μm. The distance between the anode and the topside of the extraction grid was fixed at 500 μm. The geometrical arrangement for two grid thicknesses is shown in Fig. 1 a) and b). The materials used for the emitting tips and the grid was silicon and the anode consists of copper. The surrounding vacuum was modeled with air. Hereby the background number density of molecules was set to $1 \cdot 10^{15}$ m^{-3} representing a pressure of $5 \cdot 10^{-9}$ hPa. The emitter voltage was set to -300 V, the grid voltage to 0 V and the anode voltage to 500 V.

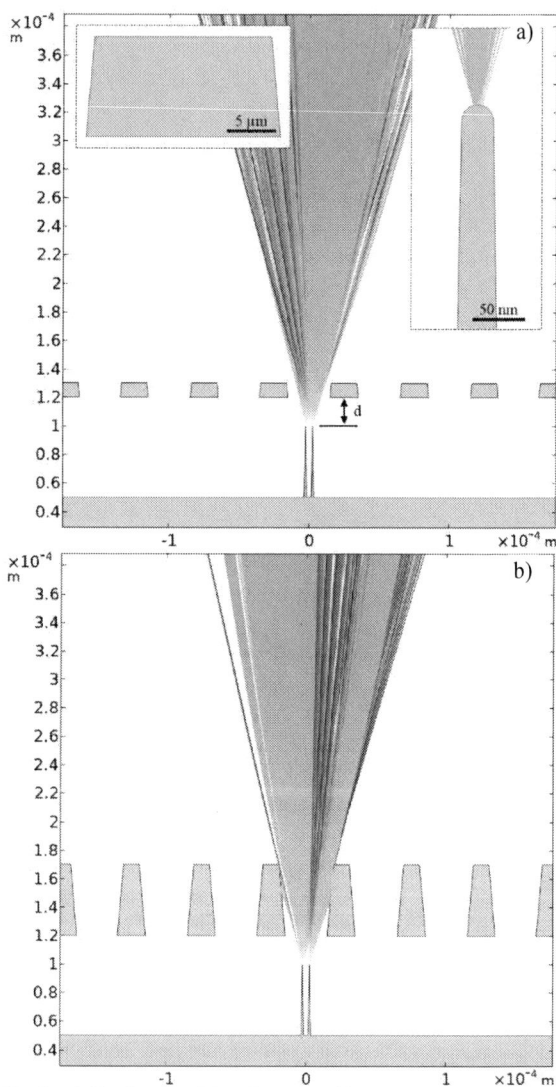

Fig. 1. a) Simulated particle trajectories with a tip to grid distance d of 20 μm and a grid thickness of 50 μm. The inset in the upper left corner shows the grid with a base width of 20 μm. The side wall angles are 5 °. The inset in the upper right corner shows the tip of an emitter pillar with a height of 50 μm, a tip radius of 15 nm and an opening angle of 1 °. b) Simulated particle trajectories with a tip to grid distance of 20 μm and a grid thickness of 50 μm.

The research work was funded by the Bavarian Research Foundation under project-number AZ-1583.

First the stationary electric field was calculated. Based on the value of the electric field at the emitter tips, particles representing a couple of electrons are released at fixed time steps of 2 ps. Their release density is proportional to the field current relationship given in [3]. Additionally, the starting velocity of the particles was calculated from the average energy of an emitted electron stated in [4]. Using the particle tracing module a transient simulation was carried out to simulate the trajectories of the individual particles. The particles that interact with the surface of the grid or the anode were counted before they disappeared. The accumulated numbers of these particles serve as an indicator for the grid current and anode current. The transmission is derived from the ratio of the accumulated numbers of particles at the anode to the number of all detected particles.

III. SIMULATION RESULTS

Fig. 2 shows the accumulated number of particles, that arrive at the grid or the anode respectively for a fixed tip to grid distance of 20 μm plotted over the time.

The accumulated number of particles increase after an initial settling time of 150 ps linearly with time. At a grid thickness of 10 μm the number of particles that hit the grid doesn´t increase and remains 0. As the grid thickness is increased the portion of particles, that hit the grid also increases resulting in a steeper slope of the accumulated number of grid particles. Meanwhile the slope of the accumulated particles of the anode decreases. The transmission can be calculated by the ratio of the accumulated particles at the anode to the sum of the accumulated particles at the anode and the grid. Fig. 3 a) shows the transmission with a grid thickness of 10 μm for various emission tip to grid distances d in the range of -10 μm to 40 μm with the negative value representing a protrusion of the tips into the grid. The transmission differs only for d = 40 μm from 100 % and reaches a plateau at 61 %. An increase of the grid thickness to 30 μm results in a decrease of the transmission for tip to grid distances greater or equal to 20 μm (Fig. 3 b)). At an even greater grid thickness of 50 μm the transmission decreases at all tip to grid distances and doesn`t reach 100 % even if the tips protrude into the grid (Fig. 3 c)).

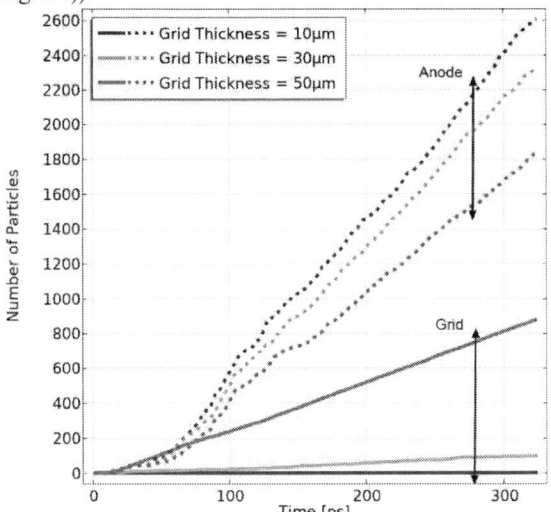

Fig. 2. Electron transmission for tip to grid distances d in the range of -10 μm to 40 μm and a grid thickness of a) 10 μm, b) 30 μm and c) 50 μm. The particles were released every 2 ps.

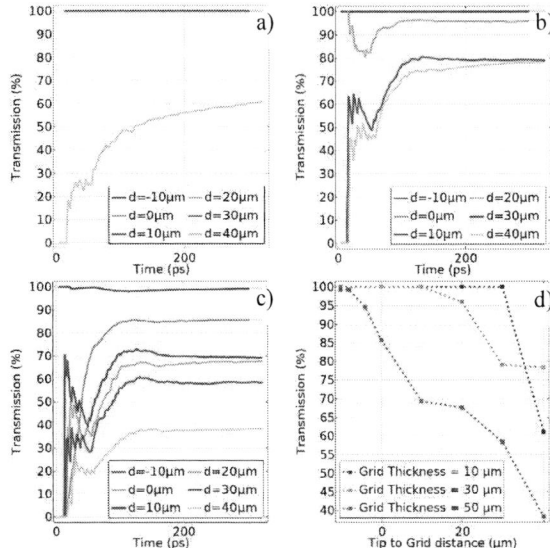

Fig. 3. Electron transmission for tip to grid distances d in the range of -10 μm to 40 μm and a grid thickness of a) 10 μm, b) 30 μm and c) 50 μm. The dependency of the transmission on the tip to grid distance at different grid thicknesses is shown in d).

Fig. 3 d) summarizes the transmission values at the end of the transient simulation of all tip to grid and grid thickness combinations. The maximal transmission can be reached with the tips protruding into the grid for all simulated grid thicknesses. The minimal transmission is at the largest gap between the tips and the grid.

IV. CONCLUSION

The transmission of particles is dependent both on the distance between the tips and the grid and the thickness of the grid. The transmission reaches its maximum when the tips protrude into the grid. An increasing distance results in a lower transmission. The thicker the grid is, the sooner and faster the transmission decreases with increasing distance. This is due to the particles hitting the inner side walls of the grid rather than the bottom. A transmission of 100 % with thick grids can only be achieved if the emitter tips protrude through the grid openings. With thin grids, the distance between the tips and the grid is less critical, so that a transmission of 100 % is possible even if a certain distance between the tips and the grid is present. These results can be used for future grid designs to optimize the electron transmission.

REFERENCES

[1] P. Buchner, M. Hausladen, M. Bartl, M. Bachmann and R. Schreiner, "High current field emission from si nanowires on pillar structures," J. Vac. Sci. Technol. B, 42, 022208, 2024, doi: 10.1116/6.0003384.

[2] P. Buchner et. al.; "An integrated silicon nanowire field emission electron source on a chip with high electron transmission,", 36th IVNC, Cambridge, MA, 2023, doi: 10.1109/IVNC57695.2023.10188878.

[3] J. Kevin, "Introduction to the Physics of Electron Emission," John Wiley & Sons, Ltd, 2017, doi: 10.1002/9781119051794.ch13.

[4] J. Paulini, T. Klein, G. Simon, "Thermo-field emission and the Nottingham effect," J. Phys. D: Appl. Phys. 26, 1310, 1993, doi: 10.1088/0022-3727/26/8/024.

Single Column Multiple Electron Beam Imaging from N-type Silicon

Jáchym Podstránský[1,4], Matthias Hausladen[3], Jakub Zlámal[1], Alexandr Knápek[2,4], Rupert Schreiner[3]

[1]Institute of Physical Engineering, FME, BUT, Technická 2896, 616 69 Brno, Czech Republic
[2]Department of Microelectronics, FEEC, BUT, Technická 10, 616 00, Brno, Czech Republic
[3]Regensburg University of Applied Sciences, Seybothstraße 2, 93053 Regensburg, Germany
[4]Institute of Scientific Instruments of the CAS, Královopolská 147, 612 00 Brno, Czech Republic
Corresponding author: podstransky@isibrno.cz

Abstract—**This work is aimed at measuring the electron emission from multiple cathodes formed by n-doped silicon and imaging the electron beams focused by an einzel lens on a CMOS camera. The experimental results are compared with computer simulation to understand the electron emission from the semiconductor cathode and the observed imaging imperfections. Finally, modifications to the experimental setup are suggested that should lead to improvement in the extraction current and spot size of focused electron beams and also better understanding of the processes taking place in the experiment.**

Keywords—Electrons, field emission, semiconductors, COMSOL Multiphysics

I. INTRODUCTION

Electron beam computed tomography (EBCT) is a technology of which the biggest advantage over conventional computed tomography (CT) is its scanning frequency is faster. Due to this, the EBCT finds its use in the scanning of moving objects, like a hearth. This comes from the fact that the EBCT machines have no moving components like the conventional CT machines. To be able to obtain the images quicker, a higher electric current of the electron beam generating X-rays is needed. In EBCT machines, the electron beam is emitted from a single tip. To increase electron beam current multiple tips are often used.

This work explores a new approach, where an array of cathodes made out of n-doped silicon would be utilized to obtain higher electron beams currents than a single tip. We describe the process of manufacturing the used cathodes and extraction grids, followed by the measurements of electron beam parameters. In the measurements, we were able to focus the electron beam onto the CMOS camera using a simple optical system consisting of electrostatic einzel lens and capture in real time electron beam image spots.

COMSOL Multiphysics and Python software for the simulation of electron emission and tracing of electron beam were used. The simulations helped to understand the system in detail, and further showed the shortcomings of the optical system design and of the executed measurements. For the calculation the extended Murphy-Good equation for the calculation of the emission by R.G. Forbes was used [1]. This work is described in more detail in a diploma thesis of Jáchym Podstránský [2].

II. MEASUREMENTS

A. Fabrication

The process of fabrication of used field emission cathodes with the extraction grid that we used was developed by M. Eng. Christoph Langer from OTH Regensburg, and it is described in depth in his work "Silicon chip field emission electron source fabricated by laser micromachining" [3].

Through the process of laser pulse ablation, we were able to create the electron source, which consisted of an extraction grid and a matrix of tips, both made of phosphorus doped silicon. By regulating the electric potential on both components, we controlled the extraction current. We were working with 2×2 and 21×21 matrices of tips. The highest measured electric current from one tip was 20 µA.

B. Image acquisition

By utilizing the einzel lens, we were able to focus the obtained emission current directly on the CMOS camera. The acquired image of one electron beam that is not fully focused by the einzel lens is shown in Figure 1. Further, the acquired image of the 21×21 electron beams, where the beams are focused, is shown in Figure 2. As we were unable to explain the "halo" effect, which is the bright ring in Figure 1, and the asymmetric geometry in the center of the spots, we made simulations of the experiment.

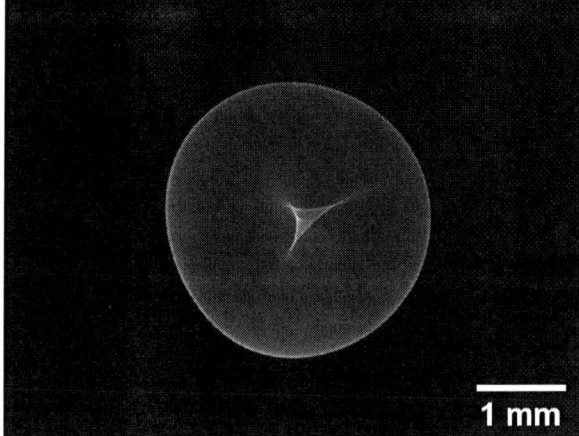

Figure 1: Image of one electron beam on CMOS camera that is not fully focused

III. SIMULATIONS OF THE EXPERIMENT

The simulations of the experiment were done in COMSOL Multiphysics version 5.4 and Python. We were able to analyze the temperature, electric potential, and more by utilizing these programs. One of the interesting results is that the temperature can suddenly jump when intrinsic conductivity occurs, which leads to the tip melting.

The biggest achievement was that we were able to reconstruct the acquired images that we observed in

979-8-3503-7977-8/24 $31.00 © 2024 IEEE

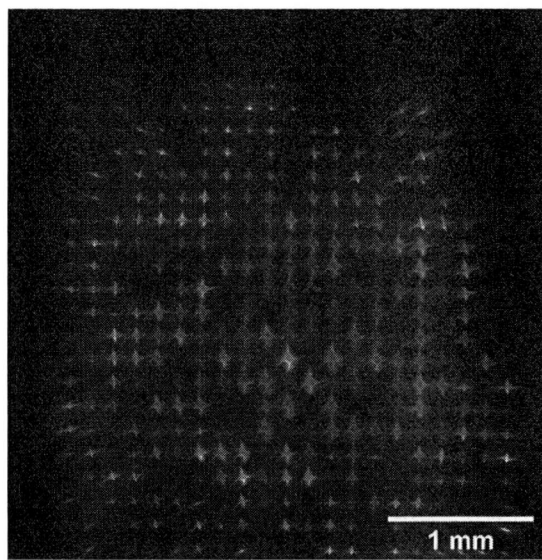

Figure 2: Electron beams from 21 × 21 tip matrix captured by CMOS camera

experiment. We found out that the mistake is in the alignment of the einzel lens. The simulated image with a displacement of 0.7 mm is shown in Figure 3. The parameters of this simulation were similar to those in the experiment. From these simulations, we were able to explain the origin of the "halo" effect and also the asymmetric geometry in the center of the spots.

From the results of the simulations, we were able to suggest optimizing the current setup. One of the most significant suggestions is to use the sapphire spheres to align the einzel lens electrodes. We also suggested measurements to understand the events of the emission in more detail, such as measurements of the conductivity and the electric current on the einzel lens electrodes, as the executed experiments did not consist of enough data to confirm some of the outcomes of the simulations.

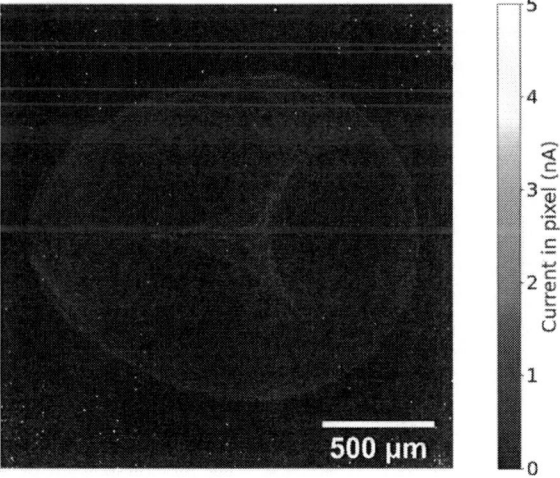

Figure 3: Simulated image with the displacement of the middle electrode of the einzel lens

REFERENCES

[1] R.G. Forbes, "The Murphy–Good plot: a better method of analysing field emission data," The Royal Society, 2019, vol. 6(12), pp. 190912,1-13

[2] J. Podstransky, "Single column multiple electron beam imaging," Master's Thesis, Vysoké učení technické v Brně, Fakulta strojního inženýrství, Institute of Physical Engineering, Brno, Supervisor Jakub Zlámal, 2024

[3] C. Langer, V. Bomke, M. Hausladen, R. Lawrowski, C. Prommesberger, "Silicon chip field emission electron source fabricated by laser micromachining," Journal of Vacuum Science & Technology B, 2020, vol. 38, pp. 013202,1-9

Slow Dynamics in Localized Heating of Carbon Nanotube Forests

Mokter M. Chowdhury[*‡], Kevin Voon[*‡], Jeff F. Young[††], George A. Sawatzky[††], Alireza Nojeh[*‡§]

[*]Department of Electrical and Computer Engineering
[†]Department of Physics and Astronomy
[‡]Quantum Matter Institute
University of British Columbia
Vancouver BC, Canada
[§]Corresponding author: alireza.nojeh@ubc.ca

Abstract—Understanding the long-term behavior of one-dimensional nanostructures at high temperatures is important for various applications ranging from thermionic emission to chemical sensing. We demonstrate unusually slow temperature decay and recovery occurring in carbon nanotube forests under strongly localized heating to above 1500 K. Through a combination of pyrometry, residual gas analysis, and computational heat transfer modeling, this phenomenon is connected to slow diffusion and desorption of adsorbed species occurring within the forest, highlighting internal mass transport properties that may be useful for nanoengineering purposes.

Index Terms—carbon nanotube, nanotube forest, nanotube array, heating, dynamics, adsorbate, diffusion

I. INTRODUCTION

Carbon nanotube (CNT) forests appear to be uniform black solids, but are in fact multiscale materials made of vertically-aligned, loosely entangled CNTs with large inter-nanotube spacing, such that over 90% of the forest is empty. The material can thus accommodate a large amount of adsorbates of various species within its volume. These can alter the properties of CNTs, such as thermal conductivity and chemical behavior [1, 2], and, by extension, they can affect the properties of the CNT forest as a bulk material.

II. METHODOLOGY AND RESULTS

We have previously shown that, despite the conductive nature of CNTs, if illuminated by focused light, the CNT forest can retain the generated heat within an area approximately the same size as the illuminated spot [3-5]. A practical outcome of this "heat trap" effect is the ability to create a hot spot with temperatures greater than 1,500 K using modest amounts of power available from simple optical sources such as laser pointers and even sunlight focused by a small lens [6]. At such elevated temperatures, significant thermal electron emission occurs, enabling devices such as compact thermionic sources and energy converters [7]. The accompanying thermal photon emission provides a spectacular visual of the localized nature of the hot spot (Fig. 1).

Funding: Natural Sciences and Engineering Research Council of Canada, Canada Foundation for Innovation, British Columbia Knowledge Development Fund, Canada First Research Excellence Fund (Quantum Materials and Future Technologies Program)

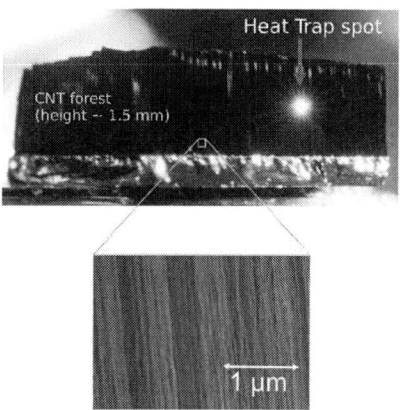

Fig. 1. Localized incandescence from a "heat trap" spot due to laser illumination of a macroscopic CNT forest. Inset is a scanning electron micrograph showing the aligned nature of the nanotubes. (Modified and reprinted with permission from [4]. Copyright 2018 by the American Physical Society.)

Here we report on surprisingly slow dynamics after the initial, rapid heating of the spot upon illumination. This is shown for two different experiments in Fig. 2, where the temperature is seen to decrease by about 200 K over the course of approximately 5 hours, only to then start a slow recovery back to its initial value after another 8 hours. Residual gas analysis using a mass spectrometer showed the presence of hydrogen, water, and nitrogen ions throughout temperature decay and recovery experiments in the vacuum chamber. These must thus originate from the CNT forest, pointing to the possible role of adsorbates within the forest in the observed slow dynamics.

Using computational heat transfer modeling, we provide a possible explanation for this effect based on the thermal removal of adsorbates from the forest through the hot spot and the accompanying transport of more adsorbates from deep within the volume of the forest towards the hot spot. The premise of the model is that adsorbates effectively act as defects leading to increased phonon scattering and a reduction in thermal conductivity, which is linked to the

979-8-3503-7977-8/24 $31.00 © 2024 IEEE

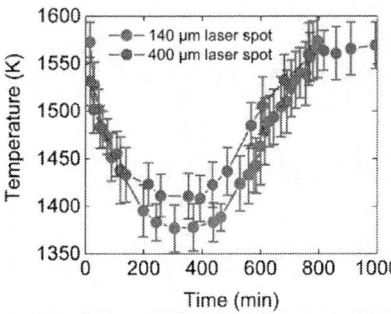

Fig. 2. The hot spot temperature, measured using an intensity comparison pyrometer, as a function of time for two different experimental runs involving different laser spot sizes and correspondingly different laser power levels.

defect concentration based on the data presented in [1]. This desorption-diffusion model qualitatively captures the observed slow dynamics and their timescale (Fig. 3). Initially, removal of adsorbates from the hot spot locally improves thermal conductivity and results in a reduction in the temperature, but this removal is then gradually mitigated and eventually countered by diffusion of further adsorbates from deep within the CNT forest towards the surface. This causes an eventual repopulation of the surface with adsorbates and its return to its starting thermal conductivity and temperature.

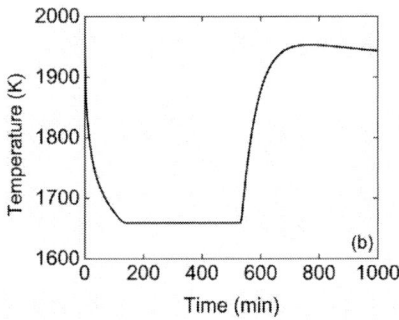

Fig. 3. The hot spot temperature simulated using a model based on desorption from the hot spot on the CNT forest into the surrounding vacuum and the diffusion of further adsorbates from deep within the forest towards the surface. The model correctly captures the experimental timescale of several hours for the decay and recovery phenomena–compare with Fig. 2. (The truncated bottom of the graph is due to the limitation in the data describing thermal conductivity based on defect concentration.)

III. SUMMARY

We observed the decay and recovery of the temperature of a spot on the side surface of a carbon nanotube forest heated by a focused laser beam. The timescales of this decay and recovery are several hours, and a model based on adsorbate removal from the surface and diffusion within the forest qualitatively captures the observed experimental behavior and the timescales involved. These results provide insight into the inner workings of the CNT forest for various applications, and may also more broadly be relevant to mass storage and

transport in other multiscale nanostructures that could be made from different nanowires and nanofibers.

REFERENCES

[1] C. Sevik, H. Sevinçli, G. Cuniberti, and T. Çağın, "Phonon engineering in carbon nanotubes by controlling defect concentration," Nano Lett., vol. 11, pp. 4971–4977, October 2011.

[2] K. Balasubramanian and M. Burghard, "Chemically functionalized carbon nanotubes," Small, vol. 1, pp. 180–192, February 2005.

[3] P. Yaghoobi, M. Vahdani Moghaddam, and A. Nojeh, ""Heat trap": Unusual light-induced-heat localization in carbon nanotube arrays," Solid State Comm., vol. 151, pp. 1105–1108, September 2011.

[4] M. Chang, H. D. E. Fan, M. M. Chowdhury, G. A. Sawatzky, and A. Nojeh "Heat localization through reduced dimensionality," Phys. Rev. B, vol. 98, p. 155422, October 2018.

[5] M. M. Chowdhury, J. F. Young, G. A. Sawatzky, and A. Nojeh, "Broadband infrared hyperspectroscopy with high spatial resolution for the study of nanoscale thermal emitters in vacuum," 2023 IEEE 36th International Vacuum Nanoelectronics Conference (IVNC), Cambridge, MA, 10-14 July 2023, pp. 20–21.

[6] P. Yaghoobi, M. Vahdani Moghaddam, and A. Nojeh, "Solar electron source and thermionic solar cell," AIP Adv., vol. 2, p. 042139, November 2012.

[7] A. Nojeh, "Carbon nanotube photo-thermionics: toward laser-pointer-driven cathodes for simple free-electron devices and systems," MRS Bullet., vol. 42, pp. 500–504, July 2017.

Study of ALD Grown Multilayers Exhibiting Vacancy Induced Conductivity for Electron Emitters

Daniel Burda[1,2,*], Mohammad Allaham[1,3], Alexandr Knápek[1], Marwan S. Mousa[4]

[1]Electron and Plasma Technologies, Institute of Scientific Instruments of the Czech Academy of Sciences, Brno, Czech Republic
[2]Department of Physics, Faculty of Electrical Engineering and Communication, Brno University of Technology, Brno, Czech Republic
[3]Central European Institute of Technology, Brno University of Technology, Brno, Czech Republic
[4]Department of Renewable Energy Engineering, Jadara University, Irbid, Jordan
*Corresponding author: burda@isibrno.cz

Abstract—**Thin oxide multilayers are prepared using low-temperature atomic layer deposition (ALD). The tungsten samples are coated with a multilayer stacks of refractory oxides: Al_2O_3, TiO_2, VO_2, and HfO_2. The properties of the multilayer oxide are controlled by the number of ALD growth cycles, which affects the thickness of the individual layers. The grown layers of dielectrics are usually amorphous. The contaminants present in the ALD chamber also affect the properties of the final multilayer. Tuning the multilayer stack thickness and composition may result in non-conventional effects on field emission from the sharp needle underneath the dielectric layer. Such effects may be oxygen-vacancy-induced conductivity, effects due to polarization of the dielectric or plamonic carrier generation in the case of photon-assisted field emission.**

Keywords—*cold field emission, atomic layer deposition, ALD, FEM, electron device*

I. INTRODUCTION

Modification of metallic and semiconductive ultrasharp emitters of several angstroms to nanometers can change the outcomes when subjected to high electric fields. Past experiments showed various dielectric materials and their contribution to protecting the emitter surface or lowering the local work function, i.e. epoxy coated tips [1,2]. Similar functional structures of nanodielectrics have been investigated in related fields of nanoelectronics, such as the design of plasmonics and transistors, i.e. the effect of dielectric nanoparticles on electron injection [3]. The growth of nanodielectrics is well documented, especially as ultra-thin insulating layers for a channel in FETs.

The effect of the thin oxide layer on the surface of the field emission tip was described in a model based on a Wentzel–Kramers–Brillouin approximation showing a correlation between layer thickness and current density of the total emission current [4]. One of the approaches to modification of a conventional surface of a (semi-)metallic emitter is atomic layer deposition (ALD). Oxygen-vacancy-induced conductivity is documented in stacks of ultrathin Al_2O_3/TiO_2 layers [5], a phenomenon in which the presence of oxygen vacancies in a material leads to increased electrical conductivity.

II. FABRICATION AND MEASUREMENTS

ALD, which is dependent on sequential self-limiting surface reactions of two vapor phase precursors, is one of the methods of precision layer growth. Using thermal ALD recipes for Al_2O_3, TiO_2, VO_2 and HfO_2 layers, we were able to explore different multilayer stacks of the aforementioned dielectrics. The thermal ALD recipe for a single layer consists of a water vapor precursor step and a metal-organic precursor step. The second precursor was one of these: Trimethyl aluminum (TMA), Tetrakis ethylmethylamino hafnium (TEMAH), tetrakis ethylmethyl amido titanium (TEMAT), and tetrakis(ethylmethylamino) vanadium (TEMAV). The process temperature was set at 250 °C. The calculated growth per cycle of ALD depositions is between 0.9 Å/cycle to 1.2 Å/cycle for different oxide recipes.

The multilayer consisted of two dielectric materials, one was always Al_2O_3, which means Al_2O_3 + HfO_2, Al_2O_3 + TiO_2, and Al_2O_3 + VO_2. The number of cycles and thus the overall multilayer thickness varied. The individual layers were stacked, alternating between the two oxides. Multilayers, which were presented during the previous 36th International Vacuum Nanoelectronics Conference [6], in Fig. 1, were thick, with individual layer thicknesses of 2.5 to 5 nm, resulting in a total multilayer thickness of 20 to 50 nm. The focus shifted towards substantially thinner multilayers with a maximum thickness of 10 to 15 nm; each layer consisted of 10 to 15 ALD cycles, thus up to 1.5 nm.

Chemical analysis using X-ray photoelectron spectroscopy found a slight contamination in the form of nitrogen atoms in several multilayer coated samples. To analyze the structure of a multilayer, we milled the coated tips using a focused-ion beam to prepare a lamella. The lamellae were analyzed using a transmission electron microscope (Talos F200i).

III. CONCLUSION

Several configurations of dielectric layers and multilayers were investigated in terms of their structure. Due to the low temperature of the ALD process, predominantly amorphous layers are formed. The research results highlight the need for further research on the subject of nanodielectrics, including the simulation of relevant phenomena.

ACKNOWLEDGMENT

We acknowledge support by the Czech Academy of Sciences (RVO:68081731) and CzechNanoLab Research Infrastructure supported by The Ministry of Education, Youth and Sports of the Czech Republic (LM2018110). D.B. thanks Martina Gálíková from IPM of CAS, Brno, for her help with TEM micrographs.

REFERENCES

[1] M. S. Mousa and T. F. Kelly, "Stabilization of carbon-fiber cold field-emission cathodes with a dielectric coating", *Ultramicroscopy*, vol. 95, pp. 125-130, 2003.

[2] M. S. Mousa and M. Al Share', "Study of the MgO-coated W emitters by field emission microscopy", *Ultramicroscopy*, vol. 79, no. 1-4, pp. 195-202, 1999.

979-8-3503-7977-8/24 $31.00 © 2024 IEEE

[3] I. Ahmed, L. Shi, H. Pasanen, P. Vivo, P. Maity, M. Hatamvand, and Y. Zhan, "There is plenty of room at the top: generation of hot charge carriers and their applications in perovskite and other semiconductor-based optoelectronic devices", *Light: Science & Applications,* vol. 10, no. 1, 2021.

[4] M. S. Mousa and R. V. Latham, "Hot-electron emission from composite metal-insulator microemitters", *Le Journal de Physique Colloques,* vol. 47, no. C7, pp. C7-139-C7-144, 1986.

[5] T. J. Seok, Y. Liu, H. J. Jung, S. B. Kim, D. H. Kim, S. M. Kim, J. H. Jang, D. -Y. Cho, S. W. Lee, and T. J. Park, "Field-Effect Device Using Quasi-Two-Dimensional Electron Gas in Mass-Producible Atomic-Layer-Deposited Al2O3/TiO2 Ultrathin (<10 nm) Film Heterostructures", *ACS Nano*, vol. 12, no. 10, pp. 10403-10409, Oct. 2018.

[6] D. Burda, M. M. Allaham, M. Horáček, and A. Knápek, "Study of Dielectric Nanolayers and Multilayer Coated Emitters", in *2023 IEEE 36th International Vacuum Nanoelectronics Conference (IVNC)*, 2023, pp. 53-54.

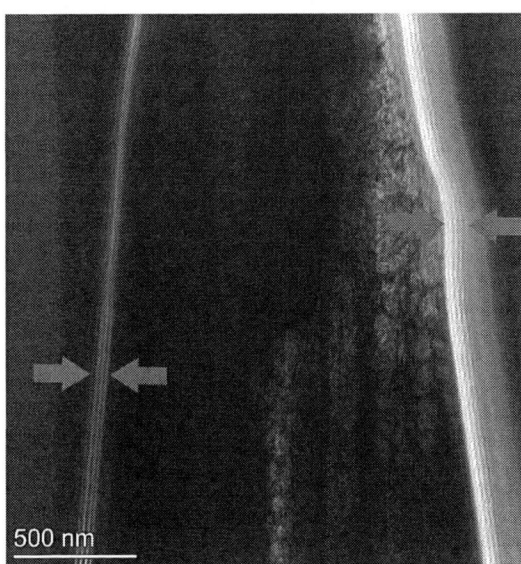

Fig. 1. Multilayer coated tungsten tip; thin lamella, TEM cross section imaging showing the multilayer structure of Al_2O_3 and HFO_2, total multilayer stack thickness is 50 nm [7].

Study of High Compression Ratio Cold Cathode Electron Gun

Qianqian Tang, Mengjie Li, Junjie Zhang, Mei Xiao[*] and Xiaobing Zhang

School of Electronic Science & Engineering, Southeast University, Nanjing 210096, China

[*]Corresponding author: xiaomei@seu.edu.cn

Abstract—**As the traveling wave tube operating frequency moves toward terahertz, the electron injection channel shrinks dramatically, which puts higher demands on the electron gun. In this work, we designed a high compression ratio cold cathode electron gun to fulfill the requirements of terahertz traveling wave tubes. Through the implementation of curved cathode for electron, as well as curved prefocusing electron, curved gate and anode configuration, CST simulation results indicate the extraction of an emission current of 10 mA from a carbon nanotube cathode with a radius of 0.4 mm at an operating voltage of 14 kV. Subsequently, the electrons undergo compression electron beam with a radius of 0.056 mm, thereby achieving an area compression ratio of approximately 51.**

Keywords—*cold cathode electron gun, terahertz, high compression ratio*

I. INTRODUCTION

Due to their exceptional transmission characteristics, traveling wave tubes find widespread application in radar, communications, electronic countermeasures and various military contexts [1-2]. As the operating frequency of traveling wave tubes develops toward terahertz, the size of the electronic channel shrinks to hundreds of microns to tens of microns, or even lower orders of magnitude, which puts higher requirements on the design of electro-optical components [3]. In contrast to hot cathodes, cold cathodes operating on the principle of field emission of electrons offer several advantages including instantaneous startup, elimination of the need for heating, high current density, and ease of miniaturize. Consequently, they hold substantial research value and promising application prospects [4]. Substituting traditional hot cathode electron guns with cold cathode counterparts to achieve the miniaturization of the traveling wave tube has emerged as a prominent research direction among scholars globally [5].

II. DESIGN OF THE COLD CATHODE ELECTRON GUN

The designed carbon nanotube cold cathode electron gun is structured akin to the Pierce electron gun, comprising a curved cathode, curved pre-focused pole, curved grid, and anode without grid control. The cathode and the prefocusing pole are spherical ensuring the equipotential contours can be more nearly spherical. This design facilitates electron flow toward the center of the cathode curvature. Near the gate aperture, an evanescent electrostatic lens is formed focusing the electron injection as the equipotential contours curves inwardly toward the anode aperture. Ultimately, the electron exit the gate and drift influenced by space charge forces.

The structural model of the electron gun is depicted in Figure 1, with detailed structural parameters provieded in Table I.

The National Natural Science Foundation of China (Grant No. 61971133)

TABLE I.   STRUCTURAL PARAMETERS OF THE ELECTRON GUN

Structures Parameters	Values
Cathode radius	0.4 mm
Cathode arc radius	2.4 mm
Gate arc radius	1.4 mm
Cathode voltage	0 kV
Prefocusing pole voltage	0 kV
Gate voltage	14 kV
Anode voltage	14 kV

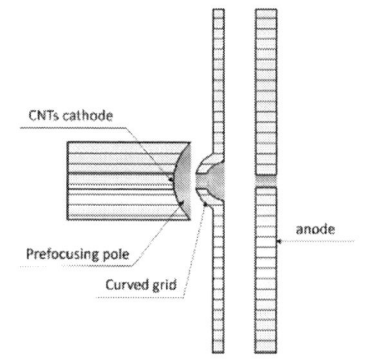

FIG1. STRUCTURAL MODEL OF CARBON NANOTUBE COLD CATHODE ELECTRON GUN

III. ELECTRON GUN SIMULATION RESULTS

The simulation results obtained from the CST demonstrate that a 10 mA electron beam is extracted from a cathode with a radius of 0.4 mm at voltage of 14 kV. Figure 2 illustrates the electron trajectory of the carbon nanotube cold cathode electron gun, showcasing well focused electron positioned after the anode. Moreover, the electron distribution diagram in Figure 3, reveals that the radius of the electron beam waist measures 56 μm, with the electron gun boasting an area compression ratio of approximately 51. These findings fulfill some of the design requirements of terahertz traveling wave tube.

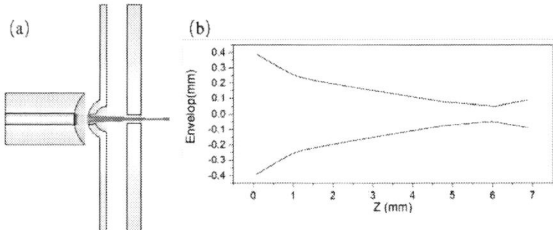

FIG2. ELECTRON BEAM TRAJECTORY DIAGRAM

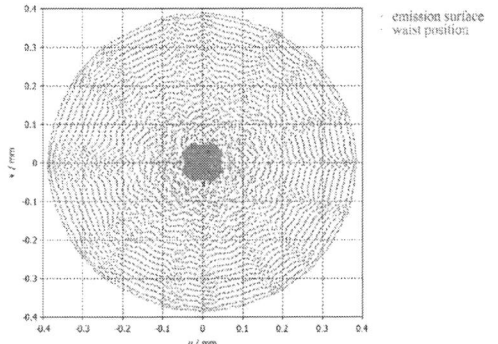

FIG3. STRUCTURAL MODEL OF CARBON NANOTUBE COLD CATHODE
ELECTRON GUN

IV. MAGNETIC FOCUS SYSTEM SETUP

The axial magnetic induction required to focus an electron injection from the cathode can be calculated by the following equation:

$$B_b(Gs) = \frac{833}{r_0} \cdot \frac{I_0^{1/2}}{V_0^{\frac{1}{4}}} \qquad (1)$$

where r_0 is the electron injection radius (cm), I_0 is the electron injection current (A), and V_0 is the electron injection voltage (V).

The calculated required magnetic induction strength for the electron gun designed in this work is 1367.5 G. However, considering the effects of high-frequency defocusing and aperture dispersion, the actual required axial magnetic induction strength should be higher, typically chosen to be between 1.5 times and 2 times the calculated value. Therefore, in this study, a line pack focusing system with an axial magnetic induction strength of 2735 G is selected to maintain the stability of the electron beam.

Figure 4 shows the magnetic field strength of the designed line pack focusing system, with Figure 5 illustrates the motion trajectory of the electron beam with the magnetic field. It is observed that the electron beam experiences minimal fluctuation within the electron channel, thus satisfying some of the design requirements for terahertz traveling wave tubes.

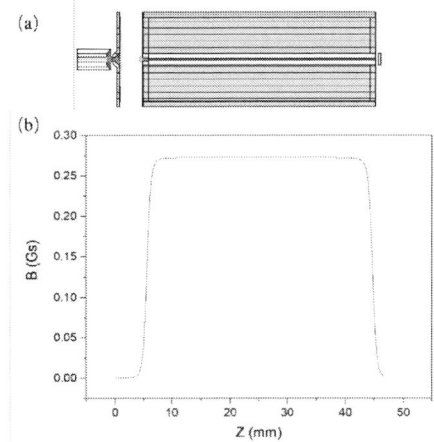

FIG4. PLOT OF THE WIRE PACKET FOCUSING SYSTEM AND ITS MAGNETIC
FIELD STRENGTH TRAJECTORY

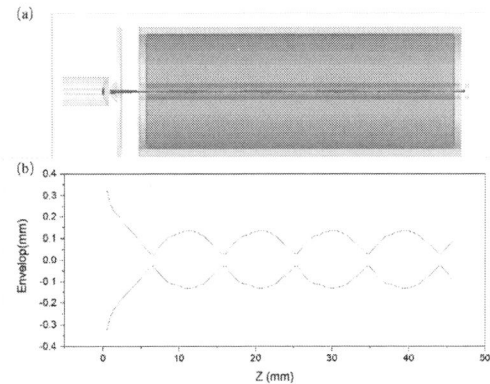

FIG5. PLOT OF ELECTRON BEAM TRAJECTORY WITH FOCUSING SYSTEM
ADDED

V. CONCLUSION AND PROSPECT

In this study, a cold cathode electron gun based on carbon nanotubes is designed to confine the electron beam using a wire-wrapped focusing system. This setup is capable of generating a microfine electron beam that satisfies the requirements of certain terahertz traveling wave tubes.

ACKNOWLEDGMENT

The author thanks the support of the National Natural Science Foundation of China (Grant No. 61971133(Xiaobing Zhang)).

REFERENCES

[1] H. J. Kim, L. B. Jang, W. B. Seo, and J. J. Choi, "Experimental investigation of broadband vaned helix traveling-wave tube," Jpn. J. Appl. Phys., vol. 45, pp. 292–299, Jan. 2006.

[2] C. K. Chong and W. L. Menninger, "Latest advancements in high-powermillimeter-wave helix TWTs," IEEE Trans. Plasma Sci., vol. 38, pp. 1227–1238, Jun. 2010.

[3] S. M. Yan, W. Su,Y. J. Wang, et al. "Theoretical and simulation study of 0.14 THzfundamental mode multi-beam folded waveguidetraveling wave tube," Acta Phys. Sin., vol. 63, pp. 1-10, Dec. 2014.

[4] J. Zhang, J. Chen, J. Xu, et al. "Development of a K-band traveling wave tube based on carbon nanotube cold cathode," Vacuum., vol. 203, pp. 1-6, Sep. 2022.

[5] C. M. Armstrong, "The quest for the ultimate vacuum tube," IEEE Spec., vol. 52, pp. 28–51, Dec. 2015.

Temperature Dependence of the Field Emission Characteristics of AlGaN/GaN Nanoscale Vacuum Diodes

Nathaniel Hernandez [1,2], Marc Cahay [1,2,*], James Hart [1,3], Jonathan O'Mara [3,4], Jonathan Ludwick [2,5], Dennis E. Walker Jr. [3], Tyson Back [5], and Harris Hall [3]

[1] *Spintronics and Vacuum Nanoelectronics Laboratory, University of Cincinnati, Cincinnati, Ohio, USA*
[2] *UES, a BlueHalo company, Dayton, Ohio, USA*
[3] *Air Force Research Laboratory, Sensors Directorate, Wright-Patterson Air Force Base, USA*
[4] *KBR, Beavercreek, Ohio, USA*
[5] *Air Force Research Laboratory, Materials and Manufacturing Directorate, Wright-Patterson Air Force Base, Ohio, USA*

*Corresponding author: cahaymm@ucmail.uc.edu

Abstract— **We investigate the temperature dependence of field emission (FE) characteristics of AlGaN/GaN heterojunction-based lateral nanoscale vacuum diodes over a temperature range varying from 77 to 373 kelvin. The vacuum diodes have a sharp (trochoidal-like) cathode and blunted anode with the shortest distance between the tip of the cathode and the anode of 210 nanometers. The DC FE characteristics were recorded using a compliance current of 100 nanoamperes. Under forward mode of operation, the vacuum diodes exhibit three regimes of operation: one associated with FE from adsorbates on the tip of the cathode at low applied bias, followed by FE from the bare cathode at larger forward bias, and finally onset of self-heating effects. The latter leads to a smaller FE current under reverse sweep. In the reverse sweep, the current-voltage characteristics suggest the onset of conduction through the substrate.**

Keywords—*Aluminum Gallium Nitride / Gallium Nitride, Field Emission from Two-Dimensional Electron Gas, Vacuum Nanodiode*

I. INTRODUCTION

Vacuum field effect transistors (VacFETs) provide a range of improvements over traditional silicon-based transistors. They offer enhancements such as quicker switching speeds, the ability to operate under extreme temperatures, and an enhanced resistance to radiation [1,2]. III-nitride semiconductor materials show promise for use in field emission vacuum electronics, thanks to their exceptional thermal and chemical stability, low electron affinity, and high breakdown fields [3,4].

Over the last three years, we have investigated the design, fabrication and experimental measurements of the DC electronic transport characteristics of AlGaN/GaN nanoscale lateral vacuum diodes. Devices with both metallic and AlGaN/GaN anodes were examined all of which had triangular cathodes and cathode to anode spacings from 50 to 600 nanometers [5]. The room temperature characterization indicated rectified FE of the AlGaN/GaN diodes forward bias FE current in the range of microamperes or milliamperes, respectively, when biased within a maximum range varying from 10 to 30 volts. Our findings represent the first successful demonstration of FE of electrons across a nanogap (created through electron beam lithography in an AlGaN/GaN heterojunction), wherein the two-dimensional electron gases (2DEG) serve as both the cathode and anode.

In this work, we focus on AlGaN/GaN vacuum diodes with trochoidal shaped cathodes facing an AlGaN/GaN anode and study their FE characteristics over a temperature range from 77 to 373 kelvin. The sharp cathodes possess two distinct features that enhance their functionality: Their shape and the presence of a confined 2DEG both contribute to the concentration of electrostatic field lines at the cathode tips, thereby increasing the field enhancement factor. Moreover, the geometric asymmetry between the cathode and anode is anticipated to result in greater rectification capabilities. The forward bias mode of operation is characterized by loop-type behavior corresponding to various modes of operation, including a regime which indicates potential FE from adsorbates on the cathode tip, FE from bare cathodes with onset of self-heating effects at large forward bias. In the reverse sweep in the forward mode of operation, an Ohmic regime is observed at low voltage which is associated to leakage current through the substrate.

II. FABRICATION AND METHODS

The wafer and epitaxy were purchased from IQE and grown by metal organic chemical vapor deposition on 4 inch 6H-SiC substrates with the following structure: AlN nucleation / 1.8 um Fe-doped GaN buffer / GaN channel / 1 nm AlN / 16 nm $Al_{0.28}Ga_{0.72}N$ / 3 nm GaN cap [6]. The vacuum field emission diodes were created by introducing a nanoscale vacuum gap, defined by electron beam lithography, into the channel of an AlGaN/GaN heterostructure. In this diode configuration, the 2DEG from the AlGaN/GaN stack act as both the cathode and the anode on either side of the nanoscale vacuum gap. A top down SEM image of a vacuum diode is shown in Fig. 1, with a gap separation of approximately 210 nanometers. The diode is tested with a DC I–V sweep, by sweeping the voltage from a negative to positive voltage and then sweeping back to the starting negative voltage. These tests were repeated four times, with a 60 milliseconds delay per voltage step, each at a temperature of 77, 150, 200, 293, and 373 kelvin, while under a 10^{-7} torr vacuum in a vacuum probe station.

III. RESULTS AND DISCUSSIONS

Fig. 2 illustrates the FE characteristics of a diode with a trochoidal-shaped cathode and blunt anode during three consecutive sweeps, forward and reverse sweeps shown as the top and bottom plot, respectively. The experimental FE data (open circles) in Fig. 2 show that the turn-on voltage rises at lower temperatures, aligning with FE theory. In forward bias, the diode displays loop-type behavior, coincident with a rapid initial current increase due to adsorbates on the cathode, followed by a slower rise from the bare cathode at larger applied bias. This pattern is similar to the one observed in the FE data from carbon nanotube fibers [7].

At low bias, the FE data for the reverse sweep can be fitted by an Ohmic model, which is attributed to leakage current through the substrate, which can be related to the etching process to create the nanogap past the 2DEG. At low applied bias, the adsorbates can re-attach to the surface leading to the reproducible loop-type behavior for successive voltage forward and reverse sweeps (not shown). The FE current in the upward swings during the second and third sweeps were found to be smaller than for the first sweep. This indicates a small degradation of the cathode because of its morphological change due to self-heating effects during the first sweep.

To fit the FE characteristics of the diode (solid lines in Fig. 2), the following modification of the AHFP mathematical form [8] to fit to FE experimental results was used:

$$\ln(I_m) = \ln(V_m^\kappa \exp(c + b/V_m) + aV_m^n) \qquad (1)$$

where b and c are fitting parameters of the FE equation (κ was set to 1.5); a and n are fitting parameters associated to the leakage current through the substrate; and I_m and V_m is the measured current and voltage, respectively. The turn-on voltage of the vacuum diode is defined as the voltage at which the FE current and leakage current from the fits are identical.

IV. Conclusions And Future Work

The results reported here are the first step towards the fabrication of vacuum field effect transistors using in-plane side gates and a thorough investigation of the temperature dependence of their FE characteristics which could find applications as sensors over a wide range of temperature.

Acknowledgment

This material is based upon work supported by the Air Force Office of Scientific Research (Program Manager, Dr. John Luginsland) under Award No. FA9550-20RXCOR027. M. Cahay and N. Hernandez acknowledge support from the Air Force Summer Research Fellowship Program (SFFP is FA9550-20-F-0005) during the Summers of 2021, 2022, and 2023 and the support by the Air Force Research Labs under contract No. FA8650-22-F-5815. J. Ludwick was supported by the Air Force Research Labs under contract No. FA8650-16-D-5408.

References

[1] H. D. Nguyen, J. S. Kang, M. Li, and Y. Hu, "High-performance field emission based on nanostructured tin selenide for nanoscale vacuum transistors," *Nanoscale*, vol. 11, no. 7, pp. 3129–3137, Feb. 2019.

[2] W. M. Jones, D. Lukin, and A. Scherer, "Practical nanoscale field emission devices for integrated circuits," *Appl. Phys. Lett.*, vol. 110, no. 26, p. 263101, Jun. 2017.

[3] K. R. Sapkota *et al.*, "Ultralow voltage GaN vacuum nanodiodes in air," *Nano Lett.*, vol. 21, no. 5, pp. 1928–1934, Mar. 2021.

[4] D.-S. Zhao *et al.*, "An Al0.25 Ga0.75N/GaN lateral field emission device with a nano void channel*," *Chinese Phys. Lett.*, vol. 35, no. 3, p. 038103, Mar. 2018.

[5] N. Hernandez *et al.*, "Field emission characteristics of AlGaN/GaN nanoscale lateral vacuum diodes," *J. App. Phys.*, vol. 135, no. 20, p. 204305, May 2024

[6] R. C. Fitch *et al.*, "Implementation of high-power-density x-band AlGaN/GaN high electron mobility transistors in a millimeter-wave

monolithic microwave integrated circuit process," *IEEE Electron Device Lett.*, vol. 36, no. 10, pp. 1004–1007, Oct. 2015.

[7] P. T. Murray *et al.*, "Evidence for adsorbate-enhanced field emission from carbon nanotube fibers," *Applied Physics Letters*, vol. 103, no. 5, p. 053113, Jul. 2013.

[8] S. V. Filippov, A. G. Kolosko, E. O. Popov, and R. G. Forbes, "Field emission: calculations supporting a new methodology of comparing theory with experiment," *R. Soc. Open Sci.*, vol. 9, no. 11, p. 220748, Nov. 2022.

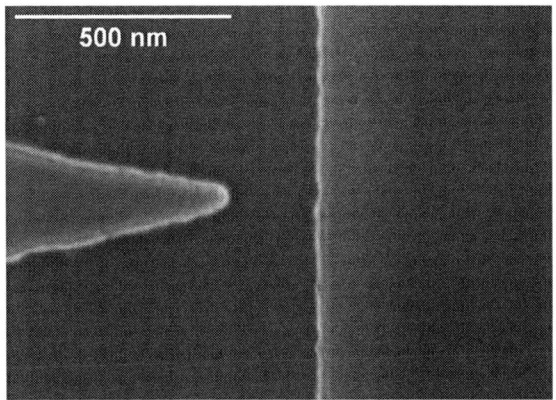

Fig. 1. SEM image of a vacuum diode with a trochoidal-shaped cathode and blunted anode. The shortest distance between the cathode and anode is 210 nanometers.

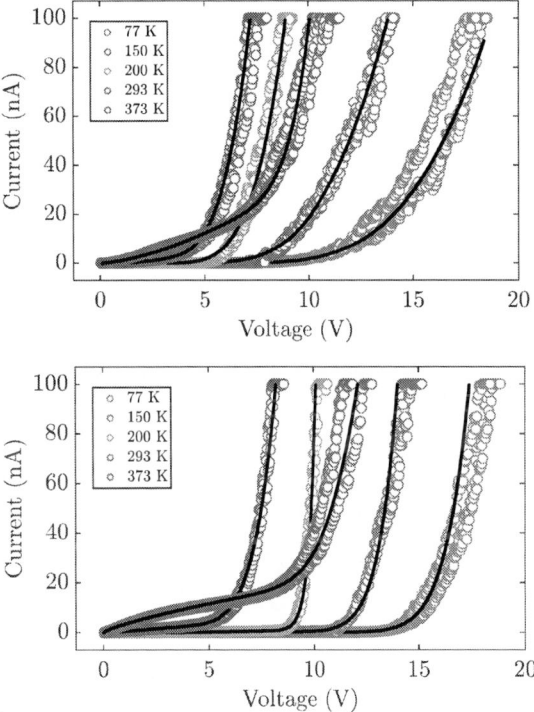

Fig. 2. Temperature dependence of the FE characteristics of a vacuum diode with have a sharp (trochoidal-like) cathode and blunted anode shown in Fig. 1 with the forward bias upsweep and the reverse bias downsweep as the top and bottom plots, respectively. The solid lines show a fit to the FE characteristic using a modified generalized Murphy-Good expression given in (1).

The Relation Between the Electron Backscattering Coefficients and Elastic Peak Intensity for C, Si, Cu, Ag, and Au at Low Energy Range

Ahmad M. D. (Assa'd) Jaber[1*], Mohamed M. El-Gomati[2,3], Marwan S. Mousa[4], Alexandr Knápek[5]

[1]Basic Medical Sciences, Aqaba Medical Sciences University, Aqaba 77110, Jordan
[2]Department of Electronic Engineering, University of York, Heslington, York YO10 5DD, United Kingdom
[3]York Probe Sources Ltd., 7 Harwood Rd, YO26 6QU, York, United Kingdom
[4]Renewable Energy Engineering, Jadara University, Irbid 21110, Jordan,
[5]Institute of Scientific Instrument of the Czech Academy of Science, Královopolská 147, 61200 Brno, Czech Republic

*Corresponding author: ahmad.jabr@amsu.edu.jo

Abstract—This study compares the backscattered electron coefficient (η) with the elastically scattered electrons coefficient (ε) for the increasingly used low electron beam energies in microscopy; low voltage electron microscopy (LVSEM) and low energy scanning electron microscopy (SLEEM/LEEM). The backscattering factor has been an extensively studied and used property of materials across the periodic table and at various beam energies. ε, however, has also been extensively used as material property but only at relatively low incident electron beam energies, and mostly in surface analysis where the material studied is under ultra-high vacuum conditions, unlike η which is mostly used or studied under high vacuum environment. Both η and ε were theoretically calculated using the Monte Carlo model based on the Geant 4 library and tools. These calculations were carried out for C, Si, Cu, Ag, and Au for electron beam energies in the range of 100 eV to 1 keV. The backscattering coefficient in this electron energy range is found to exhibit intriguing properties. Furthermore, the results indicate a direct relationship between η and ε.

Keywords—*backscattered electron coefficient, elastically scattered electrons coefficient, Monte Carlo.*

I. INTRODUCTION

When an electron beam impinges on a solid target surface, the incident electrons undergo numerous elastic and inelastic scattering events. Elastic scattering occurs when the incident electrons interacts with the nuclei of the target atoms and their energy remains unchanged, but their direction changes. This is mainly responsible for the backscattering of the incident electrons from the target surface. Inelastic scattering occurs when the incident electrons interact with the outer-shell electrons of the target atoms, resulting in energy loss and a slight change in the direction of the incident electrons. The energy loss of the incident electrons is used to excite secondary, plasmon, and Auger electrons and X-rays. The probability of an elastic scattering event increases for elements with a larger atomic number (Z) and lower incident electron energies (Ep). However, inelastic scattering events increase for elements with lower atomic number (Z) and higher incident electron energy (Ep). Backscattered electrons (BSE) are essential for studying a material's image, structure, and composition in several electron microscopy technologies [1]. The backscattering electron coefficient (η) is conventionally defined as the ratio of the number of electrons that backscatter out of the sample surface with energy greater than 50 eV to the total number of incident electrons. The elastically scattered electrons coefficient (ε) is the ratio of the intensity of the elastic peak (zero-loss energy electrons) to that of the total number of incident electrons. Several attempts have been conducted to measure η at low incident electron energies [2,3,4]. ε has also been extensively used as a material property but only at relatively low incident electron beam energies, and mostly in surface analysis where the material studied is under ultra-high vacuum conditions [5], unlike η, which is mostly used or studied in high vacuum. Both η and ε were theoretically calculated using Monte Carlo model based on the Geant 4 library and tools [6]. The present study investigates the relationship between the electron backscattering coefficients and elastic peak intensities for C, Si, Cu, Ag, and Au I the low-energy range for electron beam energies in the range of 100 eV to 1 keV.

II. MONTE CARLO MODEL

The Monte Carlo model used in the present study based on the model developed by Kieft and Bosch [7] who used the Geant4 platform of CERN [6]. The Model combines Mott cross-sections with phonon-scattering based cross-sections for the elastic scattering of electrons. These Mott (differential) cross-sections are obtained by solving the Dirac equation for electrons deflected by the positively charged nucleus of an atom, as screened by orbital electrons [8]. The model also uses discrete inelastic losses of inelastic scattering to calculate the stopping power and inelastic mean free path and to generate secondary electrons with every energy loss [9]. This model provides a spectrum of electrons that escape from the surface. The spectrum includes secondary, Auger, plasmon, and backscattered electrons (BSE). The backscattered electrons including the zero-loss energy electrons which are preforming the elastic peak with the E_p energy.

III. RESULTS AND DISCUSSION

Fig. 1 shows η as a function of primary electron energy (E_p) below 1 keV for C, Si, Cu, Ag and Au. Several observations are presented in Fig. 1. First, η increased rapidly for Cu, Ag, and Au as the incident electron energy increased. For C and Si, η increases to reach the maximum values at 150 eV, and decreases to reach lowest value at 300 eV and then remains constant for higher energies. Second, the η of Au is lower than that of all elements at an E_p less than 200 eV. At 200 eV$< E_p <$ 700 eV, η of Cu is greater than that of η for all the elements. The η of Au at E_p greater than 700 eV is greater than that of all other elements. At E_p less than 200 eV, the η

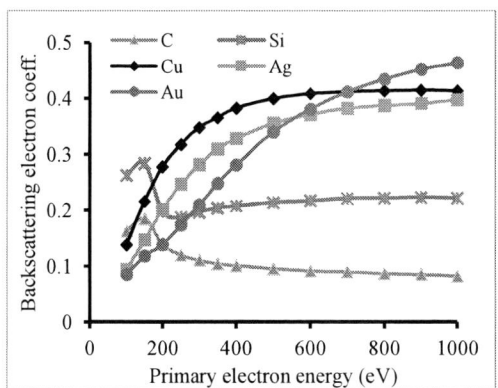

Fig. 1. Backscattering electron coefficient as a function of primary electron energy

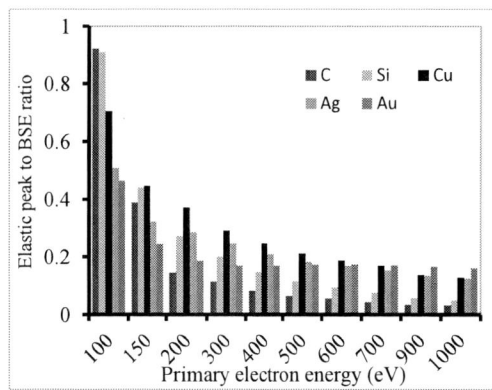

Fig. 2. The elastically scattered electrons coefficient (ε) as a function of primary electron energy

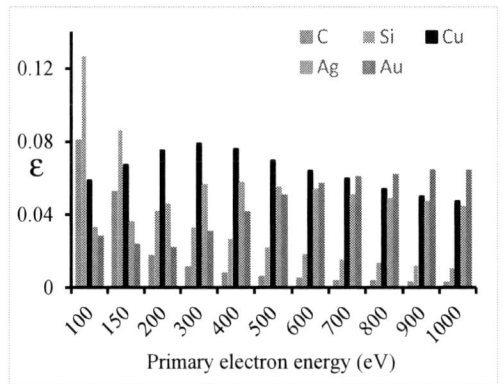

Fig. 3. Elastically scattered electrons coefficient (ε) as a function of primary electron energy

of Si is the largest among all the elements studied. Form these observations one could not obtain a general trend for η at low primary electron energies below 800 eV.

Fig. 2 shows that the elastically scattered electrons coefficient (ε) decreases as the primary electron energy increases for elements with low atomic numbers. Conversely, for heavy elements, ε increases with increasing primary electron energy. However, in the case of Cu, the intensity increases until it reaches a maximum value at 300 eV and then begins to decline for higher primary electron energies. The η results shown in Fig. 1 are strongly related to ε that are shown in Fig. 2. This could explain the behavior of η at low energies. However, El-Gomati et al [2] and Hashimoto et al [4] explained the behavior of η at low energies to be due to surface contamination. Fig. 3 shows that the ratio of the elastic peak intensity to the backscattered electron yield as a function of E_p. Generally, the ratio decreases as E_p is increased. These results show that the elastic peak electrons dominate the escaped backscattered electrons at low E_p particularly for light elements. In addition, the elastic peak electrons of Cu were larger than those of Ag and Au at Ep values less than or equal to 600 eV. This reflects that the η results of Cu were larger than those of Ag and Au. When E_p is greater than or equal to 700 eV, the Au elastic peak becomes larger than those of the other elements, thus causing η of Au to become larger than that of the other elements.

IV. CONCLUSION

The results of the present study show a strong relationship between the backscattering electron coefficient (η) and the elastically scattered electron coefficient (ε) at primary electron energies less than or equal to 1 keV. It is clear that the η behavior at a low E_p follows the elastic electron peak intensity. η and the elastic peak intensity is not proportional to the atomic number at primary electron energies below 1 keV. ε for Cu is higher than that for Ag and Au at energies lower than 700 eV. This reflects on the result for η for Cu to be higher than those for Ag and Au.

REFERENCES

[1] L. Reimer, Scanning Electron Microscopy, 2nd ed., Springer Science & Business Media Verlag Berlin Heidelberg, New York, 1998.

[2] M. M., El Gomati, C. G. Walker, A. M. D. Assa'd, M Zadrazil, "Theory experiment comparison of the electron backscattering factor from solids at low electron energy (250–5,000 eV)", Scanning: The Journal of Scanning Microscopies, vol. 20, pp. 2-15 (2008).

[3] E. Napchan, Backscattered electrons from the SEM, Microscopy and Analysis, pp. 9-11, Jan. (2001).

[4] Y. Hashimoto, A. Muto, E. Woods, T. Walters, and D.C. Joy, "Image Contrast in Energy-Filtered BSE Images at Ultra-Low Accelerating Voltages, Microscopy Today, pp.20-24, May (2015).

[5] G. Gergely,"Elastic backscattering of electrons: determination of physical parameters of electron transport processes by elastic peak electron spectroscopy". Progress in Surface Science, vol. 71, pp. 31–88, (2002).

[6] S. Agostinelli et al., "Geant4—a simulation toolkit," Nucl Instrum Methods Phys Res A, vol. 506, no. 3, pp. 250–303, Jul. (2003)

[7] E. Kieft and E. Bosch, "Refinement of Monte Carlo simulations of electron–specimen interaction in low-voltage SEM," J Phys D Appl Phys, vol. 41, no. 21, p. 215310, Nov. (2008).

[8] Z. Czyzewski, D. O'Neill MacCalium, A. Romig and D. C. Joy," Calculations of Mott scattering cross section",J. Appl. Phys. vol. 68 pp.3066–3072, (1990).

[9] W. S. M. Werner, "Electron transport in solids for quantitative surface analysis," Surface and Interface Analysis, vol. 31, no. 3, pp. 141–176, Mar. (2001).

Theoretical Analysis of Electron Emission from Hafnium Carbide Tip

Toshiaki Kusunoki[1], Noriaki Arai[2]

[1] Hitachi Ltd, R&D Group, Kokubunji, Tokyo, Japan
[2] Hitachi High-Tech, Nano-Technology Solution Business Group, Hitachi-Naka, Ibaraki, Japan
Corresponding author: toshiaki.kusunoki.hg@hitachi.com

Abstract—We analyzed electron emission characteristics of a hafnium carbide [HfC(100)] single-crystal tip at various temperatures by using Fowler-Nordheim (FN) theory and extended Schottky emission (ESE) theory. We estimated the work function of HfC(100)-CFE tip to be 3.26 eV by analyzing the FN plot and total energy distribution curve. Assuming the HfC(100) plane maintains the same work function in a wide temperature range, we calculated the total energy distribution of the HfC(100) tip. The shape and ΔE of the energy distribution curves correspond well with the experimental results at CFE, low- temperature TFE (677 K), and ESE (1706 K).

Keywords—Hafnium carbide, Fowler-Nordheim emission, Extended Schottky emission

I. INTRODUCTION

Hafnium carbide has been one of the candidates for emitter materials because of its low work function (about 3.3 eV), high melting point (about 4200 K), and high resistance to ion bombardment. In the last IVNC2023, we reported electron emission from HfC(100) tip at various temperatures[1]. To understand the emission mechanism of it, we conducted theoretical analysis of the emission properties by using Fowler-Nordheim (FN) theory[2] and extended Schottky emission (ESE) theory[3].

II. ESTIMATION OF WORKFUNCTION

First, we estimated physical parameters of the HfC(100) emitter tip—work function ϕ, field enhancement factor β, and emission area α—by fitting the experimental results of I-V measurements and energy distribution spectra at cold-field emission mode to the FN emission theory. The emission current density J (A/m^2) is expressed as

$$J = 1.54 \times 10^{-6} \frac{F^2}{\phi t^2(y)} \exp\left(-6.83 \times 10^9 \frac{\phi^{\frac{3}{2}}}{F} s(y)\right) \frac{\pi p}{sin(\pi p)} \quad (1)$$

$$y = 3.80 \times 10^{-5} \frac{\sqrt{F}}{\phi} \quad (2)$$

$$p = \frac{kT}{d} \quad (3)$$

where me is the electron mass, $h=h/2\pi$ is Dirac's constant, ϕ is the work function, e is the elementary charge, F is electric field strength, and d is the transverse enegy. The functions of $t(y)$ and $s(y)$ are correction variables due to the elliptic integral function[4,5]. The F-N plot is expressed as

$$\ln \frac{I_p}{V_e^2} = \frac{m}{V_e} + n \quad (4)$$

$$m = -6.83 \times 10^9 \frac{\phi^{\frac{3}{2}} s(y)}{\beta} \quad (5)$$

$$n = \ln\left(1.54 \times 10^6 \frac{\alpha \beta^2}{\phi t(y)^2} \frac{\pi p}{sin(\pi p)}\right) \quad (6)$$

and the transverse energy d (eV) is expressed as

$$\frac{1}{d}\left\{\frac{9.76 \times 10^{-11} V_e \beta}{\phi^{1/2}}\right\} = t\left(\frac{3.79 \times 10^{-5} (V_e \beta)^{1/2}}{\phi}\right) \quad (7)$$

As a method for experimentally obtaining d, we analyzed the slope of low energy tail ($E \ll E_F$) of energy distribution curve $J'(E)$[6].

$$J'(E) = \frac{J}{d} exp\left(\frac{E}{d}\right) f(E) \quad (8)$$

$$f(E) = \frac{1}{1 + exp\left(\frac{E - E_F}{k_B T}\right)} \quad (9)$$

$$\frac{d}{d\epsilon} \ln J'(E) = \frac{1}{d} \quad (10)$$

Figure 1 is the FN plot and Fig. 2 is the logarithm plot of energy distribution curve of HfC(100) CFE at 21 μA/sr. From these plots, we obtained m, n, and d, and by solving simultaneous equations (5) and (7), work function ϕ was calculated as 3.26 eV, and field enhancement factor β as 7.66 x 10^5 (1/m). The emission area α calculated by inputting these values to equation (6) was 33.6 nm^2. The calculated ϕ= 3.26 eV of HfC(100) CFE agrees well with the thermionic work function (ϕ = 3.34 eV) of HfC(100) at 2000 K reported by W.A. Mackie et al[7]. This suggests that the work function of HfC(100) does not depend on operating temperature.

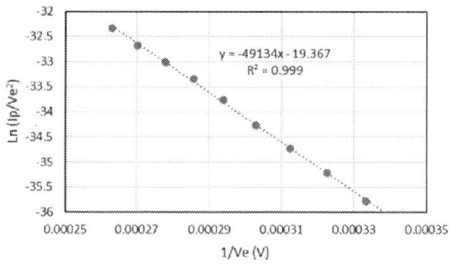

Fig. 1. FN plot of HfC(100) CFE.

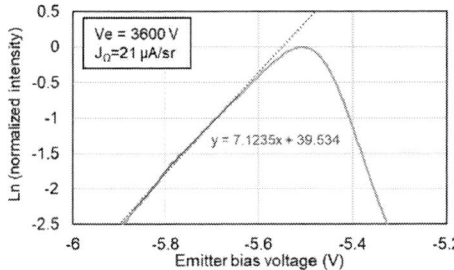

Fig. 2. Logarithm plot of energy spectrum of HfC(100) CFE.

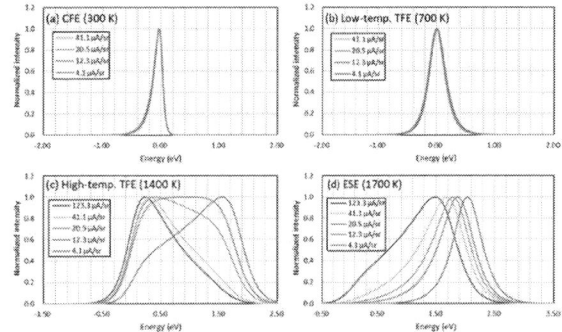

Fig. 3. Calculated energy distribution curves of HfC(100) tip at φ=3.26 eV.

Assuming 3.26 eV as the work function of the HfC(100) emission plane over the temperature range from 300 K to 2000 K, we calculated the total energy distribution of emitted electrons from FN high-temperature approximation formula (11) and ESE emission theory using parabolic function approximation (12) and compared it with the experimental results[8,3].

$$dJ_{FE}(E) = \frac{4\pi m_e ed}{h^3} exp\left(-c - \frac{\xi}{d}\right)$$
$$\times \frac{e^{E/d}}{exp[(E+\phi)/k_B T] + 1} dE \qquad (11)$$

$$dJ_{ESE}(E) = \frac{4\pi m_e e}{h^3} \frac{\kappa}{1 + exp\left(\frac{E + \phi - \Delta\phi}{k_B T}\right)}$$
$$\times ln\left\{1 + exp\left(\frac{E}{\kappa}\right)\right\} dE, \qquad (12)$$
$$\kappa = \frac{\hbar}{\pi\sqrt{m_e}}(4\pi\varepsilon_0 eF^3)^{1/4}$$

Figure 3 shows the calculated total energy distribution curves at 300, 700, 1400, and 1700 K, and figure 4 compares the energy width ΔE between calculation (open circles) and experiment(solid circles). The ΔE calculated at 300 and 700 K using the FN field emission theory coincides well with the experimental result below 20 μA/sr. The increase in the ΔE above 20 μA/sr in experiments might be due to the influence of space charge broadening of the electron beam (Boersch effect), which is not included in the emission theory. The ΔE calculated at 1700 K using ESE theory corresponds well with the experimental result. At 1400 K, there is wide gap in the ΔE value, but the trend that ΔE increases at low angular current and then decreases at high angular current is well reproduced. This might be the result of transformation of the emission mechanism from Schottky emission to thermal field emission.

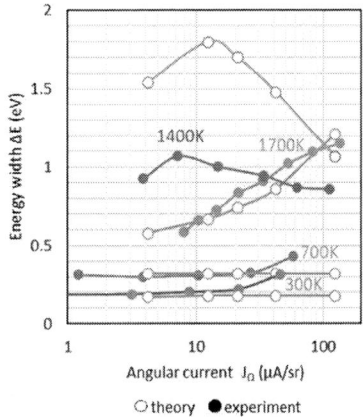

Fig. 4. Comparison of ΔE between theoretical calculation and experiment.

REFERENCES

[1] T. Kusunoki, and N.Arai, "ELECTRON EMISSION FROM HFC(100) TIP AT VARIOUS TEMPERATURES", IVNC 2023 Cambridge MA, USA, DOI: 10.1109/IVNC57695.2023.10188984

[2] P. W. Hawkes and E. Kasper, Principles of electron optics, 2, Academic press, 92 (2017)

[3] M. J. Fransen, J. S. Faber, Th. L. van Rooy, P. C. Tiemeijer, and P. Kruit, J. Vac. Sci. Technol. B 16(4), 2063 (1998)

[4] R. G. Forbes, Appl. Phys. Lett. 89, 113122, (2006)

[5] L. W. Swanson and G. A. Schwind, Adv. Imag. Ele. Phys.159,63 (2009)

[6] J. W. Gadzuk and E. W. Plummer, Rev. Mod. Phys. 45(3), 487 (1973)

[7] W. A. Mackie, J. L. Morrissey, C. H. Hinrichs, and P. R. D. Citation: J. Vac. Sci. Technol. A 10, 2852 (1992)

[8] R. D. Young and E. W. Muller, Phys. Rev. 113 115(1959)

Theoretical Analysis of the Nanovoids Influence on the Field Emission

Serhii Lebedynskyi*, Yuliia Lebedynska and Roman Kholodov

Institute of Applied Physics, National Academy of Sciences of Ukraine, Sumy, Ukraine

*Corresponding author: lebos@nas.gov.ua

Abstract— **This paper presents a study of the influence of nanopores formed in the metal surface layer on the field emission current (the first stage of the occurrence of breakdown of the accelerator structure). It is shown that the field emission current through nanopores has a resonant character. It was found that under resonance conditions, an increase in the current density from the copper surface by more than 3.5 times compared to the case of an unmodified surface can be observed. The presence of a dielectric in the nanopores expands the resonance region. When the resonance conditions are not met or the diameter of the nanopores increases, the current density decreases to zero, which can be a way to prevent breakdown.**

Keywords—accelerating structures, field emission, nanopores.

I. INTRODUCTION

The accelerator walls are irradiated during the beam passage of charged particles. One of the types of radiation defects formed during irradiation is pores. During the study of the structural materials of the future CLIC (Compact Linear Collider), it was found that the surface layer of copper (the main material of the future accelerator) at a depth of about 0.5 µm is heavily affected by pores [1].

In [2], [3], issues related to the formation of blisters, i.e., bubbles on the metal surface, were considered. Their formation is a common effect on the surface of a metal exposed to irradiation. The authors emphasize that hydrogen tends to penetrate copper and accumulate in some voids near the surface. In the simulation, the authors show atomistic mechanisms of bubble and blister growth under hydrostatic internal pressure H. All this leads to the conclusion that dielectric, in particular hydrogen, can accumulate in the near-surface layer. Therefore, the study of the effect of dielectric inclusions in the metal surface layer is important and requires further investigation.

In this paper, we theoretically investigate the effect of nanoscale pores filled with a dielectric, in particular hydrogen, on the field emission current from the structural materials of accelerators as the first stage of the occurrence of a high-vacuum, high-gradient breakdown.

II. TRANSMISSION COEFICCIENT

Let's consider a potential barrier in the case of a dielectric in the near-surface layer of a metal. The form of such a potential barrier is shown in Figure 1.

The one-dimensional time-independent Schrödinger equation is solved to calculate the probability of electron tunneling through the barrier:

$$-\frac{\hbar}{2m}\frac{\partial^2}{\partial x^2}\psi(x) + [U(x) - W]\psi(x) = 0, \quad (1)$$

where $\psi(x)$ is complex wave function of an electron, $\hbar$ is the reduced Planck constant, m is an electron mass, $U(x)$ is a potential, W is the initial longitudinal energy of electrons incident on the metal surface. For simplicity, the mass of the electron m in all three regions (i.e., in the metal, dielectric, and vacuum) is assumed to be equal to the rest mass of the electron.

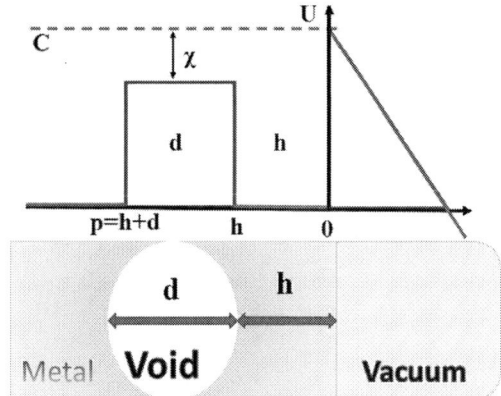

Figure 1: Schematic representation of a potential barrier for a metal-dielectric-metal-vacuum system

In the case of moderate electric fields, at which field electron emission from structural materials of accelerators usually occurs $E = 10^9 - 10^{10} \frac{B}{M}$ the transparency coefficient of a potential barrier can be written in the form:

$$D = \frac{4(C - \chi - W)W^{\frac{3}{2}}e^{-\frac{4k(C-W)^{\frac{3}{2}}}{3eE}}\sqrt{C - W}}{(C - \chi)(L + WY^2) - W^2C} \quad (2)$$

where $L = \left(\sinh(\beta d)Y\sqrt{C - \chi - W} + \cosh(\beta d)\sqrt{W}\Lambda\right)^2$

$$Y = \sqrt{C - W}\sin(\alpha h) + \cos(\alpha h)\sqrt{W},$$

$\Lambda = \sqrt{C - W}\cos(\alpha h) - \sin(\alpha h)\sqrt{W}, \alpha = k\sqrt{W}, \beta = k\sqrt{C - \chi - W}, k = \frac{\sqrt{2m}}{\hbar}, V = C - W - \chi, \eta = \xi\left(x - \frac{C-W}{eE}\right), \xi = (k^2eE)^{1/3}, C$ is a height of the potential barrier, χ is an electron affinity energy, $-e$ is an electron charge, E is an electric field strength.

From equation (2), it can be seen that the field emission current from a metal-dielectric-metal-vacuum system strongly depends on the thickness of the dielectric layer. As the thickness increases, the current will decrease exponentially. At the same time, the dependence on the depth of the dielectric layer is found in the sine and cosine argument. This means that the current must be oscillatory. It is easy to see that the maximum condition will be:

$$h = \frac{\pi}{2}\frac{1}{\alpha}(2n + 1), \quad n = 0,1,2,\dots \quad (3)$$

III. FIELD EMISSION CONSIDERING PORES FILLED WITH DIELECTRIC

Another important question is how the current density of field electron emission from accelerator construction materials will change in the case of filling the pores with a dielectric, i.e., from the metal-dielectric-metal-vacuum system. In further consideration, we chose different values of the electron affinity energy in the range of $\chi = 0 - 2$ eB. This is due to the fact that only a few atoms can be located in pores of this size at the same time. And their affinity energy will take values within these limits.

It was shown in [4] that for the parameters of accelerator structures, the influence of a generalized layer of pores with a diameter of 0.1 nm will be most noticeable. Therefore, to study the effect of pore filling with a dielectric, we chose an effective layer of this thickness to ensure the best visibility of the results. Figure 2 shows the dependence of the field emission current, expressed in units of Fowler-Nordheim current, on the depth of the defect for different affinity energies χ.

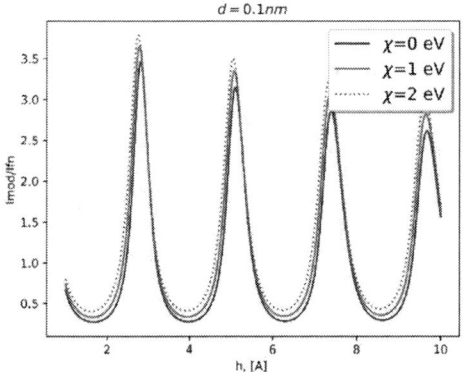

Figure 2: Comparison of field emission current densities j_{mod} *and* j_{F-N} *depending on* χ *and h for h=0÷1 nm.*

From Figure 2, it is easy to see that the current character remains unchanged when the pores are filled with a dielectric with different affinity energies. The current is still resonant. In the region of resonant amplification, the field emission current density increases by about 3.5 times for a given set of affinity energies. It is worth noting that the current enhancement associated with a decrease in the height of the step at the potential barrier (see Figure 1) due to the electron affinity does not exceed 10%.

To find the effect of both the effective layer thickness d and the depth h on the field emission current, we generated a thermal map of the current from the modified metal surface. Figure 3 shows the results of the numerical calculation of the field emission current density expressed in units of current from an ideal surface in the case when the affinity energy $\chi = 0, 1$ *and* 2 eB.

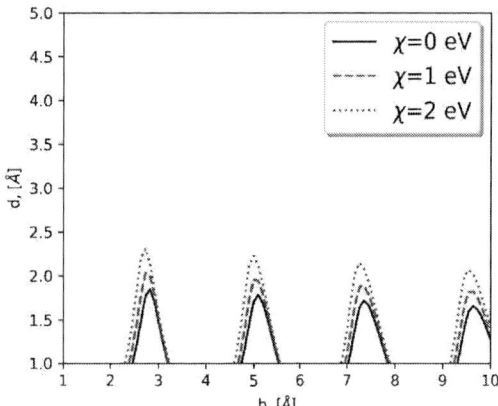

Figure 3: Graphical dependences of the current from the metal-dielectric-metal-vacuum system as a function of the pore size d & the thickness of the metal layer h

The figure shows that the current is indeed resonant. It's worth mentioning that when the size of the defect is $d > 0.2$ *nm* regardless of their depth, a reduction in the field electron emission current is anticipated. Simultaneously, the resonance region expands as the affinity energy rises.

IV. SUMMARY

Research demonstrates that the field emission current near nanopores on a metal surface exhibits an oscillatory resonant pattern. The resonance is influenced by the depth of the nanoscale cavity from the metal surface. Larger pore sizes result in a rapid decline in the field current.

It has been observed that under resonance conditions, the density of field emission current from copper surfaces surpasses that from unmodified surfaces by over 3.5 times. Introducing a dielectric expands the resonant region, visually enhancing it by approximately 20% for $\chi = 2$ eV. Failure to meet resonance conditions or an increase in nanopore diameter causes the current density to plummet close to zero, which can be a way to prevent breakdown.

REFERENCES

[1] A. Lopez-Cazalilla et al, Effect of surface orientation on blistering of copper under high fluence keV hydrogen ion irradiation, Acta Materialia, 266, 2024, 119699.

[2] Catarina Serafim, "CERN pulsed dc systems: RFQ, materials, microscopy" Presentation In-person miniMeVArc at CERN, 14-15 September 2023, Geneva, Switzerland.

[3] Alvaro Lopez Cazalilla, "Growth mechanisms of hydrogen blisters in copper" Presentation In-person miniMeVArc at CERN, 14-15 September 2023, Geneva, Switzerland.

[4] I.I. Musiienko, S.O. Lebedynskyi, R.I. Kholodov / Nanoclusters and nanoscale voids as possible sources of increasing dark current in high-gradient vacuum breakdown // The European Physical Journal D, 2022, V. 76 (4), p. 1-10

Two-stage Amplifier Using Field Emitter Array-Based Vacuum Transistor and Its Application to a Wien Bridge Oscillator

Ryosuke Hori[1], Tomoaki Osumi[1,2], Masayoshi Nagao[2], Hiromasa Murata[2], and Yasuhito Gotoh[1,*]

[1]*Department of Electronic Sience and Engineering, Kyoto University, Kyoto, 615-8510, Japan*
[2]*National Institute of Advanced Industrial Science and Technology (AIST), Tsukuba, 305-8568, Japan*
*Corresponding author: gotoh.yasuhito.5w@kyoto-u.ac.jp

Abstract—**A two-stage amplifier using FEA-based vacuum transistors was designed and its performance was investigated in view of forthcoming vacuum integrated circuits. The empirically obtained amplification factor well agreed with theoretically calculated one. A Wien bridge amplifier was designed with the two-stage amplifier. The self-driven oscillation with the frequency of 1 kHz was observed.**

Keywords—*field emitter array, vacuum transistor, vacuum integrated circuit, oscillator.*

I. INTRODUCTION

When vacuum microelectronics emerged in late '80, it was expected to produce ultrahigh frequency devices [1]. In 2000's, however, the mainstream of vacuum nanoelectronics shifted to field emission displays [2], and only a few researchers had interest to electron devices [3-5]. Recently, new movements to develop electron devices with field emitter arrays have emerged, taking advantage of possible operation in harsh environment, for example high temperature environment [6] or radiation field [7]. Developments of an oscillator [6] and a frequency mixer [5] are reported, but these reports were about a single VT. To apply VTs to practical electronic devices, circuits using multiple VTs are required. However, to the best of our knowledge, there have been no reports about such a circuit, except for a differential amplifier [3]. Thus, we designed a two-stage amplifier and investigated its performance [8]. As a consequence of the successful operation of two-stage amplifier, we designed a Wien bridge oscillator and observed preliminary results of the oscillation at the frequency of 1 kHz [9]. In this paper, some improvements of the circuit and oscillator are reported.

II. EXPERIMENTAL PROCEDURE

A. Characteristics of vacuum transistors

VTs consisted of a gated-FEA and a collector with the inter-electrode spacing of 2 mm were used. First, we measured the characteristics of two VTs, VT-A and VT-B, using 10,000-tip FEAs to evaluate transconductance g_m. For the emitter-grounded circuit, voltage amplification factor μ is expressed as

$$\mu = g_m R_{out}, \quad (1)$$

under the condition of $r_c \gg R_{out}$, where r_c is the collector resistance of the VT and R_{out} is the output load, respectively. Current-voltage ($I - V$) characteristics were measured by applying $V_G = 0$ V and $V_C = 200$ V, and sweeping V_E to negative voltage of -60 V.

B. A two-stage amplifier using VTs

A two-stage amplifier using VT-A and VT-B was operated at V_{GE} s of 51 V and 58 V, respectively. The collector resistances of VTs were both 100 kΩ. A single-stage amplifier using VT-A was operated with R_{out} of 510 kΩ at V_{GE} of 58 V for comparison.

C. A Wien bridge oscillator using a two-stage amplifier

A Wien bridge oscillator using the two-stage amplifier was designed with an improved parameters. The schematic diagram of the circuit is shown in Fig. 1. A Wien bridge oscillator requires a voltage gain of 9.5 dB or more and positive feedback. In the previous report [7], R_{out} of 100 kΩ was used. In this study, R_{out} of 51 kΩ was used to improve the cutoff frequency f_{cutoff}. The parameters shown in Fig. 1 are as follows: $C_1 = 1 \mu F$, $R_1 = 100$ kΩ, $R_2 = 1$ MΩ.

III. RESULTS AND DISCUSSION

Fig.2 shows the $I - V$ characteristics of two VTs. VT-A showed the collector current (I_C) of 510 μA or g_m of 100 μS at V_{GE} of 51.5 V, and VT-B showed I_C of 150 μA or g_m of 30 μS at V_{GE} of 57.5 V. The gate currents of VT-A and VT-B were about 1/1000 and 1/2000 of I_Cs, respectively.

Fig. 3 shows the frequency characteristics of the single stage and the two-stage amplifiers. The measured amplification factor of the two-stage amplifier at low frequency well agreed with that calculated from theory as shown in (1). The cutoff frequency of the two-stage amplifier was improved four times higher than that of the single-stage amplifier.

Fig. 4 shows the wave form of the oscillation. The oscillation was observed at V_{E1} of 56.0 V and V_{E2} of 48.8 V. The transconductances of VT-A and VT-B were 20 μS and 60 μS, respectively, so the total gain of the two-stage amplifier

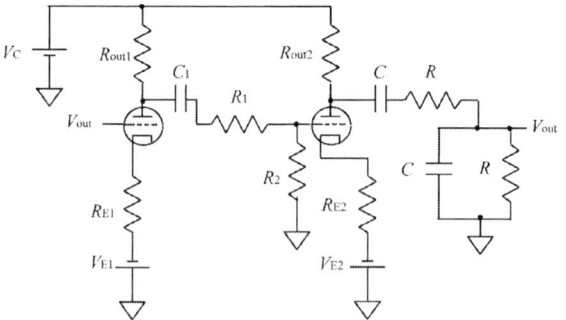

Fig. 1. Schematic diagram of a Wien bridge oscillator using a two-stage amplifier

979-8-3503-7977-8/24 $31.00 © 2024 IEEE

was 9.9 dB. The wave form showed a sinusoidal wave with little distortion, which was better than that in the previous report [9]. At this moment, however, oscillation with the frequency of 10 kHz was not observed yet.

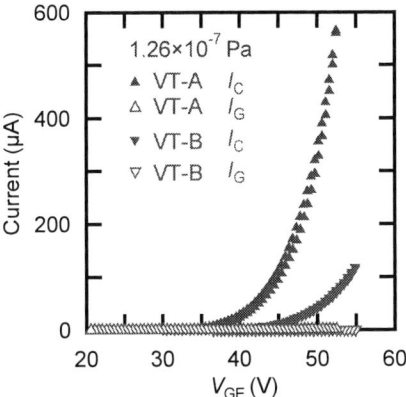

Fig. 2. $I - V$ characteristic of VTs

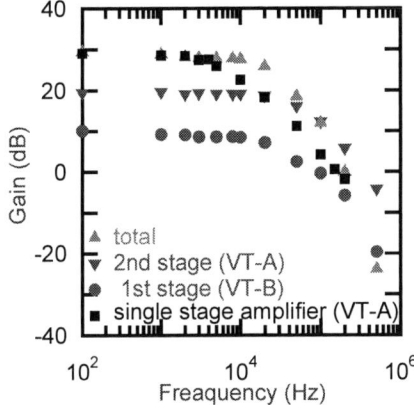

Fig. 3. Frequency characteristics of a single amplifier and a two-stage amplifier

TABLE I. CHARACTERISTICS OF A TWO-STAGE AMPLIFIER

	μ_{theory} (dB)	μ_{mes} (dB)	f_{cutoff} (Hz)
1st stage	9.5	9.1	20 k
2nd stage	20	19	50 k
total	30	29	20 k

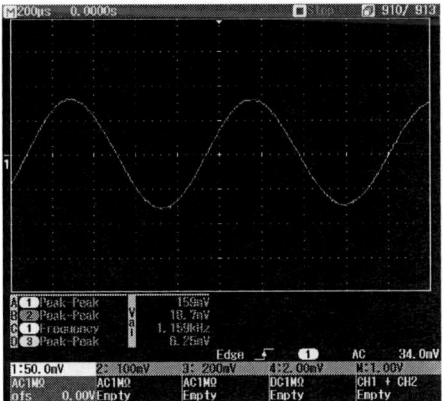

Fig. 4. The oscillation wave form

IV. SUMMARY

The characteristics of a two-stage amplifier using FEA-based VTs were investigated. It was observed that signal was amplified at each stage and μs were equal to calculated values. A Wien bridge oscillator applied it was operated and the oscillation wave with the frequency of 1 kHz was observed. It suggested that FEA was operating stably.

ACKNOWLEDGMENT

The present study was partly supported by Japan Society for the Promotion of Science through Grant-in-Aid for Scientific Research (B) KAKENHI JP21H01680.

REFERENCES

[1] K. Utsumi, "What's New and Exciting", IEEE Transactions on Electron Devices, vol. 38, no. 10, pp. 2276-2283, October, 1991.

[2] H. H. Busta, "Field Emission Flat Panel Displays", in Vacuum Microelectronics, W Zhu ed., New York, NY: John Wiley and Sons,2001, pp. 289-347.

[3] S.-H. Hsu, W. P. Kang, J. L. Davidson, J. H. Huang and D. V. Kerns, "Nanodiamond Vacuum Field Emission Integrated Differential Amplifier," in IEEE Transactions on Electron Devices, vol. 60, no. 1, pp. 487-493, Jan. 2013.

[4] K. Ikeda, W. Ohue, K. Endo, Y. Gotoh, H. Tsuji, "Development of a vacuum transistor using hafnium nitride field emitter arrays", J. Vac. Sci. Technol. B Vol. 29, pp. 02B116, March, 2011.

[5] Y. Gotoh, Y. Yasutomo, H. Tsuji, "Vacuum frequency mixer with a field emitter array", J. Vac. Sci. Technol. B, vol.31, pp. 050601, September, 2013.

[6] R. Bhattacharya, R. Hay, M. Cannon, N. Karaulac, G. Rughoobur, A. I. Akinwande, J. Browning, "Demonstration of a silicon gated field emitter array based low frequency Colpitts oscillator at 400 °C", J. Vac. Sci. Technol. B, vol. 41, pp. 023201, March 2023.

[7] Y. Gotoh, H. Tsuji, T. Masuzawa, Y. Neo, H. Mimura, T. Okamoto, T. Igari, M. Akiyoshi, N. Sato, I. Takagi, "Development of a Field Emission Image Sensor Tolerant to Gamma-Ray Irradiation", IEEE Transactions on Electron Devices, vol. 67, no. 4, pp. 1660-1665, April, 2020.

[8] R. Hori, T. Osumi, M. Nagao, H. Murata, Y. Gotoh, "Operation of FEA-based Two-Stage Amplifier for Vacuum Integrated Circuits", presented at the 84th Fall Meeting of the Japan Society of Applied Physics, 2023, 20p-A501-3 [in Japanese].

[9] R. Hori, T. Osumi, M. Nagao, H. Murata, Y. Gotoh, "Demonstration of an oscillator circuit using FEA-based vacuum transistors", presented at the 71st Spring Meeting of the Japan Society of Applied Physics, 2024, 25a-12M-1[in Japanese].

Uniform Current Supply in Gated P-Type Si-Tips for Achieving High-Performance Field Electron Emitter Array

Yang Chen, Yifeng Huang, Jun Chen, Shaozhi Deng, Ningsheng Xu, Juncong She*

State Key Laboratory of Optoelectronic Materials and Technologies, Guangdong Province Key Laboratory of Display Material and Technology, School of Electronics and Information Technology, Sun Yat-sen University, Guangzhou 510275, People's Republic of China
*Corresponding author: shejc@mail.sysu.edu.cn

Abstract—**An inversion electrons diffusion model coupled with generation electrons injection was proposed to describe the current supply for each gated p-type Si-tips in an array, which was calculated using finite element simulation. It was clarified that an equivalent position of tip-arrangement is necessary for obtaining uniform current supply. A rational design of gated p-type Si-tips in ring-arrangement with equivalent position was demonstrated for obtaining uniform current supply in the array. The ring array achieved a high-performance field electron emission, i.e., a high current intensity of 303 µA @ 179 V and spatially uniform emission sites. This work provides a new method to achieve uniform and high-performance gated p-type field emitter array.**

Keywords—*Gated p-type Si-tip, field electron emission, current supply, equivalent position of tip-arrangement.*

I. INTRODUCTION

Field electron emitter array with high current intensity and reliability is highly desired. Gated p-type Si-tip array is regarded as a promising field electron emitter with well reliability. The emission current of p-type Si-tip is typically saturated without the impact of emission surface potential fluctuation due to the limitation of electron supply from generation current in substrate depletion. In our early work, a nano-channel is on-tip integrated with the p-type Si-tip, which brings about local Joule heat effect to thermally enhance the emission current [1]. The Joule heat is initially induced by the current supplied from the substrate. It has been clarified that the average current supply to each tip can be modified by tip-numbers/gate-area ratio, thus inducing different enhancing current [2]. Accordingly, the uniformity of current supplied for each tip in an array is a crucial factor for achieving uniform emission from an array. The control on the electron supply distribution provides a new opportunity for improving the emission uniformity of each tip in the array and achieving high-performance field emitters, which has not been concerned yet. In this abstract, modelling, simulation and experiment were utilized to investigate the current supply uniformity in gated p-type Si-tip with different tip-arrangement. A rational design of gated Si-tip ring array was demonstrated for achieving uniform high current field electron emission.

II. RESULTS AND DISCUSSION

An inversion electron diffusion model coupled with generation electron injection was established to describe the current density that supply for each gated p-type Si-tips in an array. In the 2-dimentional surface inversion layer, which is overlapped with the gate pad electrode, the electrons diffuse to each tip in the array for emission. The diffusion current is expressed as: $J=qD\nabla\rho$ (Eq. 1), where J is the current line-density in A/cm, ρ is the electron sheet-concentration in cm^{-2},

q is the electron charge, and D is the electron diffusion coefficient. The generation electron in the surface depleted region keeps injecting into the inversion layer, which induces a divergence of the current density in inversion: $\nabla \cdot J=j_{gen}$ (Eq. 2), where j_{gen} is the generation current sheet-density in A/cm^2 in depletion.

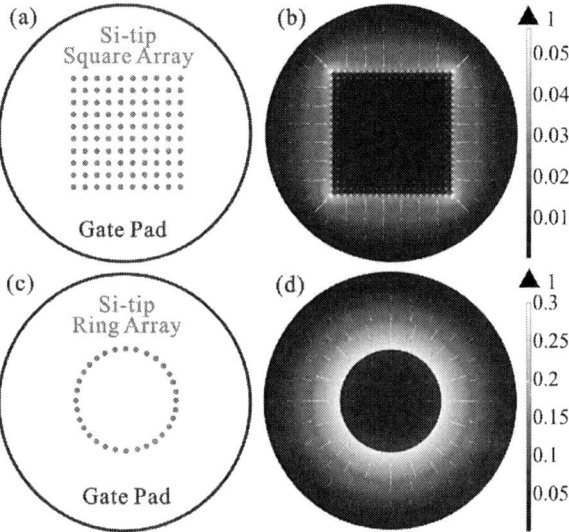

Fig. 1. The schematic illustrations of the tip array in (a) square and (c) ring arrangement, while (b) and (d) are the corresponding current density in the inversion region of the gated p-type Si-tip array. The dots represent the gated tips, while the colour-bar represents the normalized current density distribution in surface inversion. The arrows represent the local current intensity and direction.

The Eq. 1 and 2 were solved using finite element simulation and thus the current distribution in surface inversion was obtained. In a gated p-type Si-tip square array (Fig. 1 (a)), the current supplied for the tips at the edge and especially the corner of the array is much higher than that of the tips inside (Fig. 1 (b)). The tips at the edge of the array are able to obtain large quantities of electron transported from the region outside the tip-array. Conversely, the tips inside the array can only obtain the electron transported from their adjacent region, which leads to a weak current supply. Among the out-edge tips, the tips located at the corner of the array face a larger opening angle (270°) than that of other tips (180°). Accordingly, the corner-tips are able to receive electrons transported from direction of a larger angle and thus obtain more current supply. Namely, the inequivalence of geometric position of each tip in the gate pad induces non-uniform current supply. A rational design was proposed that the tips arranged as a ring at the circular gate pad (Fig. 1 (c)). The equivalent position of each tip due to the rotational symmetry induces a uniform current supply in the simulation (Fig. 1 (d)).

Both the gated p-type Si-tip array that arranged as square (20×20) and ring (400 tips) was well-controlled fabricated and characterized. Both the square array and ring array showed a current saturation tendency of ~300 nA at gate voltage ~70 V. The emission current of the ring array began to greatly enhance at a gate voltage of ~106 V and a high current of 303 μA @ 179 V was achieved. However, for the square array, the emission current began to enhance at ~126 V and a lower current of 59 μA @ 179 V was achieved. The "mark" on the ITO-anode surface that induced by electron bombardment was employed to evaluate the emission uniformity of an array. A uniform ring-shape mark was observed, suggesting a uniform emission of the ring array. However, corresponding to the square array, the marks on the out-edge is more pronounced than those inside. Namely, the emission current of the square array is spatially non-uniform, which is consistent with the simulation results.

According to the device model, the tips in the ring array get a uniform current supply from the substrate. The current brings about significant Joule-heating effect and induces high temperature at the tip-apex, which greatly promotes the local carrier generation and enhances the emission current. Due to the uniform current supply in the array, a large number of tips obtain sufficient current and work at the current enhancing regime, contributing for the high current emission together. Therefore, the current-enhance of the ring array occurs at a lower gate voltage and a higher emission current with uniform emission sites were achieved. In comparison, the non-uniform current supply for the square array only enables the tips at the out edge and especially the corner of the array to get sufficient current and work at current enhancing regime. The non-uniform field emission results in a limited current intensity for the square array.

III. SUMMARY

A device model was proposed to describe the current transport in surface inversion of gated p-type Si-tip array. Numerical simulation showed that an equivalent geometric position for each tip at the gate pad is necessary for obtaining a uniform current supply. A tip-array arranged in a ring at a circle gate pad was fabricated to achieve a high emission current and uniform emission sites. The any-angle rotational symmetry of the gated device brings about a uniform current supply in the tip array, which enables more tips to contribute for the high current emission.

ACKNOWLEDGMENTS

This work was supported in part by the National Key Research Program of China under Grant 2021YFA1200600, in part by the National Natural Science Foundation of China under Grant U22A2020.

REFERENCES

[1] Y. Huang, Y. Chen, Z. Huang, M. Zeng , Z. Gu, W. Yang, J. Chen, N. Xu, J. She, and S. Deng, "P-type Si-Tips with Integrated Nanochannels for Stable Nonsaturated High Current Density Field Electron Emission," *IEEE Trans. Electron Devices*, vol. 69, no. 7, pp. 3908-3913, Jul. 2022, doi: 10.1109/TED.2022.3172046..

[2] Z. Huang; Y. Huang, J. Chen; N. Xu; S. Deng; J. She. "Mechanism Of Non-Saturated Field Electron Emission from Gated P-Type Si-Tips," *International Vacuum Nanoelectronics Conference (IVNC)*, Kyoto, Japan, Jul. 2018, doi: 10.1109/IVNC.2018.8520279.

Vacuum Flat-Panel Solar Blind Ultraviolet Detectors Using PIN-Structure Photocathodes Formed by ZnO Nanowires on NiO-Ga₂O₃

Dunhan Mo, Zhipeng Zhang*, Zhuoran Ou, Juncong She, Shaozhi Deng, Jun Chen

State Key Laboratory of Optoelectronic Materials and Technologies, Guangdong Province Key Laboratory of Display Material and Technology, School of Electronics and Information Technology, Sun Yat-sen University, Guangzhou 510275, People's Republic of China
*Corresponding author: zhangzhp25@mail.sysu.edu.cn

Abstract—A vacuum cold cathode detector formed by a PIN-structure photocathode and a ZnS thin film is proposed to achieve highly sensitive solar blind ultraviolet detection. The PIN-structure photocathode was formed by ZnO nanowires grown on NiO-Ga₂O₃ heterojunction using a thermal oxidation technique. The electrons generated by internal photoelectric effect in PIN photocathode are accelerated in the vacuum gap and multiplied in the ZnS thin film by EBIPC effect, achieving a high ultraviolet detection responsivity.

Keywords—*Solar blind ultraviolet detectors, PIN photocathode, Field emission, Ga₂O₃, EBIPC*

I. INTRODUCTION

Solar blind ultraviolet imaging devices have important applications in space exploration, missile tracking, radiation monitoring, ultraviolet lithography and scientific research [1]. Gallium oxide (Ga_2O_3) is a promising ultra-wide bandgap semiconductor with a high absorption efficiency for solar blind ultraviolet light, a high breakdown electric field and a high-temperature stability, making it highly suitable for use in solar blind ultraviolet detection applications. For example, the reported Ga_2O_3 based photodetectors showed excellent detection performances with a dark current of 40 pA, a photo-to-dark current ratio of 10^4, a response time of 41 ms and a high responsivity of 3.23 A/W [2]. However, the sensitivity of Ga_2O_3 photoconductor based ultraviolet detectors is still needed to be improved for ultra-weak light detection.

Recently, vacuum cold cathode detectors formed by large area ZnO nanowire field emitters and ZnS photoconductor anode have demonstrated their high sensitivity detection ability due to the electron bombardment induced photoconductivity (EBIPC) mechanism [3]. When the ZnS photoconductor anode was bombarded by electrons emitted from ZnO nanowires, the impact ionization effect was generated in the photoconductor. The impact ionization effect enhanced both the conductivity of the ZnS photoconductor and the electric field at the surface of the ZnO NWs, which results in a positive feedback process. Therefore, the photomultiplier effect called EBIPC was generated to achieve high internal gain (10^4) and high responsivity (2.3×10^2 A/W) [4]. Nevertheless, achieving the vacuum cold cathode detectors with low dark current and wide dynamic range is still a challenge due to structure limitation.

In this study, the PIN-structure photocathodes formed by ZnO nanowires grown on NiO-Ga₂O₃ was proposed to achieve a vacuum flat-panel solar blind ultraviolet detector. The proposed PIN photocathodes are expected to realize a dark current, a high absorption efficiency and a high responsivity.

II. EXPERIMENTAL

Fig. 1 shows the proposed structure of vacuum flat-panel solar blind ultraviolet detector. Firstly, the glass substrate (0.3 mm) was cleaned by ultrasonication in acetone, alcohol, and deionized water for 15 minutes, respectively. The ITO electrode (540 nm) was deposited on glass substrate by magnetron sputtering with a pressure of 4.8×10^{-3} Pa and a deposition time of 35 min. Secondly, NiO thin film (125 nm, 0.5 Å/s) and Ga₂O₃ thin film (2 μm, 2 Å/s) was deposited on ITO electrode by electron beam evaporation with a pressure of 2.3×10^{-3} Pa. Lastly, arrays of patterned Zn thin film (2 μm, 0.6 Å/s) was prepared by photolithography and electron beam evaporation. Then, the sample was placed in a tubular furnace for thermal oxidation with air to grow ZnO nanowires. The oxidation temperature was 500 °C and the oxidation time was 180 min. The effective area of photocathode was 2×2 cm². The anode was formed by a 2.5 μm ZnS thin film prepared on an ITO coated glass substrate using an electron beam evaporation technique.

The morphologies of prepared PIN photocathodes were inspected by a scanning electron microscopy (SEM, SUPRA 60). The photoelectric response characteristics of PIN photocathodes were measured by a semiconductor analyzer. The field emission current of PIN photocathode was measured in a vacuum chamber with a pressure of $\sim 1.0 \times 10^{-5}$ Pa using a high-voltage power supply (Keithley 2657).

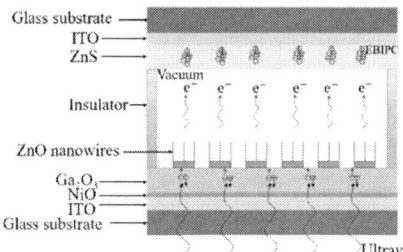

Fig. 1. Schematic diagram of the vacuum flat-panel solar blind ultraviolet detector formed by a NiO-Ga₂O₃-ZnO nanowire photocathode and a ZnS thin film anode.

III. RESULTS AND DISCUSSION

Fig. 2(a) shows the arrays of patterned ZnO nanowires prepared on NiO-Ga₂O₃ film with a single dot matrix of 25×60 μm². The ZnO nanowires were grown vertically on NiO-Ga₂O₃ film with a length of ~ 1 μm, a tip diameter of 20 nm and a population density of $\sim 10^8$ cm⁻² (Fig. 2(b)). In addition, the NiO-Ga₂O₃-ZnO photocathode exhibited a good adhesion with the glass substrate (Fig. 2(c)).

979-8-3503-7977-8/24 $31.00 © 2024 IEEE

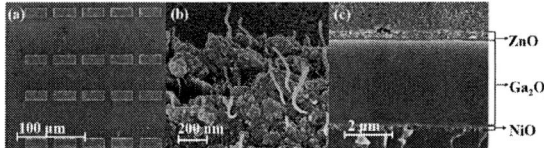

Fig. 2. (a) Arrays of patterned ZnO nanowires prepared on NiO-Ga₂O₃. (b) Morphology of ZnO nanowires. (c) Cross-view image of PIN photocathode.

Fig. 3 shows the electrical characteristic of PIN photocathode. The current increased from 2.1×10^{-4} A to 1.0×10^{-3} A with the increasing applied voltage from -3 V to 0.75 V. The current rectification ratio of photocathode was ~30.

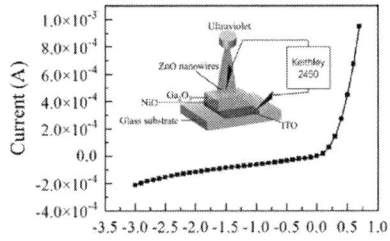

Fig. 3. Electrical characteristic of PIN photocathode with the inset showing the corresponding mearurement set-up.

The photoelectric detection characteristics of PIN photocathode were measured under a 254 nm light (Fig. 4). When the reverse bias increased from 0.1 to 5 V, the dark current increased from 15.9 μA to 1.0 mA, the photocurrent increased from 0.17 mA to 20.5 mA, and the corresponding responsivity increased from 6.9×10^{-4} to 0.08 A/W.

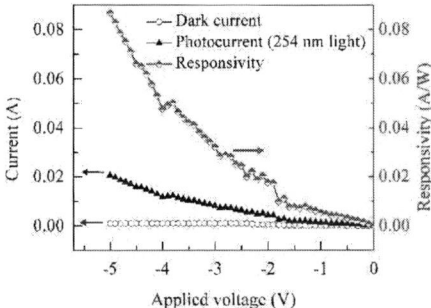

Fig. 4. Dark current, photocurrent and responsivity of the PIN photocathode.

The field emission current of fabricated PIN photocathode was further measured (Fig. 5). When the anode voltage increased from 600 to 1100 V, the field emission current increased from 1.67 nA to 4.6 μA. In order to verify the enhanced responsivity of proposed vacuum flat-panel solar blind ultraviolet detector with EBIPC effect. The photocurrent (*I*) was simulated using the F-N field emission theory [5]:

$$I = At^{-2}(y)\phi^{-1}V_1^2 \exp\left(-\frac{B\phi^{\frac{3}{2}}}{V_1}v(y)\right) \qquad (1)$$

where A and B are constants, $t(y)$ and $v(y)$ are Nordheim elliptic functions, $\emptyset$ is the work function of ZnO nanowires, $V_1 = V - V_2$ is the equivalent voltage of cold cathode field emission, V is anode voltage, $V_2 = IR$ is the equivalent voltage of ZnS thin film, R is the photoconductor resistance. The resistance of PIN photocathode was negligible compared with that of ZnS thin film. When the detector was irradiated by incident light (15 μW/cm²), the resistance of ZnS was assumed to decreased form 1 GΩ to 125 MΩ due to EBIPC effect. Therefore, the simulated photocurrent increased from 1.5 nA to 26.6 μA with the increasing anode voltage from 600 to 1100 V, achieving an enhanced responsivity of 1.63 A/W.

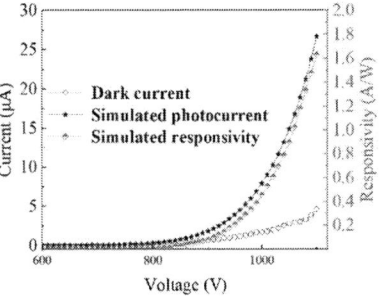

Fig. 5. Dark current, simulated photocurrent and responsivity of proposed vacuum flat-panel solar blind ultraviolet detector.

IV. CONCLUSIONS

In summary, we proposed a vacuum flat-panel solar blind ultraviolet detectors using PIN-structure photocathodes formed by ZnO nanowires on NiO-Ga₂O₃ heterostructure, and a ZnS photoconductor anode. Photoelectric measurement verified the high responsivity of PIN photocathode. Simulation results indicated that the vacuum cold cathode detectors could reach a high responsivity (1.63 A/W) owing to the internal photoelectric effect in PIN photocathode and the EBIPC effect in ZnS anode.

ACKNOWLEDGMENTS

The authors gratefully acknowledge the financial support from the National Key Research and Development Program of China (Grant No. 2022YFA1204200), Key Research and Development Program of Guangdong Province (Grant No. 2023B0101200013), National Natural Science Foundation of China (Grant No. 62271512), Natural Science Foundation of Guangdong Province (Grant No. 2024A1515012852), Guangzhou Municipal Science and Technology Bureau (Grant No. 2023A04J1664), Fundamental Research Funds for the Central Universities.

REFERENCES

[1] A. Dehzangi, J. Li, and M. Razeghi, "Band-structure-engineered high-gain LWIR photodetector based on a type-II superlattice," Light Sci. Appl., vol. 10, no. 1, p. 17, 2021.

[2] S. H. Amsterdam, A. U. Mane, and A. B. Martinson, "Ultrathin amorphous gallium oxide vacuum ultraviolet photodetectors," ACS Appl. Electron. Mater., vol. 5, no. 11, pp. 5962-5967, 2023.

[3] Z. Zhang, M. Chen, C. Wang, K. Wang, S. Deng, and J. Chen, "Highly sensitive direct-conversion vacuum flat-panel X-ray detectors formed by Ga₂O₃-ZnO heterojunction cold cathode and ZnS target and their photoelectron multiplication mechanism," Adv. Mater. Interfaces, vol. 9, no. 9, p. 2102268, 2022.

[4] Z. Zhang, K. Wang, K. Zheng, S. Deng, N. Xu, and J. Chen, "Electron bombardment induced photoconductivity and high gain in a flat panel photodetector based on a ZnS photoconductor and ZnO nanowire field emitters," ACS Photonics, vol. 5, no. 10, pp. 4147-4155, 2018.

[5] R. H. Fowler and L. W. Nordheim, "Electron emission in intense electric fields," Proc. R. Soc. London Ser. A, vol. 119, pp. 173-181, 1928.

Developing Cold Cathode Flat Panel X-Ray Source Module for Portal X-Ray Imaging System

Haonan Wei, Qi Liu, Zhuoran Ou, Song Kang, Guofu Zhang, Zhipeng Zhang, Shaozhi Deng, Ningsheng Xu, Jun Chen*

State Key Laboratory of Optoelectronic Materials and Technologies, Guangdong Province Key Laboratory of Display Material and Technology, School of Electronics and Information Technology, Sun Yat-sen University, Guangzhou 510275, Guangdong Province, People's Republic of China
*Corresponding author: stscjun@mail.sysu.edu.cn

Abstract—**Portable X-ray imaging system has important application in emergency rescue, mobile healthcare and field non-destructive testing, etc. Compared to single-focus X-ray sources, flat panel X-ray sources (FPXS) allow for a more compact imaging system due to small source-to-object distance. In this study, X-ray source module is fabricated by integrating cold cathode flat panel X-ray source, collimator and electrical driver. The effect of collimator on imaging resolution was studied. This study verified the feasibility of compact FPXS module for realizing portable and lightweight X-ray imaging system.**

Keywords—*cold cathode, flat-panel X-ray source, portable imaging system, collimator.*

I. Introduction

X-ray imaging systems equipped with thermionic cathode X-ray tubes have disadvantages such as bulky, high-power consumption and short lifetime. It is also difficult to achieve portable and lightweight imaging equipment. In recent years, with the development of one-dimensional cold cathode materials, cold cathode X-ray source has been studied extensively[1]. Furtehrmore, cold cathode flat panel X-ray source (FPXS) using ZnO nanowire cold cathode have been reported[2-4]. By using FPXS, it is possible to realized portable X-ray imaging system, which is expected to overcome the above-mentioned shortcomings.

In this study, a compact X-ray source module is designed and fabricated. FPXS was fabricated using ZnO nanowires cold cathode. X-ray source module is fabricated by integrating the FRXS, collimator and driver. The characteristics of the FPXS and the effect of collimator on imaging characteristics were studied.

II. FPXS Device and Module Fabrication

The X-ray source module of the portable X-ray imaging system is designed. As shown in Fig.1, the module includes a FPXS devices, a high-voltage power supply adapted to FPXS, a circuit board controlling the voltage output, and a multi-aperture collimator controlling the emission direction of the X-ray output from the FPXS.

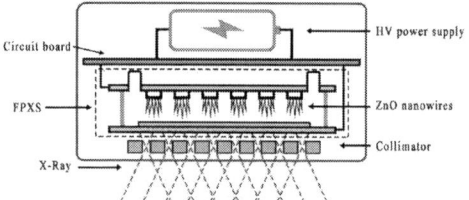

Fig. 1. Schematic of the designed FPXS module

FPXS is composed of ZnO nanowire cold cathode, transmission-type anode, and gap insulator in between. full-vacuum-encapsulated FPXS device was fabricated using the fabrication process reported in our previous work[5]. Fig.2(a)

shows a fabricated vacuum-encapsulated FPXS device used in this study. The morphology of the ZnO nanowires in the device is shown in Fig. 2(b). The electrical characteristics of FPXS were tested, and the emission uniformity and I-V characteristics were obtained as shown in Fig. 3. The field emission image recorded from the anode is relatively uniform. When the anode voltage is 34 kV, the current is ~1.5 mA. According to the electrical characteristics of FPXS, we designed a compact pulsed power supply and control circuit. The FPXS device, collimator and electronics are assembled and packed into a 3D printed housing. A finished FPXS module is shown in Fig. 4.

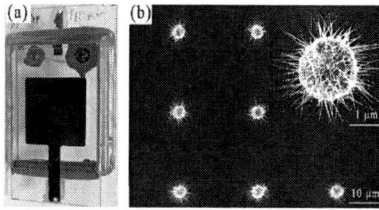

Fig. 2. (a) Picture of fabricated FPXS. (b)SEM image of ZnO nanowires in the device.

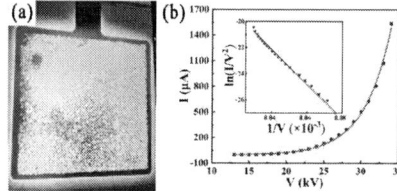

Fig. 3. (a) Field emission image and (b) I-V characteristics of the FPXS device. (inset shows the F-N plot).

Fig. 4. The picture of the assembled FPXS module.

III. Imaging Characteristics

The imaging characteristics of the FPXS module is tested using a flat panel X-ray detector. The effects of collimator parameters on the imaging properties of the FPXS module were studied.

A. Imaging characteristics of FPXS module without collimator

X-ray projection imaging experiments were carried out for different source to detector distances under the anode voltage of 30 kV using a FPXS without the collimator. In the

979-8-3503-7977-8/24 $31.00 © 2024 IEEE

measurement, line-pair sample was attached to the detector. The obtained results are shown in Fig. 3. The resolvable line-pair from the image increases with increasing source to detector distance.

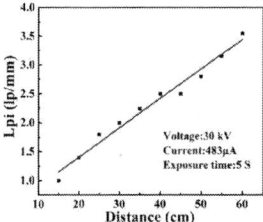

Fig. 5. The variation of the image resolution with the distance between the source and the detector.

The variation of imaging resolution of the PFXS module with the distance between the source and the detector shown in Fig.5 can be explained by the following formula for X-ray imaging resolution:

$$Resolution = \frac{1}{2}BW = \frac{\sqrt{d^2+[a(M-1)]^2}}{2M}, \quad (1)$$

where, BW represents the width of the effective radiation beam; d is the pixel size of the detector; a is the focal size; and M is the ratio of the source-to-detector distance (SDD) and the source to the object distance (SOD). Thus, in our experiment, with increasing distance, M is closer to 1, smaller BW value can be obtained, resulting in higher imaging resolution.

B. Effect of collimator on imaging quality

We first simulated the effect of the parameters of collimator on imaging resolution. The aperture size and thickness of the collimator determine the emitting angle of the X-ray. By a simulation using BEAMRC, the effect of emitting X-ray angle on imaging resolution was obtained as shown in Fig. 6. The results show that with the increase of the X-ray emitting angle, the image resolution decreases gradually, but the decreasing trend is gradually stable after the angle reaches 36 degrees. The schematic shown in Fig. 7 explained the effect of collimator on imaging resolution.

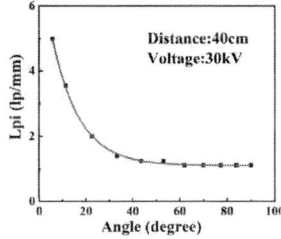

Fig. 6. Simulated results of the variation of imaging resolution with X-ray emitting angle.

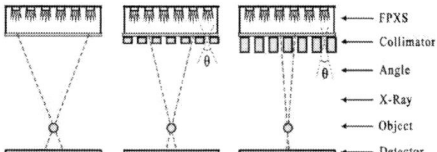

Fig. 7. The schematic showing the effect of collimator on imaging.

According to simulation, the image resolution will increase with the decrease of the angle. Therefore, we can constrain the emitting angle of FPXS to improve the imaging

resolution. However, a small emitting angle is usually realized by shrinking the diameter of the aperture or increasing the thickness of the collimator (i.e. increasing the aspect ratio). This can also number of X-ray photons reaching the detector and thus reduce the image quality. Therefore, a trade-off must be considered in the design of the collimator.

In order to verify the simulation results, collimators with aperture sizes for 18° and 53° X-ray angle were manufactured. When the distance between FPXS and detector is 40 cm and the anode voltage is 30 kV, X-ray imaging experiments were carried out using these collimators. The obtained images were presented in TABLE I and compared with the results obtained without collimator.

TABLE I. COMPARISON OF IMAGING RESULTS FOR THE DEVICE WITH AND WITHOUT COLLIMATOR

	No collimator	53° Collimator	18° Collimator
Image	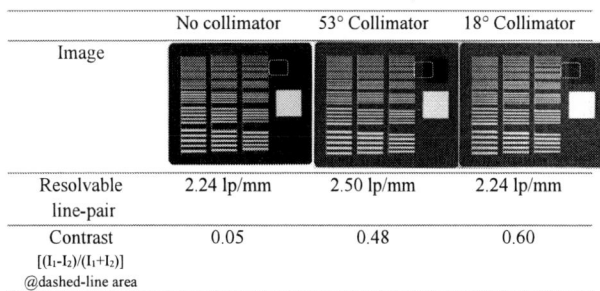		
Resolvable line-pair	2.24 lp/mm	2.50 lp/mm	2.24 lp/mm
Contrast $[(I_1-I_2)/(I_1+I_2)]$ @dashed-line area	0.05	0.48	0.60

Basically decreasing the angle will increase the resolution. However, in the results shown TABLE 1, it is found that the image quality of the PFXS module is better at 53° collimator than at 18° collimator. This is because the attenuation of the X-ray photon number caused by the small-angle collimator. Besides, it is found that the contrast at the low photon count area increases with decreasing angle of collimator. This may be due to the reduction of scattering X-ray photons when using small-angle collimator. Therefore, it is necessary to consider multiple factors such as the angle of X-ray, the number of X-ray photons and the projection area when configuring a suitable collimator for FPXS.

IV. SUMMARY

A compact FPXS module was fabricated for portable and lightweight X-ray imaging system. The effects of the parameters of the collimator on the projecting imaging characteristics were studied.

ACKNOWLEDGMENTS

Thanks for the financial support from the National Key R&D Program of China (Grant No. 2022YFA1204200), Key R& Development Program of Guangdong (Grant No. 2023B0101200013), the Science and Technology Department of Guangdong (Grant No. 2023B1212060025).

REFERENCES

[1] H. J. Kim, H. N. Kim, H. S. Raza, et al, Nuclear Engineering and Technology, vol. 48(3): pp. 799-804, 2016.

[2] D.K. Chen, X. M. Song, Z. P. Zhang, et al, Applied Physics Letters, vol. 107(24), pp. 1-5, 2015.

[3] L.B. Wang, Y.Y. Zhao, K.S. Zheng, et al, Applied Surface Science, vol. 484, pp. 966-974, 2019.

[4] Y.Y. Zhao, Y.C. Chen, G.F. Zhang, et al, Nanomaterials, vol. 11(240), pp.240, 2021.

[5] X.Q. Cao, G.F. Zhang, Y.Y. Zhao, et al, Applied Physics Letters, vol. 119(5), pp. 053501, 2021.

AUTHOR INDEX

Abuamr, Adel M. 66, 82, 84, 90
Adhikari, Bishwa Chandra 21
Akinwande, Akintunde I. 36, 138
Al-Anber, Mohammed A. 40, 108
Aljabarat, Aseel 84
Allaham, Mohammad M. 72, 134
Allaham, Mohammad 38, 152
Alqaisi, Ali F. 90
Apugade, Umesh Balaso 21, 54
Arai, Noriaki 160
Arrasheed, Enas A. 66, 84
Asgharzade, Ali 86
Asgharzadehkhorasani, Ali 146
Assis, Thiago A. De 23
Ayari, Anthony 102, 118
Bachmann, Michael 86, 140, 146
Back, Tyson 136, 156
Bammes, Fabian 126
Bartl, Mathias 86, 140, 146
Bhotkar, Ketan R. 25
Bhotkar, Ketan 114
Bialas, Marcin 98, 130
Bode, Nils 126
Bruckner, Leon 7, 31
Buchheim, Jakob 120
Buchner, Philipp 86, 140, 146
Buchta, Aleksandra M 74
Burda, Daniel 38, 72, 134, 152
Cahay, Marc 136, 156
Checoury, Xavier 102
Chen, Jun 13, 33, 42, 56, 58, 64, 68, 106, 110,
.... 112, 122, 124, 128, 166, 168, 170
Chen, Manni 124
Chen, Yang 106, 112, 166
Chen, Yinyao 48
Cheng, Yonghong 94
Chern, Winston 36
Chlouba, Tomas 31
Chowdhury, Mokter M. 150
Combrié, Sylvain 102
Cui, Tao 48
Dall'Agnol, Fernando F. 23
Dencker, Folke 74, 100
Deng, Shaozhi 3, 11, 13, 17, 33, 42, 48, 56, 58,
.... 64, 68, 106, 110, 112, 122, 124, 128, 166, 168, 170
Diekmann, Leonard Frank 100
Dienstbier, Philip 7
Dimitrakopoulos, Alexander 35

Ding, Yuyue 48
Djurabekova, Flyura 142
Dziuban, Jan 98
Eckhoff, Colin C. 1
Edgcombe, Chris 132
Edler, Simon 86
El-Gomati, Mohamed M. 158
Fan, Ruowen 58
Feng, Wenqi 106
Fohlerová, Zdenka 72
Forbes, Richard G. 23, 60, 62
Gangloff, Laurent 102
Gerner, Constanze 7
Ghotbi, Shabnam 27
Gotoh, Yasuhito 164
Gou, Mingkai 13, 68
Grzebyk, Tomasz 5, 78, 98
Guo, Dengzhu 9
Hajibaba, Soheil 29
Hall, Harris 136, 156
Han, Dong 17
Hart, James 136, 156
Hausladen, Matthias 86, 140, 146, 148
Hernandez, Nathaniel 136, 156
Holcman, Vladimir 144
Hommelhoff, Peter 7, 31, 126
Hori, Ryosuke 164
Huang, Yifeng 13, 106, 112, 166
Huang, Yiming 68
Huang, Yuan 11, 112
Huns, Janis 132
Jaber, Ahmad M D Assa'D 144
Jaber, Ahmad M D 90, 158
Jezek, Jan 44
Jiang, Jun 11
Kaiser, Alexander 140
Kandra, Mario 38
Kang, Song 33, 42, 58, 128, 170
Karande, Aniket 114
Karaoulanis, D. 19
Karaulac, Nedeljko 36, 138
Kassner, Alexander 74, 100
Kawamoto, Erina 46
Ke, Yanlin 3
Kholodov, Roman 162
Kim, Hyeonseok 1
Kim, Iksu 21, 54

Knápek, Alexandr .... 38, 72, 82, 84, 90, 134, 148, ... 152, 158
Knapkiewicz, Pawel 98
Knapp, Wolfram 52
Koch, Jannik 100
Koitermaa, Roni Aleksi 94, 142
Kokkorakis, G. C. 19
Kondo, Shun 76
Kong, Jiaquan 110
Košelová, Zuzana 72, 134
Kraus, Stefanie 31
Krysztof, Michal 78, 98
Kusunoki, Toshiaki 160
Kuzyk, Casimir 35
Kyritsakis, Andreas 94, 142
Lebed, Oleksandr 104
Lebedynska, Yuliia 162
Lebedynskyi, Serhii 104, 162
Lei, Wei 88, 92
Li, Jiaxin 3
Li, Mengjie 154
Li, Xinran 56, 64
Lin, Zufang 42
Litzel, Julian 31
Liu, Qi 33, 42, 128, 170
Lohrl, Bastian 7
Ludwick, Jonathan 136, 156
Macku, Robert 40, 108
Matejka, Milan 38
Matsunga, Soichiro 46
Meng, Guodong 94
Metzler, Luke J. 1
Mian, Md. Suruz 76
Mo, Dunhan 168
Mohammadi, Saeed 27
Mousa, Marwan S. 66, 82, 84, 90, 152, 158
Murakami, Katsuhisa 50, 76, 80, 116
Murata, Hiromasa 50, 76, 116, 164
Nagao, Masayoshi 50, 76, 80, 116, 164
Nakano, Takeo 76
Nauk, Constantin 7
Ningsheng, X. 17, 42
Nojeh, Alireza 35, 150
Novotny, Jan 44
O'Mara, Jonathan 136, 156
Orudzhev, Farid 40, 108
Osumi, Tomoaki 164
Ou, Hai 64
Ou, Zhuoran 33, 56, 64, 124, 168, 170
Ovcharenko, Artur 104
Papež, Nikola 40, 108
Park, Kyu Chang 21, 25, 54, 114

Paschen, Timo 7
Patil, Ravindra 25, 114
Pease, R. Fabian 35
Pedder, Randall E. 1
Pengbin, X. 17
Perisanu, Sorin 118
Perisanu, Sorin-Mihai 102
Plichta, Tomas 44
Podstránský, Jáchym 148
Poncharal, Philippe 118
Pu, Zhongbin 110
Purcell, Stephen T. 118
Qi, Guicai 33
Qin, Xiaoyu 17
Radtke, Lars 126
Rouillé, Goulven 102
Roumeliotis, J. A. 19
Sawant, Jaydip 25
Sawatzky, George A. 150
Schels, Andreas 86
Schmidt-Kaler, Franz 126
Schreiner, Rupert 86, 140, 146, 148
Sedlák, Petr 40, 108, 144
Seidling, Michael 126
Sery, Mojmir 44
She, Juncong 13, 33, 58, 64, 68, 106, 110, 112, 124, 166, 168
Shen, Yan 13, 17, 48, 68
Shihkgasan, Ramazanov 40, 108
Shiloh, Roy 31
Shimawaki, Hidetaka 80
Shin, Youngjin 36
Silhan, Lukas 44
Sobola, Dinara 40, 66, 82, 84, 108, 144
Song, Guichen 122
Song, Zheyu 48
Soud, Ammar Al 66, 82, 84, 144
Staron, Patrik 40, 108
Szyszka, Piotr 78, 96, 98
Tang, Qianqian 154
Tang, Shuai 13, 68
Telfah, Ahmad 84
Tiirats, Tauno 142
Trrad, Issam 90
Tsujino, Soichiro 29
Urbanski, Pawel 5, 78, 98
Vaculik, Ondrej 44
Velásquez-García, Luis Fernando 1
Velthaus, Verena 120
Vincent, Pascal 118
Voon, Kevin 150
Walker, Dennis E. 136, 156
Wang, Chengyun 64

Wang, Jiaqi 70
Wang, Zhen 112
Wang, Zhenpeng 88
Wei, Haonan 170
Wei, Xianlong 9, 15
Wen, Bin 124
Weng, Catherine 102
Wohlfartsstätter, Dominik 86
Wu, Wangjiang 33
Wurz, Marc C. 74
Wurz, Marc Christopher 100
Xanthakis, J. P. 19
Xiao, Mei 88, 92, 154
Xie, Junhang 128
Xie, Zhemiao 70
Xu, Ningsheng 3, 13, 33, 56, 106, 112, 128, 166, 170
Xu, Yuan 33
Xu, Zhaoying 11
Yeow, John T. W. 70
Yidan, H. 15
Yimeng, L. 94
Young, Jeff F. 150
Yu, Zelin 88
Zadin, Veronika 94, 142
Zeng, Wen 106
Zhai, Xin 92
Zhan, Runze 106
Zhang, Guofu 42, 56, 58, 64, 124, 128, 170
Zhang, Junjie 154
Zhang, Xiaobing 88, 92, 154
Zhang, Yu 3, 11, 13, 17, 68
Zhang, Zhipeng 56, 110, 168, 170
Zhao, Haonan 68
Zhao, Yanqing 9
Zhiwei, L. 9, 15
Zhong, Junhao 13
Zhou, Linghong 33
Zhou, Rui 70
Zhu, Zhuoya 88, 92
Zimmermann, Robert 126
Zlámal, Jakub 148

IEEE
445 Hoes Lane
Piscataway, NJ 08854-4141

ISBN 979-8-3503-7977-8

9 798350 379778